AF559988

Post Harvest Technology of Horticultural Crops

Post Harvest Technology of Horticultural Crops

R.C. Upadhyaya

ANMOL PUBLICATIONS PVT. LTD.
NEW DELHI-110 002 (INDIA)

ANMOL PUBLICATIONS PVT. LTD.
H.O.: 4374/4B, Ansari Road, Darya Ganj,
New Delhi-110 002 (India)
Ph.: 23278000, 23261597

B.O.: No. 1015, Ist Main Road, BSK IIIrd Stage
IIIrd Phase, IIIrd Block
Bangalore - 560 085 (India)
Visit us at: www.anmolpublications.com

Post Harvest Technology of Horticultural Crops

First Published, 2008

ISBN 978-81-261-3427-4

PRINTED IN INDIA

Printed at Mehra Offset Press, Delhi.

Contents

Preface

Horticulture has a special significance for the economic development. Most of the people, though hard working and face the calamities of nature and live in harsh conditions, yet produce lot of fruits, vegetables and other flowering and ornamental plants including the medicinal and aromatic plants. In order to increase the productivity of horticultural crops growers must take the advantage of the expertise of scientists. Keeping this idea in focus, the present work has been compiled which contains experiences of the experts and related people. The volumes cover advances in strategies, production, plant protection, value addition and other important areas of horticulture development.

The Encyclopaedia provides practical information on production technologies of tropical and subtropical fruits as well as plantation crops and horticultural nursery. Post-harvest management, marketing and export trends have been furnished for guidance. It also presents most realistic cost estimates and returns from cultivation of such crops. The volumes aim at viewing commercial prospects for horticulture. It has been carefully designed to meet the queries of intending investors, financial institutes, creditors, traders and growers engaged in dealing with horticultural crops. Further it will serve the purpose reference books to students, academicians and related people.

R.C. Upadhyaya

Post Harvest of Horticultural Crops

There is a growing interest in all areas for storage and transportation of horticultural products, because the products are very perishable and because of the long distances between countries and their major export market. Worldwide Post Harvest fruit and vegetables losses are as high as 30 to 40% and even much higher in some developing countries.

Reducing Post Harvest losses is very important; ensuring that sufficient food, both in quantity and in quality is available to every inhabitant in our planet. The prospects are also that the world population will grown from 5.7 billion inhabitants in 1995 to 8.3 billion in 2025. World production of vegetables amounted to 486 million tonnes, while that of fruits reached 392 million tonnes. Reduction of Post Harvest losses reduces cost of production, trade and distribution, lowers the price for the consumer and increases the farmers income.

Distribution of Vegetables and Fruit Production in the Various Regions (in million tonnes)

	Vegetables	*Fruits*
North and Central America	44.16	53.17
Europe	66.98	
Europe	66.98	65.17
Africa	35.20	54.06
Asia	292.41	142.24
Pacific	2.60	5.62
South America	16.06	60.95

(FAO)

	Production	*Edible matter*	*protein*
Vegetables	486	57	5
Fruits	388	53	2

The increase in prosperous population, the urbanization and the green revolution have enlarged the market for all kind of ornamentals. Post Harvest research has a high return.

Significant contributions, minimizing the Post Harvest losses have been made through research on the physiological changes in the product after storage, new longlife varieties, suitable the product after storage, new longlife varieties, suitable cultivation circumstances, optimum harvesting indices, recommended storage recommendations, pre-cooling refrigerated transport and careful handling.

In preserving the quality of the produce reaching the final market, the whole marketing chain from the producer to the final consumer must be examined for any weakness. The Post Harvest research topics have to be closely related to the socio-economical and financial situation, the climatical and cultivating circumstances, the market organization and the requirements of the specific commodities.

The current needs of scientific research and extension services are different between developing and high developed, industrialized countries, and the nature of the projects must be adapted, step by step, following the evolution. In this respect, the possibilities of linkages and further international research cooperation has to be stimulated by international institutions such as EC, FAO, ISHS, HR, and through initiatives such as the World Conference on Horticultural Research.

Post Harvest research is definitely a muti-disciplinary endeavour of biology chemistry, physics, pathology, biochemistry, physiology, marketing, financing, psychology and ethics in a social-economical, political and financial context.

Suitable Post Harvest Research in Developing Countries

Pole Post Harvest research in developing countries: Post Harvest research has to be not only limited to storage conditions but also to market requirements, breeding and cultivating circumstances.

Treatments to arrest deterioration, harvesting at correct maturity, pre-cooling to reduce field air, care during sorting, packaging, transportation are also determinent Post Harvest factors.

Some methods of handling, packing and storage have been developed over years, using innovative abilities and skills, also low cost and commonly available materials and technolgies.

In other methods different new techniques and disciplines are implemented.

Market situation: Roadside markets with small shops or a type of local marketing are seen commonly at the suburbs of a city and in rural districts, mostly close to farms and selling seasonal products.

Price differences according to quality are mostly too small and don't stimulate enough quality improvement.

Urban population is exploding in developing countries, rising from 35% of the total population in 1990 to 54% in 2020. With this increasing urban population more free market and wholesale markets will be required.

Product is shipped from neighbouring areas as well as from distant ones.

Comparison with imported products stimulates increase of the quality of the regional produce. Export will make clearly show what quality of the regional produce. Export will make clearly show what quality level is needed for exported produce. Continuously following consumers and traders requirements, growers have to be informed in the choice of the varieties, the cultivating techniques and the extending of the selling period. The presence of growers cooperatives or collecting centers for selling products and the requirements of exporters and supermarkets could stimulate for applying suitable varieties cultivation methods of sorting, grading and transporting, and establishing research activities and extensions services.

Cultivating Conditions: Many small growers produce a high variability of crops, in the villages for self consumption and around the cities for selling the rest. Changing the choice of the varieties and the cultivating techniques is the most important and must be oriented to a higher quality and healthy product, preferred by consumers and with a longer shelf-life required by traders. Pest control with plant extracts has to be improved.

New developments presume proper financing possibilities, not always easy to get. Exceptionally some big specialized growers produce a high quantity of nice products for export as Phaseolus beans in Senegal or mangos in the Philippines. Grading and sorting:

Quality Standardisation: Grading and sorting in quality classes is limited or not existent and this evolution can be stimulated by the development of cooperative selling organisations.

Packaging: Packaging is dependant on the disposal of packing material in the country (bags, wooden crates, woven baskets). Adaptations are necessary to reduce bruising by filling and transportation and to increase shelf life. The evolution is to use standardised reusable plastic boxes and suitable corrigated carton boxes, after wooden crates.

Transportation: Most growers have only at their disposal limited transport facilities as bicycles, three wheeled or small trucks. Some traders with bigger trucks can buy the products in the producting area and transport them to the markets in the cities, with high profit. Due to the poor condition of the roads there is a rapid deterioration of the produce in some areas. Cooperatives could place at the disposal of the growers suitable important transport vehicles.

In developing countries a few number of exporters transport proper packed and graded product by refrigerated trucks, air of boat on commissions (± 10%) to countries far away.

Storage and Processing

Storage technics are mostly limited Storage and processing: Storage technics are mostly limited to evaporative cooling or ventilated cooling with air in winter and cold water in summer (humypad cooling) or storage in caves. These technologies have to be optimized and complemented by proper ventilated or simple modified atmosphere systems.

Refrigerated warehouses in developing countries are located mainly in large cities and ports but very limited in the farmers area (garlic in Beijing).

Refrigerated storage rooms have to be developed in the production areas.

In developing countries the introduction of new expensive, not ozone degrading cool media is not easy.

Research on the optimum cooling techniques of different commodities, to improve shelf-life, have to be studied, as long-life cultivars applying biotechnology technics, better cultivating techniques and pest control, harvesting using objective maturity indices.

Reducing bruising during harvesting packing and transport, picking in the morning, putting the product in the shadow and increasing the speed of selling times, are key basic elements for quality.

To prolong storage time, heat treatment of fruit, curing of root-vegetables after harvest, chemical treatments of flowers, treatment with growth regulators or wax of fruit and vegetables, chemical drying of onions before harvest are important element.

To preserve fruit and vegetables in developing countries research on solar drying, laetables in developing countries research on solar drying, lactic acid fermentation (kimchi), fresh pack fruit added with the juice of tropical berries with very high Vit. C en low pH or anti-browning agents as grapefruit seeds must be stimulated.

Information to growers through extension services: The current emphasis of the Post Harvest extension services is to reduce losses, increase quality, to invest low cost suitable technology, using suitable financing.

Proper information is necessary to help the sometimes poor and exploited producers with technical and marketing assistance, for more market directed crops.

It is recommended that a structural system for education and training of growers will be implemented.

Suitable Post Harvest Research

Industrialized countries, research is related to more fundamental research, and characterized by the introduction of far advanced expensive technologies and the integrataion of different scientific disciplines.

Market Situation: Specialized wholesalers-exporters and retailers and supermarkets require big quantities of the same quality.

Farmers have to concentrate their production through cooperation, selling big quantities of the same quality, offering a variability in crop by remote selling, and supple same quality, offering a variability in crop by remote selling, and supply on a long period. Competitiveness is increasing, mostly with commodities from tropical countries, inducing some problems in selling final produces.

There exist a consumer preference, different from market to market with also market segments.

The relationship price-specific quality indice is studied for different markets (Northern Europe red tomatoes, Southern Europe pink).

For long storage time the elasticity supply-demand factor is important.

Export market research will make clearly show what level is needed for exported produce and will help to find new suitable markets.

Communication and information of the consumer about quality (health, nutritive) and the promotion of the product means a sophisticated knowledge and experience. Quality of food is quality of life.

Cultivating Circumstances

Few farmers produce high quantities of an high quality product. They are very specialized in one crop and a few numbers of cultivars.

Through breeding and molecular biology techniques long-life and disease resistant varieties are created. The cultivating techniques and pest control methods are modelled and optimized. In the last time most interest is given to healthy aspects by integrated fruit production and environmental friendly fertigation. The supply is extended through plastic covering and glasshouses, he supply is extended through plastic covering and glasshouses, also through storing of plants as strawberries and raspberries for later forcing.

Objective indices and apparatus indicating optimum harvest are generally available. The application of Post Harvest chemical treatments by drenching, dipping or fogging is decreasing for consumer healthy objectives. There is yet some interest for heat treatments and wax coatings.

Grading and Sorting: Quality standardisation and determination: The EC quality categories E, I, II for sorting and size (mm) or fruit-weight are generally practised. This distinction is no more sufficient for traders because external and internal quality in one quality category is changing too much.

Selling boards make own quality blocks between one category (red and ground colour for apples, morphologic aspects for vegetables). Also some supermarkets require minimum quality norms for quality (colour, brix, titric acidity, firmness).

For controlled integrated fruit production labels for quality insurance are applied. International standardization of analytical and equipment is necessary. For controlling quality and automized sorting non-destructive methods are studied based on physical, acoustic, optical, electric impedance, X-ray, chlorophyll fluorescence and nuclear magnetic resonance technics. For export to some countries or quarantine security is required.

Packaging: Standardisation of the dimensions of package (60 cm x 40 cm, 30 cm x 40 cm) and the pallets (80 cm x 120 cm, 100 cm x 120 cm) is far advanced.

After wooden boxes reusable uniform plastic boxes - sometimes pliable—are used (Europool).

Corrugated carton boxes with minimum compression and minimum 4,5% open air locks after humid air pre-cooling and long transport are available.

Modified atmosphere packaging with absoption layers for water and ethylene, and inserts are used. For expensive products, PP films with a wide range of micropores are of great interest. The presence of some pathogenic organisms need urgent research.

The disinfection of wooden boxes with fomaldehyde or azaconazole is important for long time storage in high humid circumstances.

Labels on boxes or product have to inform consumers about the origin, the quality and the welfare of the product.

Increasing attention is given to fresh product, fresh pack and tightly processed products and juices.

Transportation: To meet the nutritional needs, higher amounts of food will have to be transported over greater distances and for a further extended time. Refrigerated trucks and containers are more used.

To reduce the dependence on expensive rapid air transport, CA refrigeraP> To reduce the dependence on expensive rapid air transport, CA refrigerated trucks and containers are performed. Accordingly, maximum values for impact and vibration norms are recommended for trucks on normal roads.

Storage and Processing

On construction level there is an evolution from cheap cork or glass fibre insulations, and bitumen or paint gastightness to gastight polyurethane-metal panels.

Requirements for compressor capacity and relative humidity increase. New not ozone degrading refrigerant are used. Compared to regular storage, modified atmosphere, controlled atmosphere, ultra low oxygen storage, quick pull down of oxygen and ethylene absorption are now suitable according to costs of investments and price.

New temperature, CO_2 and O_2 sensors for scrubbers and membrane separators are favouring monitoring and lowering risks of not proper storage. Pre-cooling with humid forced air cooling and ice bank has a lot of benefits for vegetables and berries.

For each commodity specific storage conditions are dressed. CA storage for tropical fruits, vegetables and fresh pack products is recommended. To extend the supply of berries, storage of plants is applied. During storage, chemical treatment of flower with preservatives is well developed.

In processing development of juices, purees, extracts and essences significant progress is perceptible, with new technics nces significant progress is perceptible, with new technics as microwave, membrane separations, ultrahigh temperature, aseptic package.

Future Post Harvest Research Priorities

- Genetic manipulation for long life, diseases and environmental-stress resistant cultivars.
- Modelling cultivating conditions for high quality and long life (avoiding root stress by heat or drought).
- Environmental friendly pest control.
- Objective determination of suitable harvesting date.
- Post Harvest treatments (heat, UV, irradiation CO_2, chemicals) for storage ability.
- Monitoring refrigerating systems
- Optimum storage conditions as CA storage for tropical fruits, ornamentals, planting material, fresh pack and lightly processed produce.
- Possibilities of modified atmosphere packaging (MAP), absorption layers, inserts sensors, PP films with micropores
- Prevention of possible pathogenic organisms in MAP products.
- Adaptive control of storage conditions with biological sensors.
- CA storage during transport and as quarantine measures.
- Humid forced air Pre-cooling.
- Minimum impact and vibration norms for bruising during sorting and packaging.
- Objective non destructive measuring of quality and maturity

- Environmental friendly packaging.
- Consumer and traders quality preferences in each country.
- Cost and return of the investment of Post Harvest techny.
- Cost and return of the investment of Post Harvest technologies.
- Fundamental research on senescence, ripening, respiration, ethylene effect, chilling, fermentation, superficial browning.

International Post Harvest Research Cooperation

Post Harvest research has to be adapted to the specific Post Harvest problems in each countries - according to the cultivated commodities and the socio-economical, technical and financial situation. International contacts between scientists and international cooperation projects will be very important in future to accelerate the evolution in developing countries especially.

The information of existing research programmes in the world collected on occasion of the World Conference on Horticultural Research will be very helpful to determine the Post Harvest priorities for each country and to stimulate cooperation.

International organisations such as ISHS, ASHS, EC, FAO, could there after contribute considerably in extending Post Harvest networking, research contacts and cooperation and to support some priority research projects, together with the existing governmental research and extension service.

Production and Marketing Decision

Thailand's horticultural export, both fresh fruit and vegetable, has increased over the years. Trade opportunity is largely due to increasing interests of major importing countries and Thailand's relative competitiveness over other producers. To increase its presence in regional/world markets, the industry is facing increasing pressures and unfamiliar trading practice. In addition, the decade-old production system based on increasing yield and reducing production cost by the producers, as well as the traditional management style of exporters, are inadequate in the face of the challenges of an increasingly competitive and discriminating global trading environment. From the business point of view, the market operation needs to move toward improving management efficiency and effectiveness rather than developing new technology alone.

This paper provides a brief overview of international horticultural trade with special reference to tropical fruit, analyses of factors influencing production and trade and trends in market development of fresh produce, effect and implication of new emerging producers, and the future role and challenge for Post Harvest technology in response to changing market scenes.

Overview of Global Fresh Fruit Industry

Market Size

The total value of international horticultural export trades (fruit and vegetables) has increased from less than US$11 billion to US$20 billion, and to US$44 billion over the periods 1980-82, 1990-92 and 2000-02, respectively. This included fresh (70%) and processed products (30%), of which half were fruit juices.

The increase represents an impressive annual growth rate of international exports at an average of 8-10% in relation to a slower growth prospect of traditional agricultural commodities such as cereals, livestock products, oil seed, fats and oil. No country can provide all the fresh produce it consumes weekly the whole year, in turn creating trading opportunities among countries. Expansion of trade for vegetables differs from that for fruits. Suppliers vary significantly between vegetable and fruit categories. Despite the increase in international horticultural trades, a far greater amount of fresh produce produced in each country is still domestically consumed.

Major Markets

Developed countries provide the largest production and import market. Between 60% and 80% of horticultural export trades are carried out among developed countries. Demand for fresh fruit is concentrated in Europe and the United States, which absorb around 60% and 11%, respectively, of international imports. The United States is the dominant player in the international trade of horticultural commodities, ranked number one as both importer and exporter accounting for about 18% of world horticultural trade.

In general, developing countries are major producers of tropical fruits (accounting for 98% of total 60.4 million tons in 2000), but their exports remain insignificant. International trade of tropical fruit represents less than 3% of total tropical fruit production and until recently most export went to other developing countries. It is important to note that tropical fruit is mostly local fruit suited to local taste. Trade data for 1999 indicate that Latin America, the Caribbean, and Asia accounted for about 75% of world trade in fresh tropical fruits and 80% were imported by developed countries. It is generally agreed that the potential for the sustainable growth of the tropical fruit industry depends on export to developed countries and on the size of the domestic market.

Product Composition

Horticultural trade is characterized by a wide diversity of products. The product composition of horticultural exports has changed over time with an increase in the relative importance of fruit than vegetables. The more traditional varieties of fruit, namely citrus, apples, grapes, bananas, pears, pineapples and peaches, represent some 85% of world production with those of temperature climate origin (apples, grapes, pears and peaches) accounting for 42%.

These are mainly produced and consumed in the developed countries. Of the typical tropical fruit, only banana, pineapple, and mango are produced in significant quantities and these are mainly produced in developing countries and consumed domestically. The taste for exotic products is rapidly evolving in developed countries. It was in the past 10 years that exotic tropical fruits such as avocado, guava, papaya and kiwi have become mainstream trading items. Pineapple accounts for 44% of total tropical fruit trade, followed by mango at 27%, avocado at 12%, and papaya at 7%. Among fresh vegetables, the miscellaneous vegetable category had the largest share, followed by roots and tubers. Tomatoes were the most important single item in the miscellaneous category that also includes a wide range of varieties such as pumpkins, lettuce, cucumbers, garlic, chillies and peppers, cabbages, cauliflower, carrots, garlic, onions, asparagus and ginger.

Significant Trends in Production, Marketing, and Export Development

Factors influencing trends are summarized into three broad categories as a result of changing demand/supply, market environment and supply chain management, and regulatory/legal requirements, taking the recent developments in Thailand as examples.

Increased Interest in Tropical Fruit Trade: Tropical fruit is becoming increasingly available in Europe. The increasing interest in tropical fruit in the developed countries is due to increased promotional activities and better information to consumers. In addition, it is largely attributed to the fact that the food markets are now saturated, leading to intense competition, lower margins, and lower prices for food. Since 1990, many modern trade retailers have positioned their store formats and image around the fresh produce department.

In order to further attract and delight the consumer, the markets are demanding less mass customizing of food items; diversified products based on consumer food consumption patterns; and safe, environment-friendly, organic product or by introducing new products. The introduction of tropical fruit meets all these criteria. The variety and composition of produce must therefore be enhanced, so that the characteristics of tropical fruits as diverse, exotic, unique in taste, highly nutritional, have multiple uses, and based on low input production system can be fully appreciated. For Thailand, the recent worldwide increase in the popularity of Thai food also increased the demand for Thai fresh produce.

Major Export Markets: According to the Food and Agriculture Organization (FAO), international consumption of tropical fruit will increase by around 40% between 1995 and

2005, equivalent to an annual growth rate of 3.5%. Despite the increase in consumption, developing countries export little tropical fruit due to short seasonality, short shelf-life of the fresh produce, high transportation cost, and increasing strict regulations in import. Preferences of tropical produce vary among countries and also vary greatly even within localized areas in one country.

Major trades in tropical fruit, until recently, remain limited either for domestic consumption in the country, or to regional markets or other developing countries with similar taste and preference. Exports to far away markets remain constrained with the exception of bananas with some 30% of total production exported, mainly to developed countries. As more produce enter into the trade and with increasing trend in global sourcing, market prospect for tropical fruit in developed countries are becoming positive.

Off-Season Supply: Supply of tropical fresh produce during off-season, counter season and/or complementary trade will become the norm than the exception. Good examples are the shipment of Southern Hemisphere produce to meet consumer demand during the Northern Hemispheres' winter when the domestic supplies are low. Most trade is caused by differing levels of relative competitiveness between producers of the same or similar products during different seasons to avoid trade dispute. Thailand's export of asparagus, to EU and Japan, has more than twofold increase in values from 1998 to 2002 due to Thailand's natural climatic production advantage with the ability to produce almost year-round.

Domestic Market: Export performance of horticultural trade differs widely among developing countries. It has been estimated that future rate of growth of domestic demand is much higher in developing countries than in developed countries and the prospect of high intra developing-country trade may not likely be sustained because each has increased production and expected future export growth. A large domestic market remains very important for export development, more so for fruit than vegetables. Domestic markets provide both a springboard for the growth of exports and a cushion to absorb the shocks or uncertainties of export markets. Thailand's longans and durians, two top export items each with more than US$50 million in export values per year representing an exceptionally high 30% and 15% of the country's production, respectively, still require a large domestic market to avoid price fluctuation during peak season. Thailand produces approximately 800,000 tons of durian and exports 110,000 tons mainly to China (via Hong Kong), Hong Kong, Singapore and Taiwan.

Though there is also an increasing interest for producers in the developed countries to exclusively produce tropical fruit for and sell in export markets, the global production statistics of key tropical fruit fail to indicate such a trend. The relative low export indicates that tropical fruit is typically produced for domestic market and suited to local taste. A further observation is that in many cases the largest exporters of tropical fruit are not the top producing nations. This indicates that these countries are adopting a specialized export-oriented strategy, and this is largely due to a significant price differential in favour of exports and based on high cost production system, high capital investment, and a well-coordinated vertical supply chain.

Falling Price: This increasing interest in the expansion of market stimulates increase in supply and this has a consequent downward pressure on prices and tends to depress margins due to intense competition especially among exporters in the developing countries. Mangoes are a good example, with prices in Europe falling 30% since 1988, in conjunction with a 66% increase in import volumes. The increase in export market and in suppliers resulted to reduced profit margins, which discouraged many exporters, and to high cost of production. This in turn has led to a supply of relatively poor-quality produce by low-cost producers.

Concentration of Suppliers and Their Composition

The size of the tropical fruit export market relative to the output in many cases is not large enough to allow many countries/exporters to succeed and will initially result in a smaller share of the market held by each country/exporter. Thus, the enlargement in trade may eventually result in consolidation and concentration of fewer suppliers operating at a reduced profit margin offering produce at a comprised quality of larger volume. This is usually considered transitional when the small industry is at a crossroad. There is a general interrelation indicating that an increase in export trade often results in overall improvement in quality, but produce for sale in the domestic markets are often of lower quality or are export rejects. Product differentiation or specialization to meet the large sourcing need of modern trade may result in reduced level of overall country production though there may be increased export by individual firm. The long-term effect of such trend to the industry will need to be seriously assessed.

Farm to Table: Production methods for domestic consumption are simple, much less demanding, and the domestic market usually pays well when lower expenditure on cost structures and investment are taken into consideration. To meet the discerning export market requirements, farming practices and marketing system will have to change.

However, changes in the production methods, better sanitation and hygienic farm practice, and improvement in quality will not ensure that export market can be developed and maintained. Traditional fragmented small-sized producers and labour-intensive nature of tropical fruit production system needs to be optimized and to link to companies that organize the distribution of products in the export market. A developing country that wants to export a not yet established tropical produce often exports it through an agent in the importing country who handles the distribution and marketing without any financial risk but for a commission. The export bears the entire risk. The risk is shared by the importer only when trust has developed over time between the importer and the exporter.

Commission sales play an important role in fresh produce markets for new exporters and remain so even for established exporters. Regulations in some countries that prohibit commission sales and require that horticultural imports must have a prearranged buyer tend to inhibit fresh produce trade.

There is no organized system for market intelligence and trade regulations for horticultural products comparable to what exist for other agricultural products. The wide variety and small trade involved with fresh produce do not attract adequate attention for the international or national trade intelligence and statistical services. Most traders and government in the developing countries lack the leadership and vision and are ill-prepared to formulate a well-developed sustainable export strategy.

Food Safety Issues

The traditional approach to food safety was that people at each stage of the food supply chain were responsible for the handling of the food and, hence, for food safety issues. This is easily said than done as the food chains have become more complex and the initiation of food safety strategy is often difficult, if not impossible, in the fresh produce supply chain where there is no certain predominant market power, or worst, a consistently shifting market power as affected by supply and demand during the season.

Recent legislation trend in many developed countries is to put the responsibility for food safety into the hands of the retailer. This is an over-simplified approach to the complex food safety issues and the tendency is toward strict requirements for imports through an expensive time-consuming import/export inspection. There has been increasing pressure and requirement to implement third party certified food safety and quality management systems especially for import/export trade. While Good Agricultural Practices, Good Manufacturing Practices, Hazard Analysis Critical Control Point, and ISO 9000 certification have been focal points for many organizations across the world to be used for food safety and management of value chain, there have been no appropriate frameworks for these standards for fresh produce.

Some of these systems are either unrealistic and unworkable because of inherent weaknesses/uncertainty/dynamic nature of the fresh produce industry and the difficulty to provide branding and to understand certification, or too costly because such systems are still based on inspection and or regulation, or the intended holistic approaches are not meeting customer/business specifications or formalized dialogue in the fresh produce supply chain, or due to the fact that users and the bureaucracy take a simple concept and make it difficult for operators.

Without a clear purpose, too stringent/improper safety requirements and the application of different health standards for domestic and imported goods, and imposing food safety requirements often had an obvious protectionist component. Many safety factors/variables need to be considered in harmonizing national standards with the guidelines through international consensus, and mutual recognition agreement on food safety issues is expected to take time. The likelihood is high to forge increased bilateral and regional trading arrangement.

Another important consideration is the need for a dialogue between policy makers and the producers/operators. A recent bilateral agreement on fresh produce trade between Thailand and China, and the Thai government initiative on food safety project have been mistakenly interpreted as restricting trade by the producers/operators. The challenges are to provide better communications and to put in place food safety systems that are viable and effective both in terms of cost and efficiency.

Trade Liberalization and Trade Barriers

Horticultural exports of developing countries are constrained by tariff and non-tariff barriers (NTBs). It is generally agreed that the fruit sector must be prepared to confront a more liberalized trade environment. One of the most significant trade liberalization development and the new trading opportunities and threats affecting horticultural import/export industry worldwide is the recent inclusion of China and Taiwan in the World Trade Organization (WTO).

Tariffs vary by product, season, and country of origin. The restrictive effect of NTBs on horticultural imports is considerably greater than that of tariffs. A wide range of NTBs include marketing orders such as product specifications, quotas and voluntary export constraints, import licensing, variable levies, minimum price systems, countervail taxes and duties, technical specifications (especially health restrictions, and strict labelling and packaging specifications), and even bureaucratic delays or uncertainties.

Without a clear purpose or scientific justification, there is a tendency to increase the use of NTBs as a discrimination against fresh produce imports. Most of these requirements are voluntary or as recommendation at this stage, but some are soon expected to become mandatory. Among NTBs, fresh produce is particularly heavily constrained by various sanitary and phytosanitary regulation mechanisms on the basis of concerns for consumers and most notably, for the protection/security of the importing country's own production.

Sanitary and phytosanitary regulations can be characterized as trade barriers and WTO requires these to be backed by sound and verifiable scientific evidence. It should be noted that the countries with the most rigorous control procedures are precisely the major exporters, the US, EU and Japan. Almost all national standards either already have guidelines, or are in the process of harmonization with guidelines formulated through international consensus. At this transitional stage, as a result of competition, the market is at the mercy of misinformation and mud sling.

To meet these requirements, the developing exporting countries have to set up expensive, high technology laboratories. Most important of all is that exporting countries will need to provide education, training, and a sophisticated level of technological infrastructure and managerial skill to monitor conditions of production/handling processes for export and to ensure transparency, and allow consultation and dispute settlement processes between trading partners. The costly procedures for the operators to comply with regulatory requirements and restriction tends to be in favour of middle-income developing countries and give new exporting countries only very selective and limited access to enter international trade.

Changing Consumption Trend and Consumer Demand

The tropical fruit industry needs to retain the number of loyal consumers who have now moved into the middle and upper income brackets. The industry also needs to expand to new markets with a new consumer group with varying income levels in both the developed and the developing countries. In addition, recent global economic downturn, falling prices, increased competition among producers, and increased market concentration in favour of supermarkets, have all resulted in a swift of market power in a supply chain that was producer driven, to one that is driven by consumer and their proxy retailer -- a change that has a negative impact on the levels of returns to the producers and exporters.

This has created new opportunities for new companies and pressured some rethinking among existing companies to concentrate their efforts on the core business and move toward more vertical supply chain coordination, in order to improve the overall quality dimensions of the product perceived by the market. The bottom line is to capture the hearts and minds of the consumer.

Emerging New Suppliers and Markets: The expansion of the EU to integrate several countries producing fresh products (Greece, Portugal and Spain) has significantly affected the horticultural exports of developing countries because these members are substantial producers and exporters of horticultural products. The questions facing the non-EU producers are the long-run trends in export supply in and out of the EU by these countries and those that are also potential suppliers to US, Japan and other developing countries in relative competitions with other non-EU member suppliers. The enlargement of EU to include these countries has resulted in the adoption of export standards to EU from one of lenient in the past to one of the strictest.

New emerging suppliers and markets in the region are the result of China and Taiwan's entry into the WTO. China has achieved an outstanding annual horticultural growth rate of 20% per year and that has transformed the country from the fourth largest producer at the start of the decade into the world's largest producer of fruit and vegetables. China has significantly increased its production of the most widely sold varieties of fruit at the international level, namely banana

(13% growth per year), oranges (7% per year), apples (26% per year), pears (14% per year), and peaches (15% per year). This suggests that it has adopted a strategy of supplying the world, and as China's productions meet the regulatory requirements of importing countries and the quality/price expectation of the consumers, it will become an exporting giant. Opportunities still exist at present as China experiences increased purchasing power and improved standard of living.

Thailand has gained an impressive market share of fresh fruit and vegetables imported into China in recent years, mainly tropical fruits such as durian, longan and lychee. A very significant percentage of China's fruit imports are via Hong Kong's re-export trade. During the last 5 years (1996-2000), some 250,000 tons of fresh fruits were re-exported annually to China representing about 25% to 29% of fruit imported to Hong Kong.

Thai fresh fruits such as durian, longan and lychee which constitute 80%, 74%, and 58%, respectively of Hong Kong's total imports of these items in 2000, were exported to China. Demand for longans in China shows maturation for the past years with an annual growth rate of 6% as a result of China's own longan production, in comparison with durian with an annual growth rate of 94% for the past 5 years. Taiwan's recent economic downturn resulting in falling price and increasing strict sanitary and phytosanitary measures have seriously affected Thailand's export prospects into that country.

Better Demand-Driven Information Intelligence: There is an increasing trend for large majority fresh produce wherein importers source their products by purchasing them directly from foreign grower/shippers. The interest of modern trade store has resulted in the growing interrelation between domestic and international market standards and a concentration of market power in the hands of the retail stores. Another trend is the increasing demand of suppliers in the exporting countries in terms of quality, quantity, delivery times, and efficiency leading to falling in price in the importing countries and a reduction in the profit margins of intermediary agents and producers. As sourcing by chain stores become global, the potential for targeted marketing strategy to replace mass marketing with the aid of information technology based on better demand information and to share the communication will increase. The quality of tropical produce in the modern retail chains remains mixed and differs widely from poor to excellent.

Increased Share in Trade of Domestic Market

The retail sector represents about 16% of Thailand's Gross Domestic Product (GDP), and modern trade retailers, concentrating in Bangkok and suburban areas, account for more than 54% of total retail sales. Supermarkets and modern retail chains are gaining market share of fresh produce from traditional wet market distribution systems such as wholesalers, street markets and small stores in many countries' domestic market.

These new stores have emerged to challenge the traditional practice and established status quo of the existing central commodity-based wholesalers. Nowadays, almost all supermarkets place their produce department prominently in the front of the store. The large size of supermarkets permits greater produce diversity, better quality, and convenience required by the time-pressed consumers. While fruit is still sold predominantly whole, fresh vegetables for supermarkets are sold in branded small consumer packs.

An increasing attraction is the offering of more imports, more new varieties, branded products, speciality leafy greens, sprouts, the ready-to-eat hot meal to eat at home and herbal/health foods and other food services including bakery, drinks, and dairy products. Innovative marketing strategy, technology, and R&D will play an increasing important role.

Future Focus of Post Harvest Technology

The traditional goals of Post Harvest technology are to improve handling system with emphasis on Post Harvest treatments including storage, packaging and transportation to maintain quality and reduce losses of products between the field and the ultimate destination. The Post Harvest technology has increasingly become vital in the supply chain management. The medium and long-term challenges are that Post Harvest technology should be more viable to enhance the value chain and to cope with the changing market environments. Following are the new roles of Post Harvest technology:

- To meet the demand-driven legality, safety, and quality requirements;
- To provide mechanisms for combating non-tariff trade barriers and market access;
- Post Harvest technology and quality management system integration in supply chain management;
- Shift to appropriate technologies to serve the modern trade stores and global sourcing by multinational retail chain;
- Targeted/focused technologies to meet the challenge of emerging new supplies and the requirement of the new markets.

2

Technology of Fruits and Vegetables

Proper Post Harvest processing and handling is an important part of modern agricultural production. Post Harvest processes include the integrated functions of harvesting, cleaning, grading, cooling, storing, packaging, transporting and marketing. The technology of Post Harvest handling bridges the gap between the producer and the consumer—a gap often of time and distance. Post Harvest handling involves the practical application of engineering principles and knowledge of fruit and vegetable physiology to solve problems.

Utilizing improved Post Harvest practices often results in reduced food losses, improved overall quality and food safety, and higher profits for growers and marketers. It is estimated that 9 to 16 percent of the product is lost due to Post Harvest problems during shipment and handling. Mechanical injury is a major cause of losses. Many of these injuries cannot be seen at the time that the product is packed and shipped, such as internal bruising in tomatoes. Other sources of loss include over-ripening, senescence, the growth of pathogens and the development of latent mechanical injuries.

Many factors contribute to Post Harvest losses in fresh fruits and vegetables. These include environmental conditions such as heat or drouth, mechanical damage during harvesting and handling, improper Post Harvest sanitation, and poor cooling and environmental control. Efforts to control these factors are often very successful in reducing the incidence of disease. For example, reducing mechanical damage during grading and packing greatly decreases the likelihood of Post Harvest disease because many disease-causing organisms (pathogens) must enter through wounds. Chemicals have been widely used to reduce the incidence of Post Harvest disease. Although effective, many of these materials have been removed from the market in recent years because of economic, environmental, or health concerns.

Increased interest in the proper Post Harvest handling of fresh fruits and vegetables has prompted the widespread use of flumes, water dump tanks, spray washers, and hydrocoolers. To conserve water and energy, most Post Harvest processes that wet the produce recirculate the water after it has passed over the produce. This recirculated water picks up dirt, trash, and disease-causing organisms. If steps are not taken to prevent their spread, these organisms can infect all the produce that is subsequently processed. In the past, various fungicides and bactericides have been used (alone or in combination with chlorination) to prevent the transmission of diseases. These materials have often been favoured over chlorination because they provide some residual protection after treatment.

Chlorination

At present, chlorination is one of the few chemical options available to help manage Post Harvest diseases. When used in connection with other proper Post Harvest handling practices, chlorination is effective and relatively inexpensive. It poses little threat to health or the environment The primary objective was to reduce Post Harvest losses of fruits and vegetables during harvesting, farm handling, storage, marketing, transportation and processing.

The Impact of Post Harvest Losses

Too much of the world's food harvest is lost to spoilage and infestations on its journey to the consumer. In developing countries, where tropical weather and poorly developed infrastructure contribute to the problem, losses are sometimes of staggering proportions. Losses occur in all operations from harvesting through handling, storage, processing and marketing. They vary according to the influence of factors such as the perishability of the commodity; ambient temperature and relative humidity which determine the natural course of decay; fungal and bacterial decay; damage by pests -insects, rodents and birds; the length of time between harvesting and consumption; and practices of Post Harvest handling, storage and processing.

Post Harvest Disorders or Losses

Post Harvest disorders or losses in quality have economic impacts vastly greater than the actual losses caused by frequency and intensity of their occurrence. For example there are direct financial losses incurred by the grower from batches of fruit expressing the disorder. Direct losses Horticultural produce is alive and has to stay alive long after harvest. Like other living material it uses up oxygen and gives out carbon dioxide. It also means that it has to receive intensive care. For a plant, harvesting is a kind of amputation. In the field it is connected to roots that give it water and leaves which provide it with the food energy it needs to live. Once harvested and separated from its sources of water and nourishment it must inevitably die. The role of Post Harvest handling is to delay that death for as long as possible.

Effect of Temperature on Post Harvest

The ideal temperature often depends on the geographic origin of the product. Tropical plants have evolved in warmer climates and therefore cannot tolerate low temperatures during storage. Plants from tropical origins must be stored above 12°C. This is in contrast to plants which have evolved in temperate, cooler climates which can be stored at 0°C. Table gives a few examples of products and their recommended storage temperature.

Recommended Storage Temperature a Selection of Fruits and Vegetables

1 - 4 °C	*5 - 9 °C*	*> 10 °C*
Apple	Avocado (temperate origin)	Avocado (sub-tropical)
Asparagus	Zucchini	Pawpaw
Berry fruits	French Bean	Grape fruit and Lemon

Contd...

1 - 4 °C	*5 - 9 °C*	*> 10 °C*
Broccoli	Passion fruit	Mangoes
Peach and Plum	Egg plant	Banana
Cherry	Capsicum	Pineapple
Grapes	Cucumber	Sweet potato
Lettuce	Mandarin orange	Tomato
Mushroom	Potato	Pumpkin
Carnation	Protea	Ginger

The rapid cooling after harvest is so important. If the temperature is lowered and the harvested products are put in refrigerated storage, water and quality loss can be reduced. Fresh produce is alive, living and breathing. The general term for all the processes going on inside a living organism is called metabolism. Temperature has a big effect on the rate of metabolism of the product. When the temperature of the product rises, so too does the rate of metabolism.

One of the main processes of metabolism is respiration which is the process of Temperature management for fresh produce is the key to quality. Lowering the temperature as quickly as possible after harvest will slow the rate of metabolism and therefore extend the product's shelf life. For some flowers, sugar can be added (always with a biocide) to the vase solution to provide carbohydrate, or food for metabolism. As a result, for these types of produce respiration is not limited and the vase life is extended. At extreme of temperature product are damaged. Some suffer chilling injury, some suffer damage at very high temperatures and all products are damaged if they freeze.

Temperature is the most important influence on the rate of deterioration in the quality of produce. High temperatures accelerate ripening and the speed at which rots develop. A 10°C increase in temperature will cause fruit and vegetables to deteriorate twice as fast, as well as encouraging disease organisms to grow twice as fast as well. This is why it is important to remove field heat from the produce as quickly as possible after harvest. Important as it regulates the quantity of refrigerant entering the evaporator coils. When the liquid moves into the evaporator coils the liquid changes into a gas because it absorbs heat energy from the surrounding air (latent heat of vaporisation). As a result the air in the cool room is cooled.

Post Harvest Losses

Most losses of fresh produce occur between leaving the farm and reaching the consumer. Losses during this period have been estimated to be about 20% of the total crop. These losses may be caused by complete wastage of the product or by lower prices due to a reduction in quality. The cost of these losses is also important as the value of the product increases several fold from the farm gate to the final consumer, so in dollar terms Post Harvest losses are even more significant.

There are generally three main causes of Post Harvest losses:

1. Disease caused by fungi and/or bacteria
2. Physical injuries due to insects, mechanical force, chemicals, heat or freezing
3. Non-disease disorders resulting from storage conditions that upset normal metabolism

when the product is rejected further down the marketing chain. Other factors such as cultivar, weather and crop management also influence the development of disease. Some cultivars are more susceptible to diseases than others and wet weather can promote the development of disease and hamper efforts to control outbreaks.

Essential Oils for the Control of Post Harvest

These plant extracts are generally assumed to be more acceptable and less hazardous than synthetic compounds. This means that essential oils that are registered food grade materials, could be used as alternative anti-fungal and anti-bacterial treatments for fresh produce. The potential for these types of plant extracts is considerable. It is a resource that has not been fully explored.

Post Harvest Laboratory looking at the use of essential oils for the control of Post Harvest pathogens. Essential oils are made up of many different volatile compounds and the make up of the oil quite often varies between species. It seems that the anti-fungal and anti-microbial effects are the result of many compounds acting synergistically.

These means that the individual components by themselves are not as effective. Quite a lot of preliminary work has been done to demonstrate the potential of essential oils for use against Post Harvest pathogens. One work showed that essential oils from red thyme (Thymus zygis), clove buds (Eugenia caryophyllata) and cinnamon leaf (Cinnamomum zeylanicum) prevented the growth of Botrytis cinerea. Other researchers have shown that the essential oil of Monarda citrodora and Melaleuca alternifolia also exhibit antifungal activity against a wide range of common Post Harvest pathogens.

We have looked at the effect of tea tree oil as a vapour on the growth of Botrytis from grapes Tea tree oil has antibacterial and antifungal properties that have secured it a place in the commercial pharmaceutical market. We wanted to determine if these properties may also be useful for the Post Harvest control of fungi on grapes. Our work showed that concentrations of between 100 and 500 ppm were able to prevent the growth of this fungi when it was grown in the laboratory.

Role of Ethylene in the Post Harvest Life

Ethylene plays a role in the Post Harvest life of many horticultural crops. The important role of ethylene as a plant growth regulator has only been established over the last 50 years but its effects have been known for centuries. The most commonly know use of ethylene is to trigger ripening is some crops, such as bananas and avocados.

The application of ethylene at a controlled rate means that these products can be presented to the customer as "ready to eat". The concentration of ethylene required for the ripening of different products varies. The concentration applied is within the range of 1 and 100 ppm. The time and temperature of treatment also influences the rate of ripening with fruit being ripened at temperatures between 15 to 21°C and relative humidity of 85 - 90 %. Although controlled ripening is the major

Post Harvest use of ethylene it can also be applied pre-harvest to promote Post Harvest benefits. The chemical Ethephon produces ethylene and is applied in the field. Ethephon can promote several benefits such as fruit thinning (apples, cherries), fruit loosening prior to harvest (nuts), colour development (apples), degreening (citrus), flower induction (pineapples).

Other Factors of Post Harvest Losses

In addition to genetic traits, environmental factors such as soil type, temperature, wind during fruit set, frost, and rainy weather at harvest can have adverse effects on storage life, suitability for shipping, and quality. Cultural practices may have dramatic impacts on Post Harvest quality. Good Agricultural Practices during harvest operations and any subsequent Post Harvest handling, minimal or fresh-cut processing, and distribution to consumers must be developed.

Common Post Harvest Diseases

*Commodity and Disease - Pathogena**

Apples

Blue mold Penicillium expansum (f)

Gray mold Botrytis cinerea (f)

Black rot Physalospora obtusa (f)

Bitter rot Glomerella cingulata (f)

Grapes and small fruit

Blue mold Penicillium sp. (f)

Gray mold Botrytis cinerea (f)

Rhizopus rot Rhizopus stolonifer (f)

Potatoes

Fusarium tuber rot Fusarium spp. (f)

Wet rot Pythium sp. (f)

Bacterial soft rot Erwinia spp. (b)

Slimy soft rot Clostridium spp. (b)

Peaches and plums

Brown rot Monilinia fructicola (f)

Rhizopus rot Rhizopus stolonifer (f)

Gray mold Botrytis cinerea (f)

Blue mold Penicillium sp. (f)

Alternaria rot Alternaria sp. (f)

Gilbertella rot Gilbertella persicaria (f)

Sweet Potatoes

Bacterial soft rot Erwinia chrysanthemi (b)

Black rot Ceratocystis fimbriata (f)

Ring rot Pythium spp. (f)

Java black rot Diplodia gossypina (f)

Fusarium surface rot Fusarium oxysporum (f)

Fusarium root and stem rot Fusarium solani (f)

Rhizopus soft rot Rhizopus nigricans (f)

Charcoal rot Marcrophomina sp. (f)

Tomatoes and Peppers

Alternaria rot Alternaria alternata (f)

Buckeye rot Phytophthora sp. (f)

Gray mold Botrytis cinerea (f)

Soft rot Rhizopus stolonifer (f)

Sour rot Geotrichum candidum (f)

Bacterial soft rot Erwinia spp. (b) or

Pseudomonas spp. (b)

Ripe rot Colletotrichum sp. (b)

Vegetables in General

Watery soft rot Sclerotinia sp. (f)

Cottony leak Pythium butleri (f)

Fusarium rot Fusarium sp. (f)

Bacterial soft rot Erwinia sp. (b) or

Pseudomonas spp. (b)

- f = fungus
- b = bacterium

Post Harvest Disorders

Many types of Post Harvest disorders and infectious diseases affect fresh fruits and vegetables (Table). Disorders are the results of stresses related to excessive heat, cold, or improper mixtures of environmental gases such as oxygen, carbon dioxide, and ethylene. Some disorders may be caused by mechanical damage, but all are abiotic in origin (not caused by disease organisms) and cannot be controlled by chlorination or most other Post Harvest chemicals. However, abiotic disorders often weaken the natural defences of fresh produce, making it more susceptible to biotic diseases those that are caused by disease organisms. Further, in many cases injuries caused by chilling, bruising, sunburn, senescence, poor nutrition, and other factors can mimic biotic diseases.

The control of biotic Post Harvest diseases depends on understanding the nature of disease organisms, the conditions that promote their occurrence, and the factors that affect their capacity to cause losses. Post Harvest diseases may be caused by either fungi or bacteria, although fungi are more common than bacteria in both fruits and vegetables. Post Harvest diseases caused by bacteria are rare in fruits and berries but somewhat more common in vegetables. Viruses seldom cause Post Harvest diseases, although they, like Post Harvest disorders, may weaken the produce.

Chilling Injury

Tropical fruits are susceptible to many Post Harvest diseases. They are sensitive to chilling injury, and cannot be kept at low temperatures. It is very important to harvest tropical fruits at the right stage of maturity. Treatments with sulphur dioxide, fungicides and antioxidants help to reduce losses of tropical fruits from Post Harvest diseases, as does careful attention to humidity and temperature. Most important is careful handling to minimize injuries. Bruised fruit, or fruit with a damaged skin, is more vulnerable to diseases, spoils more quickly and sells more slowly.

Development of the Post Harvest sector commonly requires inputs such as pesticides, cement, galvanized sheeting, wire netting, adequate packaging for the safe handling and transportation of perishables, sprays, boxes, and simple machines. Practical technical interventions are researched and promoted to prevent losses. For example, storage bins for grain must be cleaned out completely between seasons and disinfected before refuse; shade must be provided for holding perishables together with appropriate containers for their transportation and marketing. Improved technologies for drying fruits, vegetables and root crops have also been introduced to reduce losses arising from seasonal gluts.

Biological Control

Biological control of Post Harvest diseases (BCPD) has emerged as an effective alternative. Because wound-invading necrotrophic pathogens are vulnerable to bio-control, antagonists can be applied directly to the targeted area (fruit wounds), and a single application using existing delivery systems (drenches, line sprayers, on-line dips) can significantly reduce fruit decays. The pioneering bio-control products Bio-save and Aspire were registered by EPA in 1995 for control of Post Harvest rots of pome and citrus fruit, respectively, and are commercially available. The limitations of these bio-control products can be addressed by enhancing bio-control through manipulation of the environment, using mixtures of beneficial organisms, physiological and genetic enhancement of the bio-control mechanisms, manipulation of formulations, and integration of bio-control with other alternative methods that alone do not provide adequate protection but in combination with bio-control provide additive or synergistic effects.

Another critical factor for reducing food losses is quality standards and incentives for delivery of better quality produce through the introduction of a fair and practical grading system. The operation of such a system often calls for training and extension to improve handling, storage, packing, sorting and grading practices.

Irradiation

In both domestic and international agricultural markets, expanding the use of irradiation can help to reduce the need for methyl bromide for the Post Harvest control of insect pests. Currently, irradiation treatments have been approved for a variety of food use applications by the U.S. Food and Drug Administration (FDA).

The United States Department of Agriculture (USDA) Animal and Plant Health Inspection Service (APHIS) Plant Protection and Quarantine Service (PPQ) has outlined policy positions regarding the development and use of irradiation treatments for quarantine pest control, and is actively seeking ways to incorporate additional irradiation uses into their plant protection programme (USDA 1995, USDA 1996a, 1996b, 1996c).

Food irradiation is a process by which products are exposed to ionizing radiation to sterilize or kill insects and microbial pests by damaging their DNA. The FDA permits three types of ionizing radiation to be used on foods: gamma rays from radioactive cobalt-60 and cesium-137, high energy electrons, and x-rays. Although all three have similar effects, gamma rays are most commonly used in food irradiation because of their ability to deeply penetrate pallet loads of food (Forsythe and Evangelou 1993, Morrison 1989).

Gamma irradiation equipment irradiates packaged or bulk commodities by exposing the product to gamma energy from cobalt-60 in closed chambers, which range in size from single modular pallet irradiators to large research or contract irradiation facilities. Absorbed dose is measured as the quantity of radiation imparted per unit of mass of a specified material. The unit of absorbed dose is the gray (Gy) where 1 gray is equivalent to 1 joule per kilogram (ICGFI 1991, NAPPO 1996).

In addition, by interfering with cell division, irradiation inhibits sprouting in tubers, bulbs, and root vegetables (potatoes, onions) and can delay ripening of some tropical fruits, resulting in an extended shelf life for many foods.

In turn, longer shelf lives will enhance trade opportunities between nations by extending time constraints under which fresh produce must be delivered to more distant geographic markets or by allowing the use of slower and less expensive modes of transportation (Kader 1986, Moy 1991, OTA 1985).

Irradiation is used as a pest control tool in over 40 countries, including the United States, Russia, Great Britian and Brazil (Nordion 1995).

The disinfestation of grain as it enters the Soviet Union at the Black Sea Port of Odessa, estimated at over 500,000 metric tons per year, is one of the largest documented commercial industrial applications (Giddings 1991). In the United States, the FDA approved low-doses irradiation for wheat, wheat flour, and potatoes in the early 1960s.

In 1984 and 1985, the FDA approved irradiation of spices and pork, and in the following year, approved low-dose irradiation (up to 1 kGy) to control insects in foods and extend the shelf life of fresh fruits and vegetables (Kader 1986, Morrison 1989). Irradiation has also been used to sterilize food for U.S. hospital patients and astronauts (Morrison 1992). Further, irradiation disinfestation has been found to be effective for treatment of dried fruits, spices, nuts, cut flowers, lumber, and wood chips (ICGFI 1994, Marcotte 1992, Morrison 1989, OTA 1985).

At doses below 1 kGy, irradiation is an effective treatment against various species of fruit flies, mango seed weevils, naval orange worms, potato tuber moths, coddling moths, and other insect species of significance to quarantine situations (Kader 1986). For irradiation to be approved as a quarantine treatment in the United States, either as a single treatment, or as part of a combined approach (*e.g.*, systems approach), USDA/APHIS/PPQ will require that the level of efficacy be scientifically demonstrated, and that efficacy be demonstrated under commercial settings (USDA 1996).

Factors influencing the response of fresh fruits and vegetables to irradiation include the type of commodity and cultivar, production area and season, maturity at harvest, initial quality, and Post Harvest handling procedures.

Similarly, environmental conditions during irradiation (temperature and atmospheric composition), and dose rates are also influencing factors (ICGFI 1994, Kader 1986, OTA 1985, Morrison 1992).

Relative Tolerance of Fresh Fruits and Vegetables to Irradiation below 1 kGy

High	Apple, cherry, date, guava, longan, muskmelon, nectarine, papaya, peach, rumbutan, raspberry, strawberry, tamarillo, tomato
Medium	Apricot, banana, cherimoya, fig, grapefruit, kumquat, loquat, lychee, orange, passion fruit, pear, pineapple, plum, tangelo, tangerine
Low	Avocado, cucumber, grape, green bean, lemon, lime, olive, pepper, sapodilla, soursop, summer squash, leafy vegetables, broccoli, cauliflower

Source: Kader 1986.

Comparison of Estimated Post Harvest Treatment Costs for Selected Crops

Crop	*Methyl Bromide (cents per pound)*	*Irradiation (cents per pound)*
Strawberries	0.88 to 0.94	2.5 to 8.1
Papaya	0.88 to 0.94	0.9 to 4.2
Mango	0.88 to 0.94	Data not available

Sources: Forsythe and Evalgelou 1993 and 1994, Morrison 1989.

Fumigation with Methyl Bromide

The fumigant methyl bromide is generally preferred for the disinfection of both durable and perishable agricultural commodities. This is because of its low phytotoxicity and its wide-ranging action against a large number of pests. Furthermore, large, bulky shipments can be rapidly and easily treated. However, methyl bromide has now been identified as one of the chemicals that are damaging the world's ozone layer. Efforts are being made to develop alternatives, and treatment systems which recycle methyl bromide.

VHT (Vapour heat treatment)

Fruitfly is a major problem in the import and export of tropical fruits and vegetables. VHT is an effective, non-chemical method of treatment, widely used to treat mangoes and other tropical fruits grown for export. VHT treatment maintains produce at a certain temperature for a fixed period of time, using a hot water/steam vapour system. This destroys any larvae of the fruitfly which might be present inside the fruit, as well as the eggs and pupae.

Packaging: The packaging of fruits and vegetables should protect them from injury and water loss, and be convenient for handling and marketing. Packages should also provide information about the product, including the grade, handling instructions, and appropriate storage temperatures when the product is on display. The cost of the packaging is important, including whether the container can be recycled or reused.

Transport: Fresh horticultural products should be cooled after harvest and during transport. It is very important that the cold chain is continuous. Once fruit and vegetables have been cooled, they must stay cool. Trucks for road transport may be refrigerated, or may sometimes just be insulated. It is difficult to control the temperature of air shipments, but produce shipped by air should be covered and pre-cooled.

The transportation and storage of fresh fruit and vegetables is an international operation for which the available technology must be used to ensure that produce reaches the consumer in the best possible condition. The use of controlled atmospheric conditions, as a way of reducing the use of chemical preservatives and pesticides, has great potential for the reduction of Post Harvest losses and the maintenance of nutritive value and organoleptic characteristics.

The proper application of controlled atmosphere storage is likely to have as great an impact as the introduction of refrigeration technology a century earlier, yet its potential is only just becoming appreciated, despite its use for apples for many years.

Quality Considerations

Several factors are important in determining the quality of fruits and vegetables: the appearance, the flavour, the texture, the nutritional value and the safety. Only the first three can be easily identified by consumers.

Successful Post Harvest handling depends partly on the initial quality of the crop at harvest, including the degree of maturity. It also depends on careful handling to minimize mechanical damage, proper management of the environmental conditions, and good sanitation.

Most often, Post Harvest losses are a symptom rather than the problem. Knowledge of their cause is, therefore, essential for deciding measures to prevent them. Such measures may have to be taken by the small farmer, the private trader, a cooperative, the marketing board or other operator, handlers and transporters, wholesale and retail markets, etc.

Future Post Harvest research priorities

- Genetic manipulation for long life, diseases and environmental-stress resistant cultivars.
- Modelling cultivating conditions for high quality and long life (avoiding root stress by heat or drought).
- Environmental friendly pest control.
- Objective determination of suitable harvesting date.
- Post Harvest treatments (heat, UV, irradiation CO_2, chemicals) for storage ability.
- Monitoring refrigerating systems
- Optimum storage conditions as storage for tropical fruits, ornamentals, planting material, fresh pack and lightly processed produce.
- Possibilities of modified atmosphere packaging (MAP), absorption layers, inserts sensors, PP films with micropores
- Prevention of possible pathogenic organisms in MAP products.
- Adaptive control of storage conditions with biological sensors.
- Storage during transport and as quarantine measures.
- Humid forced air pre-cooling.
- Minimum impact and vibration norms for bruising during sorting and packaging.
- Objective non destructive measuring of quality and maturity

- Environmental friendly packaging.
- Consumer and traders quality preferences in each country.
- Cost and return of the investment of Post Harvest techny.
- Cost and return of the investment of Post Harvest technologies.
- Fundamental research on senescence, ripening, respiration, ethylene effect, chilling, fermentation, superficial browning.

Major Research Programmes:

- Development of technology for packaging and storage for reduction in Post Harvest losses of fruits and vegetables.
- Development of technology for value added products from fruits, vegetables and food grains.
- Development of machines/equipment such as groundnut pod grader, evaporative cooler, colour sorter, lac scrapper and fruit grader.
- Design development and evaluation of structures for covered crop cultivation of high value crops for varied agro-climatic zones.
- Survey of agro industries, to evaluate the technological status and associated problems.
- Studies on Post Harvest management of fruits and vegetables.
- Adaptive trials on agro-processing centres.

Research in horticulture focuses in the areas of:

- Post Harvest Physiology of horticultural produce
- Impact of pre-harvest and Post Harvest factors on shelf life and Post Harvest quality
- Physiology of fruit ripening
- Ethylene management during ripening and storage of horticultural crops
- Regulation of fruit colour and aroma
- Controlled atmosphere (CA) storage of fruits
- Modified atmosphere packaging of fruits
- Production Technology of fruit crops

- Chemical control of suckers in olives
- Mechanism of Fruitlet Abscission and Fruit Ripening of Mango- with a Particular Reference to Polyamines.
- Regulation of Fruit colour development in apple.
- Genetic transformation of fruit and other horticultural crops

Conclusion

Proper evaluation of Post Harvest technologies includes technical, economic and social components. It involves beneficiary participation throughout. Decisions to adopt new or improved technologies are made by individual farmers but their decisions are often strongly influenced by incentives and credit schemes, access to reliable sources of inputs, extension advice, training opportunities, and market information all of which are generally the responsibility of national governments or their agents.

3

Existing Post Harvest Technologies

Post Harvest technologies which greatly influence the level of Post Harvest losses and the quality of produce include grading, packaging, Pre-cooling, storage and transportation. Some products also require one or more of the following treatments: trimming, cleaning, curing, disease or insect control, waxing, and ripening. A general description of those technologies adopted in the United States is available in Kader's (1992) book. More detailed but less up-to-date information is also seen in books by Ryall and Lipton (1979), Ryall and Pentzer (1982), and Pantastico (1975). Storage requirements for a wide variety of products are summarized by Hardenburg et al. (1986).

Special requirements for each major product are often published in monographs, such as that for citrus fruits by Wardowski et al. (1986), and for banana by Stover and Simmonds (1987). Kader (1992) also provides long lists of scientific journals, journals with reviews or abstracts and other periodicals which contain up-to-date information of scientific findings and new technologies related to Post Harvest physiology and handling. This Bulletin is not intended to make a comprehensive review of literature, but will give a brief summary of current technology in common use.

Grading

Essentially all fruits and vegetables sold in modern markets are graded and sized. Sophisticated marketing systems require precise grading standards for each kind of product. More primitive markets may not use written grade standards, but the products are sorted and sized to some extent. Typical grading facilities in large packinghouses include dumpers and conveyors.

Produce is graded by human eyes and hands, while moving along conveyor belts or rollers. "Electric eyes" are sometimes used to sort produce by colour. In small-scale packing operations, one or a few grading tables may be enough. Dumping, conveying and grasping can cause mechanical injury to some products. Equipment should have a smooth, soft surface, and dumping and grading operations should be gentle, to minimize injuries.

Many products are sized according to their weight. Automated weight sizers of various capacity are used in packinghouses. Round or nearly round fruits are often sized according to their diameter, using automated chain or roller sizers or hand carried ring sizers. An inefficient sizing operation can also cause significant injuries.

Packaging

There are two very different types of packaging. The first is when produce is packed in containers for transportation and wholesale. The second is when produce is packed into small retail units.

Ideal containers for packing fruits and vegetables should have the following attributes. They are easy to handle, they provide good protection from mechanical damage, they have adequate ventilation, and they are convenient for merchandising (*i.e.* they can easily carry printed information and advertising about the product etc.). They should also be inexpensive, and easily degradable or recyclable. Many kinds of containers have been used, but the "ideal" is yet to be found. Users often put economic considerations first in selecting containers. Fancy containers such as fibreboard boxes, or wooden or plastic crates, are often used for high-value products. Inexpensive containers such as bamboo baskets or nylon net sacs are used for low-priced produce.

Methods of packing can affect the stability of products in the container during shipping, and influence how much the container protects their quality. In fibreboard boxes, for example, delicate and high-priced products are often packed in trays, while other products are simply all put in the box together.

Prepackaging or consumer packaging generally provides additional protection for the products. It is also convenient for retailers as well as customers, and therefore adds value to produce. However, overuse of non-biodegradable plastic trays and wrapping materials, as often seen in modern supermarkets, creates an extra burden of waste disposal and damages the environment.

Pre-cooling

Good temperature management is the most effective way to reduce Post Harvest losses and preserve the quality of fruits and vegetables. Desirable storage and transportation temperatures for major fruits and vegetables have been identified and published (*e.g.* Kader 1992, Hardenburg et al. 1986, McGregor 1987). Temperatives which are low, but not low enough to cause chilling-injury, slow down physiological activity and hence the rates of senescence of the products. Low temperatures also reduce microbial growth rates and the rate of decay.

Products harvested from hot fields often carry field heat and have high rates of respiration. Rapid removal of field heat by Pre-cooling is so effective in quality preservation that this procedure is widely used for highly perishable fruits and vegetables. Currently used Pre-cooling methods include room cooling, forced-air cooling, watercooling, vacuum cooling, and package icing. Cooling equipment and technologies for large-scale operations in the United States are well described in Kader's (1992) book. Smaller capacity equipment can be made using the same principles. Liu (1991) has made a brief summary of cooling principles and methods.

Room cooling is a relatively simple method which needs only a refrigerated room with adequate cooling capacity. The produce is packed in containers which are loosely stacked in the cooling room, leaving enough space between containers for each one to be exposed to circulating cold air. The rate of cooling is rather slow compared to other methods of cooling, because the heat inside each container needs to be transferred to the surface of the container by means of conduction before being carried away by the refrigerated air. It may take hours or even days to cool a product, depending on what kind of product it is, the size and nature of the container, and the temperature and velocity of the circulating air.

Forced-air cooling is a more rapid way of using air to cool produce. Cold air is forced to flow through the inside of each container, so that it carries away heat directly from the surface of the produce rather than from the surface of the container. The air flow is produced by creating a pressure difference between the two perforated sides of each container. The containers are stacked inside a covered tunnel with an exhaust fan at one end. Highly perishable and high-value products such as grapes, strawberries and raspberries may be cooled in less than an hour using this method.

Water cooling (also known as hydrocooling), is a rapid and less expensive method. Produce is exposed to cold water by means of showering or dipping. The required cooling time is often a matter of minutes. However, not all kinds of products tolerate hydrocooling. Hydrocooled products inevitably have a wet surface, which may encourage decay in some kinds of produce.

Vacuum cooling is the most efficient way to cool leafy vegetables, particularly headed ones such as head lettuce, cabbage and Chinese cabbage. The produce is placed inside a vacuum tube in which air pressure is reduced. When the pressure is lowered to 4.6 mm Hg, water "boils" off at 0°C from all over the leaf surface. The boiling effect draws heat for vaporization, and hence cools the produce. The cooling time is usually in the order of 20 to 30 minutes. Unfortunately, the equipment needed for vacuum cooling is very expensive, and may not be a good choice for small-scale farming systems.

Package-icing or top-icing is the simplest way of cooling. Adding crushed ice, flake-ice or slurry of ice in containers can cool the produce. However, this method is not suitable for produce which is very sensitive to ice-cold temperatures. Cooling by ice also inevitably wets both the produce and container, and generates water which needs to be drained. Fancy equipment is used in large-scale package-icing operations in the United States (Kader 1992), but many vegetable packers in Taiwan simply add ice flakes to each container, either by hand or by shovel.

Storage

Many horticultural crops have a relatively short harvesting season. Storage is needed to extend the marketing period. Various storage methods have been used on a commercial scale (Liu 1991). Air-cooled common storage houses are often used, or underground or cave storage using natural cold air. Storage humidity is sometimes regulated by controlled ventilation and dehumidifiers. Refrigerated storage (cold storage) controls temperature and humidity precisely by mechanical means. Controlled atmosphere (CA) storage controls the concentrations of oxygen and carbon dioxide, in addition to temperature and humidity.

Modified atmosphere (MA) storage also controls oxygen and carbon dioxide concentrations, although not as precisely as in CA, by using semipermeable polymeric films. Ethylene may be scrubbed for products responsive to it, regardless of the storage system used. A good control of temperature, humidity and atmospheric composition maximizes the storage life span of a product.

Air-cooled Common Storage

This is widely used for storing horticultural products, particularly those which have good keeping quality even without a precise low temperature. However, its use is generally limited to cool seasons in temperate and sub-tropical regions, or high altitude areas where there are low ambient temperatures at night.

An ideal storage room is adequately insulated and has a good ventilation control system which pulls cool air in at night and keeps warm air out during the day. Booth and Shaw (1981) and Hallee and Hunter (undated) have given comprehensive descriptions of how to construct and ventilate air-cooled common storage, but this technology has not been adopted in many parts of the world. For instance, hundreds of thousands of tons of citrus fruits are stored in common storage in Taiwan each winter, but ideal common storage facilities are still at their trial stage. Lack of technical knowledge, and the need for a large investment to build new facilities, delay adoption of this new technology.

Large commercial-scale storage of horticultural products in caves and underground cellars are seen only in northern and central China. The major crops stored are apples, pears and citrus fruits. Underground or cave storage often provides better temperature control than above-ground air-cooled storage. This method is most useful in areas where there is a long, cold winter, a dry climate, and a thick layer of fine heavy soil. The construction and management technologies developed in China do not yet seem to have spread to other countries. Some modification of Chinese technologies to suit different climatic and soil conditions might well make this technology useful in many parts of the world.

SA Storage (Controlled Atmosphere)

Commercial application of CA storage is limited to only a few crops, apples and pears being the most popular ones. It is not used for other crops because the benefit is too slight to cover the cost. The technologies involved (Bartsch and Blanpied 1984) are complicated and sophisticated. The cost of building, facilities, and management for CA storage is considerably higher than for refrigerated storage. It should not be recommended for any crop without a thorough cost and benefit analysis.

MA Storage

This is a much simpler method than CA. Theoretically, it is possible for MA storage to generate a result similar to that of CA. However, practical experience seems to show that MA storage is usually less reliable and less effective than CA, in terms of extending storage life and preserving quality. Its biggest advantages are that it is easy to use and not expensive. Therefore, commercial trials and application are becoming more common. Although many research papers have been published on trials using different polymeric films for wrapping or bagging different products, many new experiments are still conducted each year to try and get better results. This method is unlikely to replace CA for crops which are stored in large quantities over long periods, but it is often useful for small-scale and short-period storage of products which benefit from an atmosphere with a low oxygen and high carbon dioxide content.

Transportation

Inland transportation of horticultural crops is usually by rail or way truck. Overseas transportation is by sea or by air. A limited amount of high-valued produce is sometimes transported overland by air. The basic requirements for conditions during transportation are similar to those needed for storage, including proper control of temperature and humidity and adequate ventilation (except in the case of MA). In addition, the produce should be immobilized by proper packaging and stacking, to avoid excessive movement or vibration. Vibration and impact during transportation may cause severe bruising or other types of mechanical injury.

CA containers are not used much for transporting fruit and vegetables. Only a small number of CA containers are used for long-distance shipping of highly perishable, high-value products in the United States and for international trade. Refrigerated containers and trailers are more often used for long-distance shipping, whether by sea, rail or truck. Modern technologies for the refrigerated transportation of horticultural produce have been well described by McGregor (1987).

Shipping by refrigerated trucks is not only convenient, but also effective in preserving the quality of product. However, both the initial investment and the operating costs are very high. Another possibility is insulated or properly ventilated trailer trucks. Pre-cooled products can be transported by well insulated non-refrigerated trucks for up to several hours without any significant rise in product temperature. There are considerable cost savings without any sacrifice of quality if trucks are only insulated, rather than refrigerated, for short-distance shipping. If the product is not Pre-cooled or if the shipping distance is long, a ventilated truck is a better choice than an insulated truck without ventilation and without refrigeration. Ventilation alone does not usually provide a uniform cool temperature, but it may help dissipate excessive field heat and respiration heat, and thus avoid high temperature injury.

Difficulties in Applying Western Technology

Modern Post Harvest technologies developed over the past twenty or thirty years are now widely used in North America and Europe. Japan is the only Asian country which uses technologies of comparable sophistication.

Although there are many Asian Post Harvest horticulturists who have been trained in the United States, Europe, or Japan and returned to their home countries, technical improvement in many Asian countries has been slow. There are several reasons why Western technologies cannot be applied quickly in Asia. Firstly, Western technologies use sophisticated equipment and facilities which are too large for the small-scale farming systems of Asia. There is some small-scale equipment manufactured in Europe or Japan, but it is too expensive for most Asian countries. Secondly, advantages of applying Western technologies are less obvious in traditional marketing systems than in the supermarkets which dominate retail in western countries.

Large capacity coolers, waxers, grading and packing line facilities, etc. manufactured in the United States are designed for use in large packinghouses. Asian farmers not only have small-scale operations, but also grow diverse crops. Their products are either packed by the growers themselves, or in small packingsheds operated by farmers' cooperatives. The volume each grower or cooperative packs each day is too small for large equipment. Although smaller equipment is often made in European countries and Japan, it is always sold at high prices.

The manufacturers have good reason to ask high prices, because the number of units sold is so small. Reputable manufacturers of farm machinery are reluctant to compromise the quality of their products for the sake of lower prices, so they make only sophisticated, expensive equipment. Since user countries cannot even afford to buy the first unit for trial, it is not possible for them to order many units at once for a bargain price.

Modern supermarkets have refrigerated display cabinets and temporary cold storage. They can keep produce cool continuously, provided it has been properly cooled before arrival. In contrast, traditional markets do not usually have refrigeration. They make every effort to sell produce quickly to reduce display time and avoid quality loss. Even if the produce has been cooled before and during shipping, the "cold chain" breaks at the retail markets and hence reduces the benefit of cooling. Even worse, many consumers mistakenly believe that cold produce has been stored, while warm produce means it is "fresh". This misconception further discourages cooling, which is the most effective way to prevent rapid deterioration of the produce.

Modification of High-Technology

If the capacity of a packing shed is small, only small-scale equipment is needed. If a country cannot afford to buy expensive equipment, cheaper equipment should be developed. In order to lower the price, it may be necessary to sacrifice quality in terms of appearance, convenience, mobility, durability and/or user friendliness. Sacrificing one or more of these attributes might be acceptable at the initial stage of development, provided that the equipment still functions well. In order to encourage users to purchase, the equipment must be sold at an affordable price. Therefore, sophisticated equipment and facilities developed in Western countries may need substantial modification to suit local needs in Asian countries.

This is best done in the user countries. The equipment and facilities needed for Post Harvest handling of horticultural crops are relatively simple. The principles involved have been widely published, and are readily available to farm machinery specialists, while few or no patents are involved. The fundamental need for developing modified equipment of this kind is good cooperation among Post Harvest horticulturists, farm machinery specialists and end users, plus sufficient support from government policies. Several Asian countries have made significant progress in this area in recent years. For example, locally made small-scale forced-air coolers, crushed-ice makers, and small refrigerated storage facilities are often seen at cooperative shipping points in Taiwan today.

Integration: The Key to Success

A typical route for a horticultural product to follow from harvest to arrival in the consumer's hands includes grading, packaging, transportation, wholesaling and retailing. Additional processes which might be added en route are trimming, curing, Pre-cooling, storage, waxing, disease and insect control treatments, and prepackaging. Improper procedures at any step en route may ruin all the efforts made during other steps to preserve the quality of the product. In other words, the final quality of the product is often determined by the worst procedure, which acts as a limiting factor.

Since no Post Harvest technology can remedy quality which is already damaged, a mistake at one point may not be compensated by extra care in other steps. Therefore, an integrated programme covering all Post Harvest procedures is necessary to guarantee success. For instance, if three steps in the Post Harvest handling system need to be improved, they all have to be improved simultaneously. Improving only one or two steps may not be any improvement at all.

Taiwan had at least two bad experiences in the past, when it made fragmentary effort to improve Post Harvest technology. One was to import two hydrocoolers from the United States in the 1970s for cooling vegetables produced in central Taiwan. The government subsidized more than half the cost of the equipment, but the users (Farmers' Associations) still could not justify the operational costs for the very limited benefits from using the coolers.

There were no suitable transportation and storage facilities to protect hydrocooled products at that time. The products rewarmed rapidly. Pre-cooled and then rewarmed wet vegetables often decay faster than those which have not been cooled at all. The ultimate fate of the hycrocoolers was to be dismantled and sold as waste metal.

The second failure was the subsidizing of small refrigerated trucks to transport fresh vegetables from eastern Taiwan to Taipei in the 1980s. Since no Pre-cooling facilities were available at the shipping point, the warm vegetables remained warm or became warmer inside the "refrigerated" truck, because the cooling capacity of the truck was enough to keep cool vegetables from warming but not enough to precool warm products.

After these costly lessons of failure, Post Harvest workers realized that improvement programmes should be integrated rather than fragmentary. The people who formulate and supervise a programme must have a comprehensive knowledge of the subject. They should understand all procedures involved, and be able to identify all problems and choose remedial strategies.

The first successful programme in Taiwan began with vegetables in the suburbs. Harvested vegetables were trimmed, hydrocooled, prepackaged, stored in refrigerated rooms until needed, and then shipped to supermarkets in Taipei by insulated small trucks, with or without refrigeration. The hydrocoolers, refrigerated storage, and shipping trucks used were all small and were locally made.

Encouraged by the favorable response of growers and consumers, the programme was extended to major vegetable production areas in central and southern Taiwan. As the volume of vegetables handled under the project increased and shipping distance lengthened, larger coolers, refrigeration storage rooms and shipping trucks came into use.

In order to avoid wetting vegetables, forced-air cooling gradually replaced hydrocooling for some kinds of vegetables and fruit. More and more shipping points adopted the new technology, and Pre-cooled products were not only shipped to supermarkets, but also to large institutional users, and sometimes wholesale markets. Post Harvest losses were reduced, and product value increased.

In order to provide shippers, extension staff and marketing personnel with the essential technical knowledge, a number of short training courses were offered. At the beginning of the programme, Government agencies begged users to try the new technology. Nowadays, it is the users who beg government agencies for newer technology and more technical advice.

Research into New Technologies

Existing technologies for cooling, storage, and transportation of horticultural crops are generally adequate at present. New research should aim at improving equipment, facilities and methods, to make them more efficient and less costly. Applied research for solving specific problems is likely to be more fruitful than basic research in developing countries.

Different crops often need different Post Harvest treatments. Post Harvest horticulturists must establish the requirements for each major crop. Even if optimum conditions and best treatments are known, it is often worthwhile to study the crop's response to less than optimum conditions, since the best treatment is not necessarily the most economic one. For instance, a study was made of optimum storage conditions for Taiwan citrus fruits. This was followed by experiments in refrigerated storage, improved common storage, and "traditional" common storage. Cave storage is also going to be studied. When these experiments are completed, we shall be able to make recommendations for growers, depending on their specific location and the duration of the storage.

Since improved handling methods often rely heavily on improvements in equipment and facilities, close cooperation between Post Harvest horticulturists and agricultural engineers is necessary.

An area which is urgently in need of new technology is Post Harvest disease and insect control. Effective fungicides for Post Harvest application have become fewer and fewer, since many previously used fungicides are now banned for fear of health hazards. Control of Post Harvest decay must rely on new, safe chemical treatments and practical physical methods, in addition to improved sanitary conditions. Recent findings, such as the use of acetic acid fumigation, (Moyls et al. 1996) and UV-C light (Wilson et al. 1997) indicate the feasibility of finding new chemical and physical methods.

Many countries wish to export their quality products, but have difficulty in fulfilling the quarantine requirements imposed by importing countries. Current quarantine procedures are either chemical or physical (Paull and Armstrong 1994). Many tropical products cannot tolerate any form of cold treatment. Heat treatment is expensive, and often damages the quality of products. Irradiation is a possibility (Lalaguna 1998 Miller and McDonald 1998), but still needs further development. New and safe chemical treatments are yet to be found.

Coordination and Cooperation

More about Taiwan's Experience

Although our past experiences of making improving Post Harvest handling of horticultural crops in Taiwan have not all been successful, significant progress has been made in recent years. In the early stages, even a small success could be realized only after many trials and frustrations. Projects which began in the 1970s ended in almost total failure, because every programme tackled only part of the Post Harvest system programmes were also technically unsound, according to current knowledge. More projects, were launched in the 1980s and some small progress made, but many projects still failed. Eventually the stage was reached in the late 1980s when there were enough high-caliber Post Harvest scientists, while both industry and government agencies were aware of the importance of Post Harvest handling. Taiwan was then able to carry out more effective programmes.

Recent progress has been based on good cooperation between Post Harvest scientists, government coordinators, extension staff, and decision makers at shipping points and terminal markets. Full coordination and cooperation were made possible after repeated discussions and debates. Once the major funding agencies were convinced that an integrated national programme should replace individual fragmentary small projects, the programme really took off.

This type of integrated programmes quickly generated fruitful results. Each year we were able to see some improvement in the quality of fresh produce at the markets, and growers and shippers began to enjoy the benefits of reduced Post Harvest losses and increased profit margins for higher value-added products.

Several steps were crucial in changing programmes from initial failure to recent success:

- A few high-level Post Harvest scientists were trained to serve as programme leaders, while more than a hundred middle-level technical workers were trained to implement projects.
- Key persons were organized into a strong working group.
- Problems, priorities, and programmes were discussed, to generate a consensus within the group.
- Fragmentary projects were integrated into a national programme.

Post Harvest Handling of Horticulture Produce

Horticultural crops not only provide us with nutritional and healthy foods, but also generate a cash income to growers. Appropriate production practices, careful harvesting and proper packaging, storage and transport all contribute to the good produce quality. Once a crop is harvested it is impossible to improve its quality.

The horticultural crops, because of their high moisture content are inherently more liable to deteriorate especially under tropical conditions. Moreover, they are biologically active and carry out transpiration, respiration, ripening and other biochemical activities, which deteriorate the quality of the produce.

Losses during Post Harvest operations due to improper storage and handling are enormous and can range from 10-40 percent. Post Harvest losses can occur in the field, in packing areas, in storage, during transportation and in the wholesale and retail market. Severe losses occur because of poor facilities, lack of know-how, poor management, market dysfunction or simply the carelessness of farmers. Proper storage conditions, temperature and humidity are needed to lengthen the storage life and maintain quality once the crop has been cooled to the optimum storage temperature.

Causes of Post Harvest Losses

Physiological and Biochemical Aspects

The quality of the harvested fruits and vegetables depend on the condition of growth as well as physiological and biochemical changes they undergo after harvest. Fruits and vegetable cells are still alive after harvest and continue their physiological activity. The Post Harvest quality and storage life of fruits appear to be controlled by the maturity. If the fruits are harvested at a proper stage of maturity the quality of the fruits is excellent. Poor quality and uneven ripening are due to early harvesting and late harvesting which results in extremely poor shelf life.

Respiration plays a very significant role in the Post Harvest life of the fruits. In most of the fruits, the rate of respiration increases rapidly with ripening. The sudden upsurge in respiration is called the 'climacteric rise', which is considered to be the turning point in the life of the fruit. After this the senescence and deterioration of the fruit begin. The fruits such as banana, papaya, mango, guava, jackfruit, fig, sapota, etc. belong to the category of climacteric fruits. While litchi, pineapple, grapes, pomegranate, lemon, orange, lime, etc. belong to the non-climacteric group. To extend the Post Harvest life of the fruits its respiration rate should be reduced as far as possible. Thus an understanding of the factors, which influence the rate of respiration, is indispensable to Post Harvest technologies for manipulating the storage behaviour of fruits.

Mechanical Injury

Owing to their tender texture and high moisture content, fresh fruits and vegetables are very susceptible to mechanical injury. Poor handling, unsuitable containers, improper packaging and transportation can easily cause bruising, cutting, breaking, impact wounding and other forms of injury.

Parasitic Diseases

High Post Harvest losses are caused by the invasion of fungi, bacteria, insects and other organisms. Microorganisms attack fresh produce easily and spread quickly, because the produce does not have much of a natural defence mechanism and has plenty of nutrients and moisture to support microbial growth. Post Harvest decay control is becoming a more difficult task, because the number of pesticides available is falling rapidly as consumer concern for food safety increases.

Post Harvest Technologies

Pre-cooling: Good temperature management is the most effective way to reduce Post Harvest losses and preserve the quality of fruits and vegetables. Products harvested from hot fields often carry field heat and have high rates of respiration. Rapid removal of field heat by Pre-cooling is so effective in quality preservation that this procedure is widely used for highly perishable fruits and vegetables. Currently used Pre-cooling methods include room cooling, forced-air cooling, water cooling, vacuum cooling and package icing

Room cooling is a relatively simple method, which needs only a refrigerated room with adequate cooling capacity. The produce is packed in containers, which are loosely stacked in the cooling room, leaving enough space between containers for each one to be exposed to circulating cold air.

The rate of cooling is rather slow compared to other methods of cooling, because the heat inside each container needs to be transferred to the surface of the container by means of conduction before being carried away by the refrigerated air. It may take hours or even days to cool a product, depending on what kind of product it is, the size and nature of the container, and the temperature and velocity of the circulating air.

Forced-air cooling is a more rapid way of using air to cool produce. Cold air is forced to flow through the inside of each container, so that it carries away heat directly from the surface of the produce rather than from the surface of the container. The airflow is produced by creating a pressure difference between the two perforated sides of each container. The containers are stacked inside a covered tunnel with an exhaust fan at one end. Highly perishable and high-value products such as grapes, strawberries and raspberries may be cooled in less than an hour using this method.

Hydro cooling is a rapid and less expensive method. Produce is exposed to cold water by means of showering or dipping. The required cooling time is often a matter of minutes. However, not all kinds of products tolerate hydro cooling. Hydro cooled products inevitably have a wet surface, which may encourage decay in some kinds of produce.

Vacuum cooling is the most efficient way to cool leafy vegetables, particularly headed ones such as head lettuce, cabbage and Chinese cabbage. The produce is placed inside a vacuum tube in which air pressure is reduced. When the pressure is lowered to 4.6 mm Hg, water "boils" off at 0°C from all over the leaf surface. The boiling effect draws heat for vaporization, and hence cools the produce. The cooling time is usually in the order of 20 to 30 minutes. Unfortunately, the equipment needed for vacuum cooling is very expensive, and may not be a good choice for small-scale farming systems.

Ice bank cooler is a new development in refrigeration with positive ventilation. In this system ice cool air is passed through the boxes containing horticultural produce. This facilitates quicker cooling and large amount of heat is removed in a relatively shorter period. The store maintains a temperature of 0.5 - 0.8°C and relative humidity of 98 percent.

Package icing or top icing is the simplest way of cooling. Adding crushed ice, flake-ice or slurry of ice in containers can cool the produce. However, this method is not suitable for produce, which is very sensitive to ice-cold temperatures. Cooling by ice also inevitably wets both the produce and container, and generates water, which needs to be drained.

Sanitation

Sanitation is of great concern to produce handlers, not only to protect produce against Post Harvest diseases, but also to protect consumers from food borne illnesses. E.coli 157:H_7,

Salmonella, Chryptosporidium, Hepatitis and Cyclospera are among the disease causing organisms that have been transferred via fresh fruits and vegetables. Use of a disinfectant in wash water can help to prevent both Post Harvest diseases and food borne illnesses.

Chlorine in the form of a sodium hypochlorite solution or as a dry powdered calcium hypochlorite can be used in hydro-cooling or wash water as a disinfectant. For the majority of vegetables, chlorine in wash water should be maintained in the range of 75-150 ppm (parts per million).

The antimicrobial form, hypochlorous acid, is mostly available in water with a neutral pH (6.5 to 7.5). Organic growers must use chlorine with caution, as it is classified as a restricted material.

Ozonation is another technology that can be used to sanitize produce. A naturally occurring molecule, ozone is a powerful disinfectant. Fruit and vegetable growers have begun using it in dump tanks as well, where it can be thousands of times more effective than chlorine. Ozone not only kills whatever food borne pathogens might be present, it also destroys microbes responsible for spoilage. A basic system consists of an ozone generator, a monitor to gauge and adjust the levels of ozone being produced and a device to dissolve the ozone gas into the water.

Hydrogen peroxide can also be used as a disinfectant. Concentrations of 0.5% or less are effective for inhibiting development of Post Harvest decay caused by a number of fungi. Hydrogen peroxide has a low toxicity rating and is generally recognized as having little potential for environmental damage.

Presizing and Storage

For many commodities fruits below a certain size are eliminated manually or mechanically by presizing belt. Undersized fruits are diverted for processing. The sorting process eliminates cull, overripe, misshapen and otherwise defective fruit and separates produce by colour, maturity and ripeness classes.

Waxing

Food grade waxes are commonly applied to replace some of the natural waxes removed in the washing and cleaning operations to reduce water loss and to improve appearance. It also provides protection against decay organisms. Waxing may be done after grading and fungicides may be added to the wax. Application of wax and Post Harvest fungicides must be indicated on each container where the refrigerated storage facilities are not available Protective skin coating with wax is one of the methods for increasing the storage life of fresh fruits.

Packaging

Packaging of fresh fruits and vegetables has a great significance in reducing the wastage. Packaging provides protection from physical damage during storage, transportation and marketing. There are variety of packages, packaging materials and inserts available.

There are two types of packaging. The first is when produce is packed in containers for transportation and wholesale. The second is when produce is packed into small retail units. Ideal containers for packing fruits and vegetables should have the following attributes.

They are easy to handle, they provide good protection from mechanical damage, they have adequate ventilation and they are convenient for merchandising. They should also be inexpensive and easily degradable or recyclable. Many kinds of containers have been used but the "ideal" is yet to be found. Users often put economic considerations first is selecting containers.

Fancy containers such as fibreboard boxes or wooden or plastic crates, are often used for high-value products. Inexpensive containers such as bamboo baskets or nylon net sacs are used for low-priced produce. Methods of packaging can affect the stability of products in the container during shipping, and influence how much the container protects their quality. In fibreboard boxes, for example, delicate and high-priced products are often packed in trays, while other products are simply put in the box in groups.

Prepackaging or consumer packaging generally provides additional protection for the products. It is also convenient for retailers as well as customers, and therefore adds value to produce. However, overuse of non-biodegradable plastic trays and wrapping materials, as often seen in modern supermarkets, which creates an extra burden of waste disposal and damages the environment.

Factors Affecting Storage Life

Relative Humidity

Transpiration rates (water loss from produce) are determine by the moisture content of the air, which is usually expressed as relative humidity. At high relative humidity, produce maintains salable weight, appearance, nutritional quality and flavour, while wilting, softening and juiciness are reduced. Leafy vegetables with high surface-to-volume ratios; injured produce and immature fruits and vegetables have higher transpiration rates. High temperatures, low relative humidity and high air velocity increase transpiration rates.

Relative humidity needs to be monitored and controlled in storage. Control can be achieved by a variety of methods:

1. Operating a humidifier in the storage area.
2. Regulating air movement and ventilation in relation to storage room load.
3. Maintaining refrigeration coil temperature within the storage room.
4. Using moisture barriers in the insulation of the storage room or transport vehicle.
5. Wetting the storage room floor.
6. Using crushed ice to pack produce for shipment.
7. Sprinkling leafy vegetables, cool-season root vegetables and immature fruits and vegetables with water.

Temperature

Respiration and metabolic rates are directly related to room temperatures within a given range. The higher the rate of respiration, the faster the produce deteriorates. Lower temperatures reduce respiration rates and the ripening and senescence processes, which prolong the storage life of fruits and vegetables. Low temperatures also slow the growth of pathogenic fungi, which cause spoilage of fruits and vegetables in storage.

Producers should give special care and attention to proper storage conditions for produce with high to extremely high respiration rates, as these crops will deteriorate much more quickly.

It is impossible to make a single recommendation for cool storage of all fruits and vegetables. Climate of the area where the crop originated, the plant part, the season of harvest and crop maturity at harvest are important factors in determining the optimum temperature. A general rule for vegetables is that cool-season crops should be stored at cooler temperatures (0 to 1.7°C) and warm-season crops should be stored at warmer temperatures (7 to 13°C).

Freezing Injury

Temperatures that are too low can be just as damaging as those too high. Freezing will occur in all commodities below 0°C. Whether injury occurs depends on the commodity. Some can be repeatedly frozen and thawed without damage, while others are ruined by one freezing.

Injury from freezing temperatures can appear in plant tissues as loss of rigidity, softening and water soaking. Injury can be reduced if the produce is allowed to warm up slowly to optimum storage temperatures and if it is not handled during the thawing period. Injured produce should be marketed immediately, as freezing shortens its storage life.

Chilling Injury

Fruits and vegetables that require warmer storage temperatures (4.5 to 13°C) can be damaged if they are subjected to near-freezing temperatures (0°C). Cooler temperatures interfere with normal metabolic processes. Injury symptoms are varied and often do not develop until the produce has been returned to warmer temperatures for several days. Besides physical damage, chilled produce is often more susceptible to disease infection.

Ethylene

Ethylene, a natural hormone produced by some fruits as they ripen, promotes additional ripening of produce exposed to it. The old age saying that one bad apple spoils the whole bushel is true. The damaged or diseased fruits produce high levels of ethylene and stimulate the other apples to ripen too quickly. As the fruits ripen, they become more susceptible to disease. Ethylene "producers" should not be stored with fruits, vegetables, or flowers that are sensitive to it. The result could be loss of quality, reduced shelf life and specific symptoms of injury. Ethylene producers include apples, apricots, avocados, ripening bananas, honeydew melons, papayas, peaches, pears, plums and tomatoes.

Storage Facilities

Crops that require different storage conditions will need three different storage facilities.

- Cold Storage (temperatures 0 to 2.2°C)
- Cool Storage (temperatures 4.5 to 13°F)
- Warmer storage (temperatures 13 to 15.6°F)

A recording thermometer can be helpful in determining whether storage facilities are maintaining ideal conditions and are not fluctuating. A maximum/minimum thermometer could be substituted. Relative humidity also should be monitored with a hygrometer.

Controlling and monitoring temperature and relative humidity will enable a grower to maintain optimum storage conditions for maximum storage life of the crop and to minimize crop damage from chilling, freezing or high temperature injuries and water loss from the crop.

Air-Cooled Common Storage

This is widely used for storing horticultural products, particularly those that have good keeping quality even without a precise low temperature. However, its use is generally limited to cool seasons in temperate and sub-tropical regions, or high altitude areas where there are low ambient temperatures at night. An ideal storage room is adequately insulated and has a good ventilation control system, which draws cool air inside during night and keeps warm air out during the day.

Refrigerated Storage

Refrigerated storage is a well-established technology widely used for storing horticultural crops all over the world. Its application is limited only by cost and benefit considerations. Essentially, all crops can benefit by being stored at a suitable low temperature, which extends the storage life and preserves quality.

Many horticultural crops have storage life spans ranging from less than one month to several months when refrigerated. Therefore, refrigerated storage can be used continuously only if different crops with different harvesting seasons can share the facility. There are other important reasons why this method is not used in many tropical and sub-tropical countries, where refrigeration is needed most. The initial investment cost is too high and its energy consumption too large for many countries.

Hydrobaric Storage

In this system the horticultural produce is kept in a vacuum-tight and refrigerated container and the air is evacuated by vacuum pump to achieve desired low pressure. The low pressure retards ripening by decreasing respiration. However this is more expensive method.

Transportation

Inland transportation of horticultural crops is usually by rail or by truck. Overseas transportation is by sea or by air. A limited amount of high-valued produce is sometimes transported overland by air. The basic requirements for conditions during transportation are proper control of temperature and humidity and adequate ventilation. In addition, the produce should be immobilized by proper packaging and stacking, to avoid excessive movement or vibration. Vibration and impact during transportation may cause severe bruising or other types of mechanical injury. Refrigerated containers and trailers are more often used for long distance shipping, whether by sea, rail or truck.

Shipping by refrigerated trucks is not only convenient, but also effective in preserving the quality of product. However, both the initial investment and the operating costs are very high. Another possibility is insulated or properly ventilated trailer trucks. Pre-cooled products can be transported through well-insulated non-refrigerated trucks for up to several hours without any significant rise in product temperature.

There are considerable cost savings without any sacrifice of quality if trucks are only insulated, rather than refrigerated, for short-distance shipping. If the product is not Pre-cooled or if the shipping distance is long, a ventilated truck is a better choice than an insulated truck without ventilation and without refrigeration. Ventilation alone does not usually provide a uniform cool temperature, but it may help dissipate excessive field heat and respiration heat, and thus avoid high temperature injury.

Post Harvest Technology of Horticulture

In developing countries agriculture is the mainstay of the economy. As such, it should be no surprise that agricultural industries and related activities can account for a considerable proportion of their output. Of the various types of activities that can be termed as agriculturally based, fruit and vegetable processing are among the most important.

Both established and planned fruit and vegetable processing projects aim at solving a very clearly identified development problem. This is that due to insufficient demand, weak infrastructure, poor transportation and perishable nature of the crops, the grower sustains substantial losses. During the Post Harvest glut, the loss is considerable and often some of the produce has to be fed to animals or allowed to rot.

Even established fruit and vegetable canning factories or small/medium scale processing centres suffer huge loss due to erratic supplies. The grower may like to sell his produce in the open market directly to the consumer, or the produce may not be of high enough quality to process even though it might be good enough for the table. This means that processing capacities will be seriously underexploited.

The main objective of fruit and vegetable processing is to supply wholesome, safe, nutritious and acceptable food to consumers throughout the year.

Fruit and vegetable processing projects also aim to replace imported products like squash, yams, tomato sauces, pickles, etc., besides earning foreign exchange by exporting finished or semi-processed products.

The fruit and vegetable processing activities have been set up, or have to be established in developing countries for one or other of the following reasons:

- diversification of the economy, in order to reduce present dependence on one export commodity;
- government industrialisation policy;
- reduction of imports and meeting export demands;
- stimulate agricultural production by obtaining marketable products;
- generate both rural and urban employment;
- reduce fruit and vegetable losses;
- improve farmers' nutrition by allowing them to consume their own processed fruit and vegetables during the off-season;
- generate new sources of income for farmers/artisans;
- develop new value-added products.

What Fruit and Vegetables can be Processed?

Practically any fruit and vegetable can be processed, but some important factors which determine whether it is worthwhile are:

a. The demand for a particular fruit or vegetable in the processed form;

b. The quality of the raw material, *i.e.* whether it can withstand processing;

c. Regular supplies of the raw material.

For example, a particular variety of fruit which may be excellent to eat fresh is not necessarily good for processing. Processing requires frequent handling, high temperature and pressure.

Many of the ordinary table varieties of tomatoes, for instance, are not suitable for making paste or other processed products. A particular mango or pineapple may be very tasty eaten fresh, but when it goes to the processing centre it may fail to stand up to the processing requirements due to variations in its quality, size, maturity, variety and so on.

Even when a variety can be processed, it is not suitable unless large and regular supplies are made available. An important processing centre or a factory cannot be planned just to rely on seasonal gluts; although it can take care of the gluts it will not run economically unless regular supplies are guaranteed.

To operate a fruit and vegetable processing centre efficiently it is of utmost importance to pre-organise growth, collection and transport of suitable raw material, either on the nucleus farm basis or using outgrowers.

Processing Planning

The secret of a well planned fruit and vegetable processing centre is that it must be designed to operate for as many months of the year as possible. This means the facilities, the buildings, the material handling and the equipment itself must be inter-linked and coordinated properly to allow as many products as possible to be handled at the same time, and yet the equipment must be versatile enough to be able to handle many products without major alterations.

A typical processing centre or factory should process four or five types of fruits harvested at different times of the year and two or three vegetables. This processing unit must also be capable of handling dried/dehydrated finished products, juices, pickles, tomato juice, ketchup and paste, jams, jellies and marmalades, semi-processed fruit products.

Advanced planning is necessary to process a large range of products in varied weather and temperature conditions, each requiring a special set of manufacturing and packaging formulae. The end result of the efforts should be a well-managed processing unit with lower initial investment.

A unit which is sensibly laid out and where one requirement co-relates to another, with a sound costing analysis, leads to an integrated operation.

Instead of over-sophisticated machinery, a sensible simple processing unit may be required when planned production is not very large and is geared mainly to meet the demand of the domestic market.

Location

The basic objective is to choose the location which minimises the average production cost, including transport and handling.

It is an advantage, all other things being equal, to locate a processing unit near the fresh raw material supply. It is a necessity for proper handling of the perishable raw materials, it allows the processing unit to allow the product to reach its best stage of maturation and lessens injury from handling and deterioration from changes during long transportation after harvesting.

An adequate supply of good water, availability of manpower, proximity to rail or road transport facilities and adequate markets are other important requirements.

Processing Systems

a. *Small-Scale Processing:* This is done by small-scale farmers for personal subsistence or for sale in nearby markets. In this system, processing requires little investment: however, it is time consuming and tedious. Until recently, small-scale processing satisfied the needs of rural and urban populations. However, with the rising rates of population and urbanisation growth and their more diversified food demands, there is need for more processed and diversified types of food.

b. *Intermediate-Scale Processing:* In this scale of processing, a group of small-scale processors pool their resources. This can also be done by individuals. Processing is based on the technology used by small-scale processors with differences in the type and capacity of equipment used. The raw materials are usually grown by the processors themselves or are purchased on contract from other farmers. These operations are usually located on the production site of in order to assure raw materials availability and reduce cost of transport. This system of processing can provide quantities of processed products to urban areas.

c. *Large-Scale Processing:* Processing in this system is highly mechanised and requires a substantial supply of raw materials for economical operation. This system requires a large capital investment and high technical and managerial skills. Because of the high demand for foods in recent years many large-scale factories were established in developing countries. Some succeeded, but the majority failed, especially in West Africa. Most of the failures were related to high labour inputs and relatively high cost, lack of managerial skills, high cost and supply instability of raw materials and changing governmental policies. Perhaps the most important reason for failure was lack of adequate quantity and regularity of raw material supply to factories. Despite the failure of these commercial operations, they should be able to succeed with better planning and management, along with the undertaking of more in-depth feasibility studies.

It can be concluded that all three types of processing systems have a place in developing countries to complement crop production to meet food demand. Historically, however, small and intermediate scale processing proved to be more successful than large-scale processing in developing countries.

Choice of Processing Technologies for Developing Countries

FAO maintains (in FAO, 1992c), that the basis for choosing a processing technology for developing countries ought to be to combine labour, material resources and capital so that not only the type and quantity of goods and services produced are taken into account, but also the distribution of their benefits and the prospects of overall growth. These should include:

a. Increasing farmer/artisan income by the full utilisation of available indigenous raw material and local manufacturing of part or all processing equipment;

b. Cutting production costs by better utilisation of local natural resources (solar energy) and reducing transport costs;

c. Generating and distributing income by decentralising processing activities and involving different beneficiaries in processing activities (investors, newly employed, farmers and small-scale industry);

d. maximising national output by reducing capital expenditure and royalty payments, more effectively developing balance-of-payments deficits through minimising imports (equipment, packing material, additives), and maximising export-oriented production;

e. Maximising availability of consumer goods by maximisation of high-quality, standard processed produce for internal and export markets, reducing Post Harvest losses, giving added value to indigenous crops and increasing the volume and quality of agricultural output.

Knowledge and control of the means of production, local manufacturing of processing equipment and development of appropriate/new technologies and more suitable raw material for processing must all be better researched.

Decentralisation of activities must be maintained and coordinated. The introduction of more sophisticated processing equipment and packaging material must be subordinated to internal and export marketing references.

Choosing a technology solely to maximise profits can actually work against true development. Choice should also be based on a solid, long-term market opportunity to ensure viability.

The internal market should be given greater consideration, safeguarded and supported.

Training courses, at all levels, in processing and preservation of indigenous crops, must be expanded.

Chemical Composition and Nutritional Aspects

General Properties

Fruit and vegetables have many similarities with respect to their compositions, methods of cultivation and harvesting, storage properties and processing. In fact, many vegetables may be considered fruit in the true botanical sense. Botanically, fruits are those portions of the plant which house seeds. Therefore such items as tomatoes, cucumbers, eggplant, peppers, and others would be classified as fruits on this basis.

However, the important distinction between fruit and vegetables has come to be made on an usage basis. Those plant items that are generally eaten with the main course of a meal are considered to be vegetables. Those that are commonly eaten as dessert are considered fruits. That is the distinction made by the food processor, certain marketing laws and the consuming public, and this distinction will be followed in this document.

Vegetables are derived from various parts of plants and it is sometimes useful to associate different vegetables with the parts of the plant they represent since this provides clues to some of the characteristics we may expect in these items.

Fruit as a dessert item, is the mature ovaries of plants with their seeds. The edible portion of most fruit is the fleshy part of the pericarp or vessel surrounding the seeds. Fruit in general is acidic and sugary. They commonly are grouped into several major divisions, depending principally upon botanical structure, chemical composition and climatic requirements.

Berries are fruit which are generally small and quite fragile. Grapes are also physically fragile and grow in clusters. Melons, on the other hand, are large and have a tough outer rind. Drupes (stone fruit) contain single pits and include such items as apricots, cherries, peaches and plums. Pomes contain many pits, and are represented by apples, quince and pears.

Citrus fruit like oranges, grapefruit and lemons are high in citric acid. Tropical and subtropical fruits include bananas, dates, figs, pineapples, mangoes, and others which require warm climates, but exclude the separate group of citrus fruits.

The compositions of representative vegetables and fruits in comparison with a few of the cereal grains are seen in Table.

Typical Percentage Composition of Foods from Plant Origin Percentage Composition- Edible Portion

Food	*Carbo-hydrate*	*Protein*	*Fat*	*Ash*	*Water*
Cereals					
wheat flour, white	73.9	10.5	1.9	1.7	12
rice, milled, white	78.9	6.7	0.7	0.7	13
maize, whole grain	72.9	9.5	4.3	1.3	12
Earth vegetables					
potatoes, white	18.9	2.0	0.1	1.0	78
sweet potatoes	27.3	1.3	0.4	1.0	70
Vegetables					
carrots	9.1	1.1	0.2	1.0	88.6
radishes	4.2	1.1	0.1	0.9	93.7
asparagus	4.1	2.1	0.2	0.7	92.9
beans, snap, green	7.6	2.4	0.2	0.7	89.1
peas, fresh	17.0	6.7	0.4	0.9	75.0
lettuce	2.8	1.3	0.2	0.9	94.8
Fruit					
banana	24.0	1.3	0.4	0.8	73.5
orange	11.3	0.9	0.2	0.5	87.1
apple	15.0	0.3	0.4	0.3	84.0
strawberries	8.3	0.8	0.5	0.5	89.9

Source: *Anon. (196O)*

Compositions of vegetables and fruit not only vary for a given kind in according to botanical variety, cultivation practices, and weather, but change with the degree of maturity prior to harvest, and the condition of ripeness, which is progressive after harvest and is further influenced by storage conditions. Nevertheless, some generalisations can be made.

Most fresh vegetables and fruit are high in water content, low in protein, and low in fat. In these cases water contents will generally be greater than 70% and frequently greater than 85%.

Commonly protein content will not be greater than 3.5% or fat content greater than 0.5 %. Exceptions exist in the case of dates and raisins which are substantially lower in moisture but cannot be considered fresh in the same sense as other fruit. Legumes such as peas and certain beans are higher in protein; a few vegetables such as sweet corn which are slightly higher in fat and avocados which are substantially higher in fat.

Vegetables and fruit are important sources of both digestible and indigestible carbohydrates. The digestible carbohydrates are present largely in the form of sugars and starches while indigestible cellulose provides roughage which is important to normal digestion.

Fruit and vegetables are also important sources of minerals and certain vitamins, especially vitamins A and C. The precursors of vitamin A, including beta-carotene and certain other carotenoids, are to be found particularly in the yellow-orange fruit and vegetables and in the green leafy vegetables.

Citrus fruit are excellent sources of vitamin C, as are green leafy vegetables and tomatoes. Potatoes also provide an important source of vitamin C for the diets of many countries. This is not so much due to the level of vitamin C in potatoes which is not especially high but rather to the large quantities of potatoes consumed.

Chemical Composition

Water: Vegetal cells contain important quantities of water. Water plays a vital role in the evolution and reproduction cycle and in physiological processes. It has effects on the storage period length and on the consumption of tissue reserve substances.

In vegetal cells, water is present in following forms:

- bound water or dilution water which is present in the cell and forms true solutions with mineral or organic substances;
- colloidal bound water which is present in the membrane, cytoplasm and nucleus and acts as a swelling agent for these colloidal structure substances; it is very difficult to remove during drying/dehydration processes;

- constitution water, directly bound on the chemical component molecules and which is also removed with difficulty.

Vegetables contain generally 90-96% water while for fruit normal water content is between 80 and 90%.

Mineral substances: Mineral substances are present as salts of organic or inorganic acids or as complex organic combinations (chlorophyll, lecithin, etc.); they are in many cases dissolved in cellular juice.

Vegetables are more rich in mineral substances as compared with fruits. The mineral substance content is normally between 0.60 and 1.80% and more than 60 elements are present; the major elements are: K, Na, Ca, Mg, Fe, Mn, Al, P. Cl, S.

Among the vegetables which are especially rich in mineral substances are: spinach, carrots, cabbage and tomatoes. Mineral rich fruit includes: strawberries, cherries, peaches and raspberries. Important quantities of potassium (K) and absence of sodium chloride (NaCl) give a high dietetic value to fruit and to their processed products. Phosphorus is supplied mainly by vegetables.

Vegetables usually contain more calcium than fruit; green beans, cabbage, onions and beans contain more than 0.1% calcium. The calcium/phosphorus or Ca/P ratio is essential for calcium fixation in the human body; this value is considered normal at 0.7 for adults and at 1.0 for children. Some fruit are important for their Ca/P ratio above 1.0: pears, lemons, oranges and some temperate climate mountain fruits and wild berries.

Even if its content in the human body is very low, iron (Fe) has an important role as a constituent of haemoglobin. Main iron sources are apples and spinach.

Salts from fruit have a basic reaction; for this reason fruit consumption facilitates the neutralisation of noxious uric acid reactions and contributes to the acid-basic equilibrium in the blood.

Carbohydrates: Carbohydrates are the main component of fruit and vegetables and represent more than 90% of their dry matter. From an energy point of view carbohydrates represent the most valuable of the food components; daily adult intake should contain about 500 g carbohydrates.

Carbohydrates play a major role in biological systems and in foods. They are produced by the process of photosynthesis in green plants. They may serve as structural components as in the case of cellulose; they may be stored as energy reserves as in the case of starch in plants; they may function as essential components of nucleic acids as in the case of ribose; and as components of vitamins such as ribose and riboflavin.

Carbohydrates can be oxidised to furnish energy, and glucose in the blood is a ready source of energy for the human body. Fermentation of carbohydrates by yeast and other microorganisms can yield carbon dioxide, alcohol, organic acids and other compounds.

Some properties of sugars. Sugars such as glucose, fructose, maltose and sucrose all share the following characteristics in varying degrees, related to fruit and vegetable technology:

- they supply energy for nutrition;
- they are readily fermented by micro-organisms;
- in high concentrations they prevent the growth of micro-organisms, so they may be used as a preservative;
- on heating they darken in colour or caramelise;
- some of them combine with proteins to give dark colours known as the browning reaction.

Some properties of starches:

- They provide a reserve energy source in plants and supply energy in nutrition;
- they occur in seeds and tubers as characteristic starch granules.

Some properties of celluloses and hemicelluloses:

- They are abundant in the plant kingdom and act primarily as supporting structures in the plant tissues;
- they are insoluble in cold and hot water;
- they are not digested by man and so do not yield energy for nutrition;

- the fibre in food which produces necessary roughage is largely cellulose.

Some properties of pectins and carbohydrate gums.

- Pectins are common in fruits and vegetables and are gumlike (they are found in and between cell walls) and help hold the plant cells together;
- pectins in colloidal solution contribute to viscosity of the tomato paste;
- pectins in solution form gels when sugar and acid are added; this is the basis of jelly manufacture.

Fats: Generally fruit and vegetables contain very low level of fats, below 0.5%. However, significant quantities are found in nuts (55%), apricot kernel (40%), grapes seeds (16%), apple seeds (20%) and tomato seeds (18%).

Organic Acids

Fruit contains natural acids, such as citric acid in oranges and lemons, malic acid of apples, and tartaric acid of grapes. These acids give the fruits tartness and slow down bacterial spoilage.

We deliberately ferment some foods with desirable bacteria to produce acids and this give the food flavour and keeping quality. Examples are fermentation of cabbage to produce lactic acid and yield sauerkraut and fermentation of apple juice to produce first alcohol and then acetic acid to obtain vinegar.

Organic acids influence the colour of foods since many plant pigments are natural pH indicators.

With respect to bacterial spoilage, a most important contribution of organic acids is in lowering a food's pH. Under anaerobic conditions and slightly above a pH of 4.6, Clostridium botulinum can grow and produce lethal toxins. This hazard is absent from foods high in organic acids resulting in a pH of 4.6 and less.

Acidity and sugars are two main elements which determine the taste of fruit. The sugar/acid ratio is very often used in order to give a technological characterisation of fruits and of some vegetables.

Nitrogen-containing Substances

These substances are found in plants as different combinations: proteins, amino acids, amides, amines, nitrates, etc. Vegetables contain between 1.0 and 5.5 % while in fruit nitrogen-containing substances are less than 1% in most cases.

Among nitrogen containing substances the most important are proteins; they have a colloidal structure and, by heating, their water solution above 50°C an one-way reaction makes them insoluble. This behaviour has to be taken into account in heat processing of fruits and vegetables.

From a biological point of view vegetal proteins are less valuable then animal ones because in their composition all essential amino-acids are not present.

Vitamins

Vitamins are defined as organic materials which must be supplied to the human body in small amounts apart from the essential amino-acids or fatty acids.

Vitamins function as enzyme systems which facilitate the metabolism of proteins, carbohydrates and fats but there is growing evidence that their roles in maintaining health may extend yet further.

The vitamins are conveniently divided into two major groups, those that are fat-soluble and those that are water-soluble. Fat-soluble vitamins are A, D, E and K. Their absorption by the body depends upon the normal absorption of fat from the diet. Water-soluble vitamins include vitamin C and several members of the vitamin B complex.

Vitamin-A or Retinol

This vitamin is found as such only in animal materials - meat, milk, eggs and the like. Plants contain no vitamin A but contain its precursor, beta-carotene. Man needs either vitamin A or beta-carotene which he can easily convert to vitamin A. Beta-carotene is found in the orange and yellow vegetables as well as the green leafy vegetables, mainly carrots, squash, sweet potatoes, spinach and kale.

A deficiency of vitamin A leads to night blindness, failure of normal bone and tooth development in the young and diseases of epithelial cells and membrane of the nose, throat and eyes which decrease the body's resistance to infection.

Vitamin C

Vitamin C is the anti-scurvy vitamin. Lack of it causes fragile capillary walls, easy bleeding of the gums, loosening of teeth and bone joint diseases. It is necessary for the normal formation of the protein collagen, which is an important constituent of skin and connective tissue. Like vitamin E, vitamin C favours the absorption of iron. Vitamin C, also known as ascorbic acid, is easily destroyed by oxidation especially at high temperatures and is the vitamin most easily lost during processing, storage and cooking.

Excellent sources of vitamin C are citrus fruits, tomatoes, cabbage and green peppers. Potatoes also are a fair source (although the content of vitamin C is relatively low) because we consume large quantities of potatoes.

Enzymes

Enzymes are biological catalysts that promote most of the biochemical reactions which occur in vegetable cells.

Some properties of enzymes important in fruit and vegetable technology are the following:

- in living fruit and vegetables enzymes control the reactions associated with ripening;
- after harvest, unless destroyed by heat, chemicals or some other means, enzymes continue the ripening process, in many cases to the point of spoilage - such as soft melons or overripe bananas;
- because enzymes enter into a vast number of biochemical reactions in fruits and vegetable, they may be responsible for changes in flavour, colour, texture and nutritional properties;
- the heating processes in fruit and vegetables manufacturing/processing are designed not only to destroy micro-organisms but also to deactivate enzymes and so improve the fruit and vegetables' storage stability.

Enzymes have an optimal temperature - around +50°C where their activity is at maximum. Heating beyond this optimal temperature deactivates the enzyme. Activity of each enzyme is also characterised by an optimal pH.

In fruit and vegetable storage and processing the most important roles are played by the enzymes classes of hydrolases (lipase, invertase, tannase, chlorophylase, amylase, cellulase) and oxidoreductases (peroxidase, tyrosinase, catalase, ascorbinase, polyphenoloxidase).

Turgidity and Texture

The range of textures that are encountered in fresh and cooked vegetables and fruit is indeed great, and to a large extent can be explained in terms of changes in specific cellular components. Since plants tissues generally contain more than two-thirds water, the relationships between these components and water further determine textural differences.

Cell Turgidity: Quite apart from other contributing factors, the state of turgidity, determined by osmotic forces, plays a paramount role in the texture of fruit and vegetables. The cell walls of plant tissues have varying degrees of elasticity and are largely permeable to water and ions as well as to small molecules.

The membranes of the living protoplast are semi-permeable, that is they allow passage of water but are selective with respect to transfer of dissolved and suspended materials.

The cell vacuoles contain most of the water in plant cells and sugars, acids, salts, amino acids, some water-soluble pigments and vitamins, and other low molecular weight constituents are dissolved in this water.

In the living plant, water taken up by the roots passes through the cell walls and membranes into the cytoplasm of the protoplasts and into the vacuoles to establish a state of osmotic equilibrium within the cells.

The osmotic pressure within the cell vacuoles and within the protoplasts pushes the protoplasts against the cell walls and causes them to stretch slightly in accordance with their elastic properties. This is the situation in the growing plant and the harvested live fruit or vegetable which is responsible for desired plumpness, succulence, and much of the crispness.

When plant tissues are damaged or killed by storage, freezing, cooking, or other causes, an important major change that results is denaturation of the proteins of cell membranes resulting in the loss of perm-selectivity. Without perm-selectivity the state of osmotic pressure in cell vacuoles and protoplasts cannot exist, and water and dissolved substances are free to diffuse out of the cells and leave the remaining tissue in a soft and wilted condition.

Other Factors Affecting Texture. The existence of a high degree of turgidity in live fruit and vegetables or whether a relative state of flabbiness develops from loss of osmotic pressure as well as final texture depends on several cell constituents.

Cellulose, Hemicellulose, and Lignin. Cell walls in young plants are very thin and are composed largely of cellulose. As the plant ages cell walls tend to thicken and become higher in hemicellulose and in lignin. These materials are fibrous and tough and are not significantly softened by cooking.

Pectic Substances. The complex polymers of sugar acid derivatives include pectin and closely related substances. The cement-like substance found especially in the middle lamella which helps hold plant cells to one another is a water-insoluble pectic substance.

On mild hydrolysis it yields water-soluble pectin which can form gels or viscous colloidal suspensions with sugar and acid. Certain water-soluble pectic substances also react with metal ions, particularly calcium, to form water-insoluble salts such as calcium pectates. The various pectic substances may influence texture of vegetables and fruits in several ways.

When vegetables or fruit are cooked, some of the water-insoluble pectic substance is hydrolysed into water-soluble pectin. This results in a degree of cell separation in the tissues and contributes to tenderness. Since many fruits and vegetables are somewhat acidic and contain sugars the soluble pectin also tends to form colloidal suspensions which will thicken the juice or pulp of these products.

Fruit and vegetables also contain a natural enzyme which can further hydrolyse pectin to the point where the pectin loses much of its gel forming property. This enzyme is known as pectin methyl esterase. Materials such as tomato juice or tomato paste will contain both pectin and pectin methyl esterase.

If freshly prepared tomato juice or paste is allowed to stand the original viscosity gradually decreases due to the action of pectin methyl esterase on pectin gel.

This can be prevented if the tomato products are quickly heated to a temperature of about 82°C (180 F°) to deactivate the pectin methyl esterase liberated from broken cells before it has a chance to hydrolyse the pectin. Such a treatment is commonly practiced in the manufacture of tomato juice products. This is known as the "hot-break process" and yields products of high viscosity.

In contrast, where low viscosity products are desired no heat is used and enzyme activity is allowed to proceed. This is "cold-break" process. After sufficient decrease in viscosity is achieved the product can be heat treated, as in canning, to preserve it for long term storage. It is often also desirable to firm the texture of fruit and vegetables, especially when products are normally softened by processing. In this case advantage is taken of the reaction between soluble pectic substances and calcium ions which form calcium pectates. These calcium pectates are water insoluble and when they are produced within the tissues of fruit and vegetables they increase structural rigidity. Thus, it is common commercial practice to add low levels of calcium salts to tomatoes, apples, and other vegetables and fruits prior to canning or freezing.

Sources of Colour and Colour Changes

In addition to a great range of textures, much of the interest that fruits and vegetables add to our diets is due to their delightful and variable colours. The pigments and colour precursors of fruit and vegetables occur for the most part in the cellular plastic inclusions such as the chloroplasts and other chromoplasts, and to a lesser extent dissolved in fat droplets or water within the cell protoplast and vacuoles.

These pigments are classified into four major groups which include the chlorophylls, carotenoids, anthocyanins, and anthoanthins. Pigments belonging to the latter two groups also are referred to as flavonoids, and include the tannins.

The Chlorophylls. The chlorophylls are contained mainly within the chloroplasts and have a primary role in the photosynthetic production of carbohydrates from carbon dioxide and water. The bright green colour of leaves and other parts of plants is largely due to the oilsoluble chlorophylls, which in nature are bound to protein molecules in highly organised complexes.

When the plant cells are killed by ageing, processing, or cooking, the protein of these complexes is denatured and the chlorophyll may be released. Such chlorophyll is highly unstable and rapidly changes in colour to olive green and brown. This colour change is believed to be due to the conversion of chlorophyll to the compound pheophytin.

Conversion to pheophytin is favoured by acid pH but does not occur readily under alkaline conditions. For this reason peas, beans, spinach, and other green vegetables which tend to lose their bright green colours on heating can be largely protected against such colour changes by the addition of sodium bicarbonate or other alkali to the cooking or canning water.

However, this practice is not looked upon favourably nor used commercially because alkaline pH also has a softening effect on cellulose and vegetable texture and also destroys vitamin C and thiamin at cooking temperatures.

The Carotenoids: Pigments belonging to this group are fat-soluble and range in colour from yellow through orange to red. They often occur along with the chlorophylls in the chloroplasts, but also are present in other chromoplasts and may occur free in fat droplets. Important carotenoids include the orange carotenes of carrot, maize, apricot, peach, citrus fruits, and squash; the red lycopene of tomato, watermelon, and apricot; the yellow-orange xanthophyll of maize, peach, paprika and squash; and the yellow-orange crocetin of the spice saffron. These and other carotenoids seldom occur singly within plant cells.

A major importance of some of the carotenoids is their relationship to vitamin A. A molecule of orange beta-carotene is converted into two molecules of colourless vitamin A in the animal body. Other carotenoids like alpha-carotene, gamma-carotene, and cryptoxanthin also are precursors of vitamin A, but because of minor differences in chemical structure one molecule of each of these yields only one molecule of vitamin A.

In food processing the carotenoids are fairly resistant to heat, changes in pH, and water leaching since they are fat-soluble. However, they are very sensitive to oxidation, which results in both colour loss and destruction of vitamin A activity.

The Flavonoids: Pigments and colour precursors belonging to this class are water-soluble and commonly are present in the juices of fruit and vegetables. The flavonoids include the purple, blue, and red anthocyanins of grapes, berries, plump, eggplant, and cherry; the yellow anthoxanthins of light coloured fruit and vegetables such as apple, onion, potato, and cauliflower, and the colourless catechins and leucoanthocyanins which are food tannins and are found in apples, grapes, tea, and other plant tissues. These colourless tannin compounds are easily converted to brown pigments upon reaction with metal ions.

Properties of the anthocyanins include a shifting of colours with pH. Thus many of the anthocyanins which are violet or blue in alkaline media become red upon addition of acid.

Cooking of beets with vinegar tends to shift the colour from a purplish red to a brighter red, while alkaline water can influence the colour of red fruits and vegetables toward violet and gray-blue.

The anthocyanins also tend toward the violet and blue hues upon reaction with metal ions, which is one reason for lacquering the inside of metal cans when the true colour of anthocyanin-containing fruits and vegetables is to be preserved.

The water-soluble property of anthocyanins also results in easy leaching of these pigments from cut fruit and vegetables during processing and cooking.

The yellow anthoxanthins also are pH sensitive tending toward a deeper yellow in alkaline media. Thus potatoes or apples become somewhat yellow when cooked in water with a pH of 8 or higher, which is common in many areas. Acidification of the water to pH 6 or lower favours a whiter colour.

The colourless tannin compounds upon reaction with metal ions form a range of dark coloured complexes which may be red, brown, green, grey, or black. The various shades of these coloured complexes depend upon the particular tannin, the specific metal ion, pH, concentration of the complex, and other factors not yet fully understood.

Water-soluble tannins appear in the juices squeezed from grapes, apples, and other fruits as well as the brews from extraction of tea and coffee. The colour and clarity of tea are influenced by the hardness and pH of the brewing water. Alkaline waters that contain calcium and magnesium favour the formation of dark brown tannin complexes which precipitate when the tea is cooled.

If acid in the form of lemon juice is added to such tea its colour lightens and the precipitate tends to dissolve. Iron from equipment or from pitted tin cans has caused a number of unexpected colours to develop in products containing tannins, such as coffee, cocoa and foods flavoured with these.

The tannins are also important because they have an astringency which influences flavour and contributes body to such beverages as tea, wine, apple cider, etc.

Fruit and vegetables are in a live state after harvest. Continued respiration gives off carbon dioxide, moisture, and heat which influence storage, packaging, and refrigeration requirements. Continued transpiration adds to moisture evolved and further influences packaging requirements.

Further activities of fruit and vegetables, before and after harvest, include changes in carbohydrates, pectins, organic acids, and the effects these have on various quality attributes of the products.

As for changes in carbohydrates, few generalizations can be given with respect to starches and sugars. In some plant products sugars quickly decrease and starch increases in amount soon after harvest. This is the case for ripe sweet corn which can suffer flavour and texture quality losses in a very few hours after harvest.

Unripe fruit, in contrast, is frequently high in starch and low in sugars. Continued ripening after harvest generally results in a decrease in starch and a increase in sugars as in the case of apples and pears. However, this does not necessarily mean that the starch is the source of the newly formed sugars.

Further, the courses of change in starch and sugars are markedly influenced by Post Harvest storage temperatures. Thus potatoes stored below about 10 C° (50 F°) continue to build up high levels of sugars, while the same potatoes stored above 10 C° do not.

This property is used to help the dehydration process in potato storage. Here potatoes should have a low reducing sugar content so as to minimise Maillard browning reactions during drying and subsequent storage of the dried product. In this case potatoes are stored above 10°C prior to being further processed.

After harvest the pectin changes in fruit and vegetables are more predictable. Generally there is decrease in water-insoluble pectic substance and a corresponding increase in water soluble pectin. This contributes to the gradual softening of fruits and vegetables during storage and ripening. Further breakdown of water-soluble pectin by pectin methyl esterase also occurs.

The organic acids of fruit generally decreases during storage and ripening. This occurs in apples and pears and is especially important in the case of oranges. Oranges have a long ripening period on the tree and time of picking is largely determined by degree of acidity and sugar content which have major effects upon juice quality.

It is important to note that the reduction of acid content on ripening influences more than just the tartness of fruit. Since many of the plant pigments are sensitive to acid, fruit colour would be expected to change. Additionally, the viscosity of pectin gel is affected by acid and sugar contents, both of which change with ripening.

Stability of Nutrients

One of the principal responsibilities of the food scientist and food technologist is to preserve food nutrients through all phases of food acquisition, processing, storage, and preparation. The key is in the specific sensitivities of the various nutrients, the principles of which are illustrated.

Specific Sensitivity and Stability of Nutrients

Nutrient	*Neutral* p^{H7}	*Acid* $< p^{H7}$	*Alkaline* $> p^{H7}$	*Air or Oxygen*	*Light*	*Heat*	*Cooking Losses, Range*
			Vitamins				
Vitamin A	S	U	S	U	U	U	0-40
Ascorbic acid(C)	U	S	U	U	U	U	0-100
Biotin	S	S	S	S	S	U	0-60
Carotenes (pro A)	S	U	S	U	U	U	0-30 0-5
Choline	S	S	S	U	S	S	0-10
Cobalamin (B_{12})	S	S	S	U	U	S	0-40
Vitamin D	S		U	U	U	U	0-10
Essential fatty acids	S	S	U	U	U	S	
Folic acid	U	U	S	U	U	U	0-100
Inositol	S	S	S	S	S	U	0-95
Vitamin K	S	U	U	S	U	S	0-5
Niacin (PP)	S	S	S	S	S	S	0-75
Pantbothenic acid	S	U	U	S	S	U	0-50
p-Amino Benzoic acid	S	S	S	U	S	S	0-5
Vitamin B_6	S	S	S	S	U	U	0-40
Riboflavin (B_2)	S	S	U	S	U	U	0-75
Tbiamin (B_1)	U	S	U	U	S	U	0-80
Tocopherols	S	S	S	U	U	U	0-55

***Source:** Harris and Karmas, 1975

(U = Unstable; S = Stable)

This shows the stability of vitamins, essential amino acids, and minerals to acid, air, light, and heat, and gives an indication of possible cooking losses. Vitamin A is highly sensitive to acid, air, light and heat; vitamin C to alkalinity, air, light and heat; vitamin D to alkalinity, air, light and heat; thiamin to alkalinity, air, and heat in alkaline solutions; etc. Cooking losses of some essential nutrients may be in excess of 75%. In modern food processing operations, however, losses are seldom in excess of 25% .

The ultimate nutritive value of a food results from the sum total of losses incurred throughout its history - from farmer to consumer. Nutrient value begins with genetics of the plant and animal. The farmland fertilization programme affects tissue composition of plants, and animals consuming these plants. The weather and degree of maturity at harvest affect tissue composition.

Storage conditions before processing affect vitamins and other nutrients. Washing, trimming, and heat treatments affect nutrient content. Canning, evaporating, drying, and freezing alter nutritional values, and the choices of times and temperatures in these operations frequently must be balanced between good bacterial destruction and minimum nutrient destruction.

Packaging and subsequent storage affect nutrients. One of the most important factors is the final preparation of the food in the home and the restaurant - the steam table can destroy much of what has been preserved through all prior manipulations.

Structural Features

The structural unit of the edible portion of most fruits and vegetables is the parenchyma cell. While parenchyma cells of different fruit and vegetables differ somewhat in gross size and appearance, all have essentially the same fundamental structure.

Parenchyma cells of plants differ from animal cells in that the actively metabolising protoplast portion of plant cells represents only a small fraction, of the order of five per cent, of the total cell volume. This protoplast is film-like and is pressed against the cell wall by the large water-filled central vacuole.

The protoplast has inner and outer semi-permeable membrane layers; the cytoplasm and its nucleus are held between them. The cytoplasm contains various inclusions, among them starch granules and plastics such as the chloroplasts and other pigment-containing chromoplasts. The cell wall, cellulose in nature, contributes rigidity to the parenchyma cell and limits the outer protoplasmic membrane. It is also the structure against which other parenchyma cells are cemented to form extensive three-dimensional tissue masses.

The layer between cell walls of adjacent parenchyma cells is referred to as the middle lamella, and is composed largely of pectic and polysaccharide cement-like materials. Air spaces also exist, especially at the angles formed where several cells come together. The relationships between these structures and their chemical compositions are further outlined below. The parenchyma cells will vary in size among plants but are quite large when compared to bacterial or yeast cells. The larger parenchyma cells may have volumes many thousand times greater than a typical bacterial cell.

There are additional types of cells other than parenchyma cells that make up the familiar structures of fruit and vegetables. These include various types of conducting cells which are tube-like and distribute water and salts throughout the plant.

Such cells produce fibrous structures toughened by the presence of cellulose and the wood-like substance lignin. Cellulose, lignin, and pectic substances also occur in specialised supporting cells which increase in importance as plants become older.

An important structural feature of all plants, including fruit and vegetables is protective tissue. This can take many forms but usually is made up of specialised parenchyma cells that are pressed compactly together to form a skin, peel or rind.

Surface cells of these protective structures on leaves, stems or fruit secrete waxy cutin and form a water impermeable cuticle. These surface tissues, especially on leaves and young stems will also contain numerous valve-like cellular structures, the stomata, through which moisture and gases can pass.

Deterioration Factors and their Control

Enzymic Changes

Enzymes which are endogenous to plant tissues can have undesirable or desirable consequences. Examples involving endogenous enzymes include (a) the Post Harvest senescence and spoilage of fruit and vegetables; (b) oxidation of phenolic substances in plant tissues by phenolase (leading to browning); (c) sugar-starch conversion in plant tissues by amylases; (d) Post Harvest demethylation of pectic substances in plant tissues (leading to softening of plant tissues during ripening, and firming of plant tissues during processing).

The major factors useful in controlling enzyme activity are: temperature, water activity, pH, chemicals which can inhibit enzyme action, alteration of substrates, alteration of products and pre-processing control.

Chemical Changes

Sensory Quality: The two major chemical changes which occur during the processing and storage of foods and lead to a deterioration in sensory quality are lipid oxidation and non-enzymatic browning. Chemical reactions are also responsible for changes in the colour and flavour of foods during processing and storage.

1. Lipid oxidation rate and course of reaction is influenced by light, local oxygen concentration, high temperature, the presence of catalysts (generally transition metals such as iron and copper) and water activity. Control of these factors can significantly reduce the extent of lipid oxidation in foods.
2. Non-enzymic browning is one of the major causes of deterioration which occurs during storage of dried and concentrated foods. The non-enzymic browning, or Maillard reaction, can be divided into three stages: (a) early Maillard reactions which are chemically well-

defined steps without browning; (b) advanced Maillard reactions which lead to the formation of volatile or soluble substances; and (c) final Maillard reactions leading to insoluble brown polymers.

Colour Changes

Chlorophylls: Almost any type of food processing or storage causes some deterioration of the chlorophyll pigments. Phenophytinisation (with consequent formation of a dull olive-brown phenophytin) is the major change; this reaction is accelerated by heat and is acid catalysed.

Other reactions are also possible. For example, dehydrated products such as green peas and beans packed in clear glass containers undergo photo-oxidation and loss of desirable colour.

Anthocyanins: These are a group of more than 150 reddish water-soluble pigments that are very widespread in the plant kingdom. The rate of anthocyanin destruction is pH dependent, being greater at higher pH values. Of interest from a packaging point of view is the ability of some anthocyanins to form complexes with metals such as Al, Fe, Cu and Sn.

These complexes generally result in a change in the colour of the pigment (for example, red sour cherries react with tin to form a purple complex) and are therefore undesirable. Since metal packaging materials such as cans could be sources of these metals, they are usually coated with special organic linings to avoid these undesirable reactions.

Carotenoids: The carotenoids are a group of mainly lipid soluble compounds responsible for many of the yellow and red colours of plant and animal products. The main cause of carotenoid degradation in foods is oxidation. The mechanism of oxidation in processed foods is complex and depends on many factors. The pigments may auto-oxidise by reaction with atmospheric oxygen at rates dependent on light, heat and the presence of pro- and antioxidants.

Flavour Changes: In fruit and vegetables, enzymically generated compounds derived from long-chain fatty acids play an extremely important role in the formation of characteristic flavours. In addition, these types of reactions can lead to significant off-flavours. Enzyme-induced oxidative breakdown of unsaturated fatty acids occurs extensively in plant tissues and this yield characteristic aromas associated with some ripening fruits and disrupted tissues.

The permeability of packaging materials is of importance in retaining desirable volatile components within packages, or in permitting undesirable components to permeate through the package from the ambient atmosphere.

Nutritional Quality

The four major factors which affect nutrient degradation and can be controlled to varying extents by packaging are light, oxygen concentration, temperature and water activity. However, because of the diverse nature of the various nutrients as well as the chemical heterogeneity within each class of compounds and the complex interactions of the above variables, generalizations about nutrient degradation in foods will inevitably be broad ones.

Vitamins: Ascorbic acid is the most sensitive vitamin in foods, its stability varying markedly as a function of environmental conditions such as pH and the concentration of trace metal ions and oxygen. The nature of the packaging material can significantly affect the stability of ascorbic acid in foods. The effectiveness of the material as a barrier to moisture and oxygen as well as the chemical nature of the surface exposed to the food are important factors.

For example, problems of ascorbic acid instability in aseptically packaged fruit juices have been encountered because of oxygen permeability of the package and the oxygen dependence of the ascorbic acid degradation reaction.

Also, because of the preferential oxidation of metallic tin, citrus juices packaged in cans with a tin contact surface exhibit greater stability of ascorbic acid than those in enamelled cans or glass containers. The aerobic and anaerobic degradation reactions of ascorbic acid in reduced-moisture foods have been shown to be highly sensitive to water activity, the reaction rate increasing in an exponential fashion over the water activity range of 0.1-0.8.

Physical Changes

One major undesirable physical change in food powders is the absorption of moisture as a consequence of an inadequate barrier provided by the package; this results in caking. It can occur either as a result of a poor selection of packaging material in the first place, or failure of the package integrity during storage. In general, moisture absorption is associated with increased cohesiveness.

Anti-caking agents are very fine powders of an inert chemical substance that are added to powders with much larger particle size in order to inhibit caking and improve flowability. Studies in onion powders showed that at ambient temperature, caking does not occur at water activities of less than about 0.4.

At higher activities, however, (aw > 0.45) the observed time to caking is inversely proportional to water activity, and at these levels anti-caking agents are completely ineffective. It appears that while they reduce inter-particle attraction and interfere with the continuity of liquid bridges, they are unable to cover moisture sorption sites.

Biological Changes

Microbiological: Micro-organisms can make both desirable and undesirable changes to the quality of foods depending on whether or not they are introduced as an essential part of the food preservation process or arise unintentionally and subsequently grow to produce food spoilage.

The two major groups of micro-organisms found in foods are bacteria and fungi, the latter consisting of yeasts and moulds. Bacteria are generally the fastest growing, so that in conditions favourable to both, bacteria will usually outgrow fungi.

Foods are frequently classified on the basis of their stability as non-perishable, semi-perishable and perishable. For example, hermetically sealed and heat processed (*e.g.* canned) foods are generally regarded as non-perishable. However, they may become perishable under certain circumstances when an opportunity for recontamination is afforded following processing.

Such an opportunity may arise if the can seams are faulty, or if there is excessive corrosion resulting in internal gas formation and eventual bursting of the can. Spoilage may also take place when the canned food is stored at unusually high temperatures: thermophilic spore-forming bacteria may multiply, causing undesirable changes such as flat sour spoilage.

Low moisture content foods such as dried fruit and vegetables are classified as semi-perishable. Frozen foods, though basically perishable, may be classified as semi-perishable provided that they are properly stored at freezer temperatures.

The majority of foods (*e.g.* meat and fish, milk, eggs and most fresh fruits and vegetables) are classified as perishable unless they have been processed in some way. Often, the only form of processing which such foods receive is to be packaged and kept under controlled temperature conditions. The species of micro-organisms which cause the spoilage of particular foods are influenced by two factors: a) the nature of the foods and b) their surroundings. These factors are referred to as intrinsic and extrinsic parameters.

The intrinsic parameters are an inherent part of the food: pH, aw, nutrient content, antimicrobial constituents and biological structures. The extrinsic parameters of foods are those properties of the storage environment that affect both the foods and their microorganisms. The growth rate of the micro-organisms responsible for spoilage primarily depends on these extrinsic parameters: temperature, relative humidity and gas compositions of the surrounding atmosphere. The protection of packaged food from contamination or attack by micro-organisms depends on the mechanical integrity of the package (*e.g.* the absence of breaks and seal imperfections), and on the resistance of the package to penetration by micro-organisms.

Metal cans which are retorted after filling can leak during cooling, admitting any microorganisms which may be present in the cooling water, even when the double seam is of a high quality. This fact is widely known in the canning industry and is the reason for the mandatory chlorination of cannery cooling water. Extensive studies on a variety of plastic films and metal foils have shown that microorganisms (including mounds, yeasts and bacteria) cannot penetrate these materials in the absence of pinholes. In practice, however, thin sheets of packaging materials such as aluminium and plastic do contain pinholes. There are several safeguards against the passage of micro-organisms through pinholes in films:

- because of surface tension effects, micro-organisms cannot pass through very small pinholes unless the micro-organisms are suspended in solutions containing wetting agents and the pressure outside the package is greater than that within;
- materials of packaging are generally used in thicknesses such that pinholes are very infrequent and small;
- for applications in which package integrity is essential (such as sterilisation of food in pouches), adequate test methods are available to assure freedom from bacterial recontamination.

Macrobiological

Insect Pests: Warm humid environments promote insect growth, although most insects will not breed if the temperature exceeds about 35 C° or falls below 10 C°. Also many insects cannot reproduce satisfactorily unless the moisture content of their food is greater than about 11%. The main categories of foods subject to pest attack are cereal grains and products derived from cereal grains, other seeds used as food (especially legumes), dairy products such as cheese and milk powders, dried fruits, dried and smoked meats and nuts.

As well as their possible health significance, the presence of insects and insect excrete in packaged foods may render products unsaleable, causing considerable economic loss, as well as reduction in nutritional quality, production of off-flavours and acceleration of decay processes due to creation of higher temperatures and moisture levels.

Early stages of infestation are often difficult to detect; however, infestation can generally be spotted not only by the presence of the insects themselves but also by the products of their activities such as webbing, clumped-together food particles and holes in packaging materials.

Unless plastic films are laminated with foil or paper, insects are able to penetrate most of them quite easily, the rate of penetration usually being directly related to film thickness. In general, thicker films are more resistant than thinner films, and oriented films tend to be more effective than cast films.

The looseness of the film has also been reported to be an important factor, loose films being more easily penetrated than tightly fitted films.

Generally, the penetration varies depending on the basic resin from which the film is made, on the combination of materials, on the package structure, and of the species and stage of insects involved. The relative resistance to insect penetration of some flexible packaging materials is as follows:

- excellent resistance: polycarbonate; poly-ethylene-terephthalate;
- good resistance: cellulose acetate; polyamide; polyethylene (0.254 mm); polypropylene (biaxially oriented); poly-vinyl-chloride (unplasticised);
- fair resistance: acrylonitrile; poly-tetra-fluoro-ethylene; polyethylene (0.123 mm);
- poor resistance: regenerated cellulose; corrugated paper board; kraft paper; polyethylene (0.0254-0.100 mm); paper/foil/polyethylene laminate pouch; poly-vinylchloride (plasticised).

Some simple methods for obtaining insect resistance of packaging materials are as following:

- select a film and a film thickness that are inherently resistant to insect penetration;
- use shrink film over-wraps to provide an additional barrier;
- seal carton flaps completely.

Rodents: Rats and mice carry disease-producing organisms on their feet and/or in their intestinal tracts and are known to harbour salmonella of sero types frequently associated with food-borne infections in humans.

In addition to the public health consequences of rodent populations in close proximity to humans, these animals also compete intensively with humans for food. Rats and mice gnaw to reach sources of food and drink and to keep their teeth short.

Their incisor teeth are so strong that rats have been known to gnaw through lead pipes and unhardened concrete, as well as sacks, wood and flexible packaging materials.

Proper sanitation in food processing and storage areas is the most effective weapon in the fight against rodents, since all packaging materials apart from metal and glass containers can be attacked by rats and mice.

Summary: Major causes of food deterioration include the following:

a. Growth and activities of micro-organisms, principally bacteria, yeasts and moulds;
b. Activities of natural food enzymes;
c. Insects, parasites and rodents;
d. temperature, both heat and cold;
e. Moisture and dryness;
f. Air and in particular oxygen;
g. Light;
h. Time.

Extrinsic factors controlling the rate of food Deterioration reactions are mainly:

a. Effect of temperature;
b. Effect of water activity (aw);
c. Effect of gas atmosphere;
d. Effect of light.

Fruit and Vegetable Preservation

Fresh Storage

Fresh Fruit and Vegetable Storage: Once fruit is harvested, any natural resistance to the action of spoiling micro-organisms is lost. Changes in enzymatic systems of the fruit also occur on harvest which may also accelerate the activity of spoilage organisms.

Means that are commonly used to prevent spoilage of fruits must include:

- care to prevent cutting or bruising of the fruit during picking or handling;
- refrigeration to minimise growth of micro-organisms and reduce enzyme activity;
- packaging or storage to control respiration rate and ripening;
- use of preservatives to kill micro-organisms on the fruit.

A principal economic loss occurring during transportation and/or storage of produce such as fresh fruit is the degradation which occurs between the field and the ultimate destination due to the effect of respiration. Methods to reduce such degradation are as follows:

- refrigerate the produce to reduce the rate of respiration;
- vacuum cooling;
- reduce the oxygen content of the environment in which the produce is kept to a value not above 5% of the atmosphere but above the value at which anaerobic respiration would begin. When the oxygen concentration is reduced within 60 minutes the deterioration is in practice negligible.

The following is a summary of some recent developments in Post Harvest technology of fresh fruits and vegetables (Source: Thompson, 1989).

Harvest Maturity: This is particularly important with fruit for export. One recent innovation is the measurement of resonant frequency of the fruit which should enable the grading out of over mature and under-mature fruit before they are packed for export.

Harvest Method: Considerable research is continuing on mechanical harvesting of perishable crops with a view to minimising damage. In fruit trees, controlling their height by use of dwarfing rootstocks, pruning and growth regulating chemicals will lead to easier, cheaper more accurate harvesting.

Handling Systems: Field packing of various vegetables for export has been carried out for many years. In the last decade or so this has been applied, in selected cases, to a few tropical fruit types. Where this system can be practiced it has considerable economic advantages in saving the cost of building, labour and equipment and can result in lower levels of damage into crops.

Pre-cooling: Little innovation has occurred in crop pre-cooling over the last decade. However high velocity, high humidity forced air systems have continued to be developed and refined. These are suitable for all types of produce and are relatively simple to build and operate and, while not providing the speed of cooling of a vacuum or hydrocooler, have the flexibility to be used with almost all crops.

Chemicals: There is a very strong health lobby whose objective is to reduce the use of chemicals in agriculture and particularly during the Post Harvest period. Every year sees the prohibition of the use of commonly used Post Harvest chemicals. New ways need to be developed to control Post Harvest diseases, pest and sprouting.

Coatings: Slowing down the metabolism of fruit and vegetables by coating them with a material which affects their gaseous exchange is being tested and used commercially on a number of products.

Controlled Environment Transport: Recent innovations in this technique have produced great progress as a result of the development and miniaturisation of equipment to measure carbon dioxide and oxygen. Several companies now offer containers where the levels of these two gases can be controlled very precisely.

4

Preservation by Reduction of Water Content

Water and Water Activity (aw) in Foods

Microorganisms in a healthy growing state may contain in excess of 80% water. They get this water from the food in which they grow. If the water is removed from the food it also will transfer out of the bacterial cell and multiplication will stop. Partial drying will be less effective than total drying, though for some microorganisms partial drying may be quite sufficient to arrest bacterial growth and multiplication.

Bacteria and yeasts generally require more moisture than moulds, and so moulds often will be found growing on semi-dry foods where bacteria and yeasts find conditions unfavourable; example are moulds growing on partially dried fruits. Slight differences in relative humidity in the environment in which the food is kept or in the food package can make great differences in the rate of microorganism multiplication. Since microorganisms can live in one part of a food that may differ in moisture and other physical and chemical conditions from the food just millimetres away, we must be concerned with conditions in the "micro-environment" of the microorganisms. Thus it is common to refer to water conditions in terms of specific activity.

The term "water activity" is related to relative humidity. Relative humidity is defined as the ratio of the partial pressure of water vapour in the air to the vapour pressure of pure water at the same temperature. Relative humidity refers to the atmosphere surrounding a material or solution.

Water activity or aw is a property of solutions and is the ratio of vapour pressure of the solution compared with the vapour pressure of pure water at the same temperature. Under equilibrium conditions water activity equals:

aw = RH / 100

When we speak of moisture requirements of microorganisms we really mean water activity in their immediate environment, whether this be in solution, in a particle of food or at a surface in contact with the atmosphere.

At the usual temperatures permitting microbial growth, most bacteria require a water activity in the range of about 0.90 to 1.00.

Some yeasts and moulds grow slowly at a water activity down to as low as about 0.65.

Qualitatively, water activity is a measure of unbound, free water in a system available to support biological and chemical reactions. Water activity, not absolute water content, is what bacteria, enzymes and chemical reactants encounter and are affected by at the micro-environmental level in food materials.

Two foods with the same water content can have very different aw values depending upon the degree to which water is free or otherwise bound to food constituents. This is a representative water absorption isotherm for a given food at a given temperature. It shows the final moisture content the food will have when it reaches moisture equilibrium with atmospheres of different relative humidities.

Thus, this food, at the temperature for which this absorption isotherm was established, will ultimately attain a moisture content of 20% at 75% RH (relative humidity). If this food was previously dehydrated to below 20% moisture and placed in an atmosphere of 75% RH, it would absorb moisture until it reached 20%. Conversely, if this food was moistened to greater than 20% water and then placed at 75% RH, it would lose moisture until it reached the equilibrium value of 20%.

Under such conditions some foods may reach moisture equilibrium in the very short time of a few hours, others may require days or even weeks. When a food is in moisture equilibrium with its environment, then the aw of the food will be quantitatively equal to the RH divided by 100.

Qualitatively, water activity is a measure of free or available water, to be distinguished from unavailable or bound water. These states of water also bear a relationship to the characteristic sigmoid shapes of water absorption isotherm curves of various foods.

Thus, according to theory, most of the water corresponding to the portion of the curve below its first inflection point (below 5% moisture) is believed to be tightly bound water, often referred to as an adsorbed mono-molecular layer of water. Moisture corresponding to the region above this point and up to the curve's second inflection point (above 20 % moisture) is thought to exist largely as multi-molecular layers of water less tightly held to food constituent surfaces.

Beyond this second inflection point moisture generally is considered to be largely free water condensed in capillaries and interstices within the food. In this latter portion of the sorption isotherm curve small changes in moisture content result in great changes in a food's aw. The illustrated the moisture sorption isotherms for various dried fruits at 25 °C.

Preservation by Drying/dehydration

The technique of drying is probably the oldest method of food preservation practiced by mankind. The removal of moisture prevents the growth and reproduction of microorganisms causing decay and minimises many of the moisture mediated deterioration reactions.

It brings about substantial reduction in weight and volume minimising packing, storage and transportation costs and enable storability of the product under ambient temperatures, features especially important for developing countries. The sharp rise in energy costs has promoted a dramatic upsurge in interest in drying worldwide over the last decade.

Heat and Mass Transfer: Dehydration involves the application of heat to vaporise water and some means of removing water vapour after its separation from the fruit/ vegetable tissues. Hence it is a combined/simultaneous (heat and mass) transfer operation for which energy must be supplied. A current of air is the most common medium for transferring heat to a drying tissue and convection is mainly involved.

The two important aspects of mass transfer are:

- the transfer of water to the surface of material being dried and
- the removal of water vapour from the surface.

In order to assure products of high quality at a reasonable cost, dehydration must occur fairly rapidly. Four main factors affect the rate and total drying time:

- the properties of the products, especially particle size and geometry;
- the geometrical arrangement of the products in relation to heat transfer medium (drying air);
- the physical properties of drying medium/ environment;
- the characteristics of the drying equipment.

It is generally observed with many products that the initial rate of drying is constant and then decreases, sometimes at two different rates. The drying curve is divided into the constant rate period and the falling rate period.

Surface area. Generally the fruit and vegetables to be dehydrated are cut into small pieces or thin layers to speed heat and mass transfer. Subdivision speeds drying for two reasons:

- large surface areas provide more surface in contact with the heating medium (air) and more surface from which moisture can escape;
- smaller particles or thinner layers reduce the distance heat must travel to the centre of the food and reduce the distance through which moisture in the centre of the food must travel to reach the surface and escape.

Temperature: The greater the temperature difference between the heating medium and the food the greater will be the rate of heat transfer into the food, which provides the driving force for moisture removal. When the heating medium is air, temperature plays a second important role.

As water is driven from the food in the form of water vapour it must be carried away, or else the moisture will create a saturated atmosphere at the food's surface which will slow down the rate of subsequent water removal. The hotter the air the more moisture it will hold before becoming saturated.

Thus, high temperature air in the vicinity of the dehydrating food will take up the moisture being driven from the food to a greater extent than will cooler air. Obviously, a greater volume of air also can take up more moisture than a lesser volume of air.

Air Velocity: Not only will heated air take up more moisture than cool air, but air in motion will be still more effective. Air in motion, that is, high velocity air, in addition to taking up moisture will sweep it away from the drying food's surface, preventing the moisture from creating a saturated atmosphere which would slow down subsequent moisture removal. This is why clothes dry more rapidly on a windy day.

Some other phenomena influence the drying process and a few elements are summarised below.

Dryness of Air: When air is the drying medium of food, the drier the air the more rapid is the rate of drying. Dry air is capable of absorbing and holding moisture. Moist air is closer to saturation and so can absorb and hold less additional moisture than if it were dry. But the dryness of the air also determines how low a moisture content the food product can be dried to.

Atmospheric Pressure and Vacuum: If food is placed in a heated vacuum chamber the moisture can be removed from the food at a lower temperature than without a vacuum. Alternatively, for a given temperature, with or without vacuum, the rate of water removal from the food will be greater in the vacuum. Lower drying temperatures and shorter drying times are especially important in the case of heat-sensitive foods.

Evaporation and Temperature: As water evaporates from a surface it cools the surface. The cooling is largely the result of absorption by the water of the latent heat of phase change from liquid to gas.

In doing this the heat is taken from the drying air or the heating surface and from the hot food, and so the food piece or droplet is cooled.

Time and Temperature: Since all important methods of food dehydration employ heat, and food constituents are sensitive to heat, compromises must be made between maximum possible drying rate and maintenance of food quality.

As is the case in the use of heat for pasteurization and sterilisation, with few exceptions drying processes which employ high temperatures for short times do less damage to food than drying processes employing lower temperatures for longer times.

Thus, vegetable pieces dried in a properly designed oven in four hours would retain greater quality than the same products sun dried over two days.

Several drying processes will achieve dehydration in a matter of minutes or even less if the food is sufficiently subdivided.

Drying Techniques: Several types of dryers and drying methods, each better suited for a particular situation, are commercially used to remove moisture from a wide variety of food products including fruit and vegetables.

While sun drying of fruit crops is still practiced for certain fruit such as prunes, figs, apricots, grapes and dates, atmospheric dehydration processes are used for apples, prunes, and several vegetables; continuous processes as tunnel, belt trough, fluidised bed and foam-mat drying are mainly used for vegetables.

Spray drying is suitable for fruit juice concentrates and vacuum dehydration processes are useful for low moisture / high sugar fruits like peaches, pears and apricots.

Factors on which the selection of a particular dryer/ drying method depends include:

- form of raw material and its properties;
- desired physical form and characteristics of dried product;
- necessary operating conditions;
- operation costs.

There are three basic types of drying process:

- sun drying and solar drying;
- atmospheric drying including batch (kiln, tower and cabinet dryers) and continuous (tunnel, belt, belt-trough, fluidised bed, explosion puff, foam-mat, spray, drum and microwave);
- sub-atmospheric dehydration (vacuum shelf/belt/drum and freeze dryers).

The scope has been expanded to include use of low temperature, low energy process like osmotic dehydration.

As far dryers are concerned, one useful division of dryer types separates them into air convection dryers, drum or roller dryers, and vacuum dryers. Using this breakdown, The applicability of the more common dryer types to liquid and solid type foods.

Common Dryer Types Used for Liquid and Solid Doods

Dryer type	*Usual food type*
Air convection dryers	
Kiln	pieces
cabinet, tray or pan	pieces, purees, liquids
Tunnel	pieces
continuous conveyor belt	purees, liquids
belt trough	pieces
air lift	small pieces, granules
fluidized bed	small pieces, granules
Spray	liquid, purees
Drum or roller dryers	
Atmospheric	purees, liquids
Vacuum	purees, liquids
Vacuum dryers	
vacuum shelf	pieces, purees, liquids
vacuum belt	purees, liquids
freeze dryers	pieces, liquids

Source: Potter, 1984

Fruit and Vegetable Natural Drying—Sun and Solar Drying

Surplus production and specifically grown crops may be preserved by natural drying for use until the next crop can be grown and harvested. Natural dried products can also be transported cheaply for distribution to areas where there are permanent shortages of fruit and vegetables.

The methods of producing sun and solar dried fruit and vegetables described here are simple to carry out and inexpensive. They can be easily employed by grower, farmer, cooperative, etc.

The best time to preserve fruits and vegetables is when there is a surplus of the product and when it is difficult to transport fresh materials to other markets. This is especially true for crops which are very easily damaged in transport and which stay in good condition for a very short time. Preservation extends the storage (shelf) life of perishable foods so that they can be available throughout the year despite their short harvesting season.

Sun and solar drying of fruits and vegetables is a cheap method of preservation because it uses the natural resource/ source of heat: sunlight. This method can be used on a commercial scale as well at the village level provided that the climate is hot, relatively dry and free of rainfall during and immediately after the normal harvesting period. The fresh crop should be of good quality and as ripe (mature) as it would need to be if it was going to be used fresh. Poor quality produce cannot be used for natural drying.

Dried fruit and vegetables have certain advantages over those preserved by other methods. They are lighter in weight than their corresponding fresh produce and, at the same time, they do not require refrigerated storage. However, if they are kept at high temperatures and have a high moisture content they will turn brown after relatively short periods of storage.

Different lots at various stages of maturity (ripeness) must NOT be mixed together; this would result in a poor dried product. Some varieties of fruit and vegetables are better for natural drying than other; they must be able to withstand natural drying without their texture becoming tough so that they are not difficult to reconstitute. Some varieties are unsuitable because they have irregular shape and there is a lot of wastage in trimming and cutting such varieties.

Damaged parts which have been attacked by insects, rodents, diseases, etc. and parts which have been discoloured or have a bad appearance or colour, must be removed. Before trimming and cutting, most fruit and vegetables must be washed in clean water. Onions are washed after they have been peeled.

Trimming includes the selection of the parts which are to be dried, cutting off and disposing of all unwanted material. After trimming, the greater part of the fruit and vegetables cut into even slices of about 3 to 7 mm thick or in halves/quarters, etc.

It is very important to have all slices/parts in one drying lot of the same thickness/size; the actual thickness will depend on the kind of material. Uneven slices or different sizes dry at different rates and this result in a poor quality end product. Onions and root crops are sliced with a hand slicer or vegetable cutter; bananas, tomatoes and other vegetables or fruit are sliced with stainless-steel knives.

As a general rule plums, grapes, figs, dates are dried as whole fruits without cutting/slicing.

Some fruit and vegetables, in particular bananas, apples and potatoes, go brown very quickly when left in the air after peeling or slicing; this discoloration is due to an active enzyme called phenoloxidase. To prevent the slices from going brown they must be kept under water until drying can be started. Salt or sulphites in solution give better protection. However, whichever method is used, further processing should follow as soon as possible after cutting or slicing.

Blanching—exposing fruit and vegetable to hot or boiling water—as a pre-treatment before drying has the following advantages:

- it helps clean the material and reduce the amount of microorganisms present on the surface;

- it preserves the natural colour in the dried products; for example, the carotenoid (orange and yellow) pigments dissolve in small intracellular oil drops during blanching and in this way they are protected from oxidative breakdown during drying;
- it shortens the soaking and/or cooking time during reconstitution.

During hot water blanching, some soluble constituents are leached out: water-soluble flavours, vitamins (vitamin C) and sugars. With potatoes this may be an advantage as leaching out of sugars makes the potatoes less prone to turning brown.

Blanching is a delicate processing step; time, temperature and the other conditions must be carefully monitored. A suitable water-blanching method in traditional processing is as follows:

- the sliced material is placed on a square piece of clean cloth; the corners of the cloth are tied together;
- a stick is put through the tied corners of the cloth;
- the cloth is dipped into a pan containing boiling water and the stick rests across the top of the pan thus providing support for the cloth bag.

The average blanching time is 6 minutes. The start of blanching has to be timed from the moment the water starts to boil again after the cloth bag has been dipped into the pan. While the material is being blanched the cloth bag should be raised and lowered in the water so that the material is heated evenly.

When the blanching time is completed the cloth bag and its content should be dipped into cold water to prevent over-blanching. If products are over-blanched (boiled for too long) they will stick together on the drying trays and they are likely to have a poor flavour.

Green beans, carrots, okra, turnip and cabbage should always be blanched. The producer can choose whether or not potatoes need blanching. Blanching is not needed for onions, leeks, tomatoes and sweet peppers. Tomatoes are dipped into hot water for one minute when they need to be peeled but this is not blanching.

Use of Preservatives

Preservatives are used to improve the colour and keeping qualities of the final product for some fruits and vegetables. Preservatives include items such as sulphur dioxide, ascorbic acid, citric acid, salt and sugar and can either be simple or compound solutions.

Treatment with preservatives takes place after blanching or, when blanching is not needed, after slicing. In traditional, simple processing the method recommended is:

- put enough preservative solution to cover the cloth bag into a container/pan;
- dip the bag containing the product into the preservative solution for the amount of time specified;
- remove the bag and put it on a clean tray while the liquid drains out. The liquid which drains out must not go back into the preservative solution because it would weaken the solution.

Care must be taken after each dip to refill the container to the original level with fresh preservative solution of correct strength. After five lots of material have been dipped, the remaining solution is thrown away; *i.e.* a fresh lot of preservative solution is needed for every 5 lots of material. The composition and strength of the preservative solution vary for different fruit and vegetables.

The strength of sulphur dioxide is expressed as "parts per million" (ppm). 1.5 grams of sodium metabisulphite in one litre of water gives 1000 ppm of sulphur dioxide. Details for solutions of different strengths are given in the following table.

Dilutions of Sodium Metabisulphite with Water to Obtain "PM" of Sulphur Dioxide (SO_2)

PPM SO_2	*Sodium Metabisulphite*	
	Grams per litre of water	*Grams per 20 litre tin of water*
1000	1.5	30.0
2000	3.0	60.0
3000	4.5	90.0

Contd...

PPM SO_2	*Sodium Metabisulphite*	
	Grams per litre of water	*Grams per 20 litre tin of water*
4000	6.0	120.0
5000	7.5	150.0
6000	9.0	180.0
7000	10.5	210.0

One level teaspoon of sodium metabisulphite = c. 5 g.

Sodium bicarbonate is added to the blanching water when okra, green peas and some other green vegetables are blanched. The chemical raises the pH of the blanching water and prevents the fresh green colour of chlorophyll being changed into pheophytin which is unattractive brownish-green.

The preservative solutions in the fruit and vegetable pretreatment can only be used in enamelled, plastic or stainless-steel containers; never use ordinary metal because solutions will corrode this type of container.

As a general rule, preservatives are not used for treating onions, garlic, leeks, chillies and herbs.

Osmotic Dehydration

In osmotic dehydration the prepared fresh material is soaked in a heavy (thick liquid sugar solution) and/or a strong salt solution and then the material is sun or solar dried. During osmotic treatment the material loses some of its moisture. The syrup or salt solution has a protective effect on colour, flavour and texture.

This protective effect remains throughout the drying process and makes it possible to produce dried products of high quality. This process makes little use of sulphur dioxide.

Sun Drying

The main problems for sun drying are dust, rain and cloudy weather. Therefore, drying areas should be dust-free and whenever there is a threat of a dust storm or rain, the drying trays should be stacked together and placed under cover.

In order to produce dust-free and hygienically clean products, fruit and vegetable material should be dried well above ground level so that they are not contaminated by dust, insects, livestock or people. All materials should be dried on trays designed for the purpose; the most common drying trays have wooden frames with a fitted base of nylon mosquito netting. Mesh made of woven grass can also be used. Metal netting must NOT be used because it discolours the product.

The trays should be placed on a framework at table height from the ground. This allows the air to circulate freely around the drying material and it also keeps the food product well away from dirt. Ideally the area should be exposed to wind and this speed up drying, but this can only be done if the wind is free of dust.

With 80 cm x 50 cm trays, the approximate load for a tray is 3 kg; the material should be spread in even layers. During the first part of the drying period, the material should be stirred and turned over at least once an hour.

This will help the material dry faster and more evenly, prevent it sticking together and improve the quality of the finished product. Products for sun drying should be prepared early in the day; this will ensure that the material enjoys the full effect of the sun during the early stages of drying.

At night the trays should be stacked in a ventilated room or covered with canvas. Plastic sheets should NEVER be used for covering individual trays during sun drying.

Dry or nearly dry products can be blown out of the tray by the wind. However, this can be protected by covering the loaded tray with an empty one, this also gives protection against insects and birds.

Shade Drying

Shade drying is carried out for products which can lose their colour and/or turn brown if put in direct sunlight. Products which have naturally vivid colours like herbs, green and red sweet peppers, chillies, green beans and okra give a more attractive end-product when they are dried in the shade.

The principles for the shade drying are the same as for sun drying. The material to be dried requires full air circulation. Therefore, shade drying is carried out under a roof or thatch which has open sides; it cannot be done either inside conventional buildings with side walls or in compounds sheltered from wind.

Under dry conditions when there is a good circulation of air, shade drying takes little more time than is normally required for drying in full sunlight.

Identification of suitable designs of solar dryers for different applications:

In the selection of appropriate solar dryers for commercial scale operation, it is imperative that economics be kept in view at all time. A total Energy System concept should be employed and due consideration be given to parasitic energy consumption.

The following features have been identified:

- large scale dryers are more promising than small scale ones. However, small scale dryers should not be neglected.
- the dryer should be designed to maximise the utilisation factor of the capital investment, *i.e.* multi-products (fruit, vegetables and other raw material) and multi-use (*e.g.* drying and heating water for domestic use).
- in general, an auxiliary heat source should be provided to assure reliability, to handle peak loads and also to provide continuous drying during periods of no sunshine.
- forced convection indirect dryers are preferred because they offer better control, more uniform drying and because of their high heat collection efficiency result in smaller collector area. However, parasitic power should be kept to a minimum.

Two dryer systems have been identified:

- a cabinet type dryer with natural convection for internal air circulation for the processing of dried fruit such as mango, banana, pineapple, apricot, pear, apple, etc. and also for potato chips and other vegetables;
- a greenhouse type dryer with forced air circulation.

Some of the barriers to the commercial development of solar dryers have been attributed to:

- initial cost—poor farmers cannot afford them
- durability—constant breakdown due to using low cost building materials
- misuse—through lack of training and technical skills
- dependability and reliability - during the wet season when drying is critical there is not enough solar energy available
- the wider use of Solar Drying Systems has been limited by other factors which are not necessarily of a technical or technological nature. Among the most important are the lack of national policies directed to promoting the drying of produce at the production site, in order to reduce losses, improve quality and increase farmers' earnings.

Construction of Solar Dryers

In the case of simple natural convection dryers it may be more appropriate to build and operate a number of small units. Multiplicity allows diversity, since more than one crop can be dried at a time. A further advantage is that if one dryer is out of operation due to damage, drying can still continue at reduced capacity using the other dryers.

On the other hand, more sophisticated dryers, such as forced convection solar dryers, benefit from economies of scale due to the investment tied up in the fan and the source of heat.

Generally speaking, one large dryer will be more cost-effective than two smaller units. However, it should be taken into consideration that an oversized unit will be operating at less than full capacity, reducing any cost advantage. The drying area required will depend on local conditions, commodity and number of trays on each rack or trolley.

Construction methods and materials: Construction methods and available materials may vary considerably from location to location. It is not within the scope of this document to discuss individual, local circumstances. Some general guidelines regarding factors which must be considered can, however, be given: dimensions of standard materials. Where possible, design should take account of the sizes of material locally available. For example, it would be poor design to specify the width of a corrugated iron collector as 1.1 m if the standard width of a corrugated iron sheet is 1 m.

Before finalising a design the commercial availability of materials must be ascertained.

Use of Rural Materials: The cost of building of solar dryer can be minimised if the producer is able to use wood cut straight from the forest rather than prepared timber.

Careful design in the development stage of a dryer can often facilitate the use of cheaper materials. Difficulties caused by these materials are in joining pieces of the structure, in sealing the structure against air leaks, and in attaching the plastic sheet to the (wooden) frame.

There is obvious scope for designs which use prepared timber for strategic points and unprepared at others.

Where the use of wood is necessary, remember to take environmental factors into consideration. For example, determine the effect of flash flooding or termites might be and take the appropriate preventive action.

Use of Plastic Sheets: For many solar dryers, the clear plastic sheet used is the major capital cost to the farmer; therefore, the type of plastic chosen is important.

A choice must be made between a relatively cheap plastic such as ordinary polyethylene which will last, at best, for one season due to photo-degradation and wear and tear; and a more expensive, better quality plastic less prone to photo-degradation; or even glass or a rigid plastic.

Attaching plastic sheet to the framework structure, so as to minimise the likelihood of the plastic being torn is, perhaps, the most difficult part of building a dryer. Listed below are some general points which should be followed to prolong the useful lifetime of plastic sheet on a solar dryer:

- when attaching plastic sheet to the framework, care should be taken to stretch the plastic at the points of attachment, but the plastic should not be so loose that it will flap about in the wind;
- rather than merely stapling or nailing the plastic directly to the framework, it is preferable to sandwich the plastic between the framework and a batten. This may not be practical when unprepared wood or other materials are being used.
- no sharp edges should come in contact with the plastic sheet since these will initiate tears;
- fold over the plastic at the point of attachment to the frame, so that there are two or more layers of plastic. This will help prevent tears;
- when fixing the sheet over the framework, sags and hollows in which water can collect should be avoided wherever possible;
- the dryer should be handled as carefully and as seldom as possible during operation and when not in use.

Technical Criteria: The following design factors must be established:

- the throughput of the dryer over the productive season; the size of batch to be dried;
- the drying period(s) under stated conditions;
- the initial and desired final moisture content of the commodity (if known);
- the drying characteristics of the commodity, such as maximum drying temperature, effect of sunlight upon the product quality, etc.;
- climatic conditions during the drying season, *i.e.* sunlight intensity and duration; air temperature and humidity; wind speed (such data may be available from local meteorological stations);
- availability and reliability of electrical power;
- the availability, quality, durability and price of potential construction materials such as:

- o glazing materials: glass, plastic sheet or film;
- o wood (prepared or unprepared);
- o nails, screw, bolts, etc.;
- o metal sheet, flat or corrugated angle iron;
- o bricks (burnt or mud), concrete blocks, stones, cement, sand, etc.
- o roofing thatch;
- o metal mesh, wire netting, etc.
- o mosquito netting, muslin, etc.
- o bamboo or fibre weave;
- o black paint, other blackening materials;
- o insulation material; sawdust, etc.;

- the type of labour available to build and operate the dryer;
- the availability of clean water at the site for preparation of the commodity prior to drying.

In any one situation there may well be other technical factors that need to be considered.

Socio-economic Criteria: From the initial considerations, estimates of the capital costs of the dryer, the price of the commodity to be dried, and the likely selling price of the dried product will have been made. Other question that need to be considered are the following:

- who will own the dryer?
- is the dryer to be constructed by the end-user (with or without advice from extension agencies), local contractors, or other organisations?
- who will operate and maintain it?
- how can the drying operation be incorporated into current practices?
- are sources of finance from local authorities or extension agencies available, etc.?

Obviously there are many other socio-economic factors, particularly those of a local nature, which must be taken into account. It cannot be stressed too highly that if such factors are not taken into account and evaluated, then is every chance that an inappropriate dryer design may result. Equal emphasis must be placed on both technical and socio-economic factors.

Summary

(a) Situations where solar dryer may be useful:

- where the cost of conventional energy is prohibitive and/or the supply is erratic, to supplement existing artificial drying systems and reduce fuel costs;
- where land is in short supply or expensive;
- where the quality of existing sun dried products can be improved upon;
- where the labour is in short supply;
- where is plenty of sunshine, but high humidity.

(b) Situations where solar dryers may not be useful:

- where conventional energy sources are abundant and cheap;
- where large amounts of combustible by-products or waste materials are freely available;
- where there is insufficient sunshine;
- where is plenty of sunshine and arid conditions (sun drying may suffice);
- where the quality of sun dried products already made cannot be improved upon;
- where local operators are insufficiently trained;
- where the ramifications of introducing a solar dryer have not been completely thought out.

Sun/solar Drying Tray

The drying tray described requires seasoned timber 22.5 mm thick x 50 mm wide.

A sun drying tray requires 6 meters of seasoned timber 22.5 mm thick x 50 mm wide.

Cut the timber into lengths of 900 mm long for the sides of the tray and 600 mm long for the ends -4 pieces of each length will be needed. The ends of each piece are cut as shown in the drawing—this is to make flush fitting joints. Join the corners using small brass screws 20 mm long. To make extra strong joints use good quality wood glue as well as the screws.

The nylon mosquito netting or grass woven mesh can be fitted between the frames as shown in the bottom drawing. Cut the mesh a little larger than the size of the frame. Using drawing pins, pin the mesh to the outside edges of one of the frames - the mesh should be pulled tight as the pins are put in around the edges.

Lay the other frame on top and drill holes about 3 mm in diameter at the points marked X in the top drawing. Use nails that are a tight fit in the holes and tap gently into place leaving a portion standing above the frame.

Cut off the standing part to leave a piece about 12 mm long which is then bent over and tapped firmly down onto the frame. When the frame has been put together tightly, the drawing pins can be removed.

Dryers

Preservation by Concentration

Foods are concentrated for many of the same reasons that they are dehydrated; concentration can be a form of preservation but this is true only for some foods. Concentration reduces weight and volume and results in immediate economic advantages.

Nearly all liquid foods which are dehydrated are concentrated before they are dried. This is because in the early stages of water removal, moisture can be more economically removed in highly efficient evaporators than in dehydration equipment. Further, increased viscosity from concentration often is needed to prevent liquids from running off drying surfaces or to facilitate foaming or puffing.

Foods are also concentrated because the concentrated forms have become desirable components of diet in their own right. Thus, fruit juices plus sugar with concentration becomes jelly. The more common concentrated fruit and vegetable products include items as fruit and vegetable juices and nectars, jams and jellies, tomato paste, many types of fruit purees used by bakers, candy makers and other food manufacturers.

Aspects of Preservation by Concentration

The level of water in virtually all concentrated foods is in itself more than enough to permit microbial growth. Yet while many concentrated foods such as non-acid fruit and vegetable purees may quickly undergo microbial spoilage unless additionally processed, such items as sugar syrups, jellies and jams are relatively "immune" to spoilage; the difference of course is in what is dissolved in the remaining water and what osmotic concentration is reached.

1. Cell walls
2. Trays on a car
3. Metal plat with burning sulphur
4. Hole for sulphur dioxide fumes
5. Exhaust hole

Tunnel type dehydration unit—tunnel dryer:

1. Control and switch-board
2. Burner
3. Platform for burner
4. Frontal plate
5. Air flow regulating plate
6. Burner's cylinder
7. Air circulating fan
8. Air direction conveying plates
9. Cars rail
10. Tunnel "feeding" door: inlet of cars with fresh product
11. Tunnel "evacuating" door: outlet of cars with dry product
12. Cars pushing device
13. Cars with trays
14. Drying trays

Typical Counter-flow Tunnel Dryer Construction

Sugar and salt in concentrated solutions have high osmotic pressure. When these are sufficient to draw water from microbial cells or prevent normal diffusion of water into these cells, a preservative condition exists.

The critical concentration of sugar in water to prevent microbial growth will vary depending upon the type of microorganisms and the presence of other food constituents, but usually 70% sucrose in solution will stop growth of all microorganisms in foods. Less than this concentration may be effective but for short periods of time unless the foods contain acid or they are refrigerated.

Salt becomes a preservative when its concentration is increased and levels of about 18% to 25% in solution generally will prevent all growth of microorganisms in foods. Except in the case of certain briny condiments, however, this level is rarely tolerated in foods. Removal of water by concentration also increases the level of food acids in solution (particularly significant in concentrated fruit juices).

Reduced weight and volume by concentration: While the preservation effects of food concentration are important, the main reason of most food concentration is to reduce food weight and bulk. Tomato pulp which is ground tomato minus the skins and seeds, has a solid content of only 6 % and so a 3.785 litre can would contain only 231 g of tomato solids.

Specific Gravity and Solids—Tomato Pulp and Commercial Tomato Concentrates

Tomato Solids, %	*Specific gravity at 68°F (20°C)*	*Dry Tomato Solids per Gal. at 68°F, lb*	*per Litre at 20°C, g*
6.0 Tomato pulp	1.025	0.51	61
10.8	1.045	0.94	113
12.0 Tomato puree	1.050	1.05	126
14.2	1.060	1.25	151
16.5	1.070	1.47	177
25.0	1.107	2.31	277
26.0	1.112	2.41	289
28.0 Tomato paste	1.120	2.61	314
30.0	1.129	2.82	339
32.0	1.138	3.03	364

Source: Adapted from National Canners Assoc. (1956)

Concentrated to 32% solids, the same can would contain 1.38 kg of tomato solids or six times the value of product. For a manufacturer needing tomato solids such a producer of soups, canned spaghetti or frozen pizza the saving from concentration are enormous in cans, transportation costs, warehousing costs and handling costs throughout his operation.

Methods of Concentration

a. *Solar Concentration:* As in food dehydration, one of the simplest methods of evaporating water is with solar energy. A typical example of this method is production at farm level in developing countries of fruit pastes/ leathers (such as apricot or plum pastes).

b. *Open Kettles:* Some foods can be satisfactorily concentrated in open kettles that are heated by steam. This is the case for jellies and jams, tomato juices and purees and for certain types of soups. High temperatures and long concentration times should be avoided in order to reduce or eliminate damage. It is also necessary to avoid thickening and burn-on of product to the kettle wall as these gradually lower the efficiency of heat transfer and slow the concentration process.

However, when the process is under control, this type of evaporation is still highly recommended for small scale operations in developing countries. It is a quite widely used system, mainly for jellies, jams and marmalades.

c. Flash evaporators.

d. Thin-film Evaporators.

e. *Vacuum Evaporators:* It is common to construct several vacuumised vessels in series so that the product moves from one vacuum chamber to the next and thereby becomes progressively more concentrated in stages.

With such an arrangement the successive stages are maintained at progressively higher degrees of vacuum, and the hot water vapour produced by the first stage is used to heat the second stage, the vapour from the second stage heats the third stage and so on. In this way maximum use of heat energy is made. Such system is called a multiple effect vacuum evaporator and is illustrated. It is a widely used system for concentrated tomato paste.

f. *Freeze Concentration*: This process has been known for many years and has been applied commercially to orange juice. However, high processing costs due largely to losses of juice occlude [unclear] to the ice crystals, have limited the number of installations to date.

g. Ultrafiltration and reverse Osmosis.

Changes from concentration: Obviously concentration that exposes food to 100° C or higher temperatures for prolonged periods can cause major changes in organoleptic and nutritional properties. Cooked foods and darkening of colour are two of the more common heat induced results which must be kept under control during a well designed process with an efficient evaporator which is still "safe".

Microbial destruction is another type of change that may occur during concentration and will be largely dependent upon temperature. Concentration at a temperature of 100 °C or slightly above will kill many microorganisms but cannot be depended on to destroy bacterial spores. When the food contains acid, such as fruit juices, the kill will be greater but again sterility is unlikely.

On the other hand, when concentration is done under vacuum many bacterial types not only survive the low temperatures but multiply in the concentrating equipment. It is therefore necessary to stop frequently and sanitise low temperature evaporators and where sterile concentrated foods are required, to resort to an additional preservation treatment.

Chemical Preservation

Many chemicals will kill microorganisms or stop their growth but most of these are not permitted in foods; chemicals that are permitted as food preservatives are listed. Chemical food preservatives are those substances which are added in very low quantities (up to 0.2%) and which do not alter the organoleptic and physico-chemical properties of the foods at or only very little.

Preservation of food products containing chemical food preservatives is usually based on the combined or synergistic activity of several additives, intrinsic product parameters (*e.g.* composition, acidity, water activity) and extrinsic factors (*e.g.* processing temperature, storage atmosphere and temperature).

This approach minimises undesirable changes in product properties and reduces concentration of additives and extent of processing treatments.

The concept of combinations of preservatives and treatments to preserve foods is frequently called the hurdle or barrier concept. Combinations of additives and preservatives systems provide unlimited preservation alternatives for applications in food products to meet consumer demands for healthy and safe foods.

Chemical food preservatives are applied to foods as direct additives during processing, or develop by themselves during processes such as fermentation. Certain preservatives have been used either accidentally or intentionally for centuries, and include sodium chloride (common salt), sugar, acids, alcohols and components of smoke. In addition to preservation, these compounds contribute to the quality and identity of the products, and are applied through processing procedures such as salting, curing, fermentation and smoking.

Traditional chemical food preservatives and their use in fruit and vegetable processing technologies could be summarised as follows:

1. Common salt: brined vegetables;
2. Sugars (sucrose, glucose, fructose and syrups):
   (i) Foods preserved by high sugar concentrations: jellies, preserves, syrups, juice concentrates;
   (ii) Interaction of sugar with other ingredients or processes such as drying and heating;
3. Indirect food preservation by sugar in products where fermentation is important (naturally acidified pickles and sauerkraut).

Acidulants and other preservatives formed in or added to fruit and vegetable products are as follows:

Lactic Acid: This acid is the main product of many food fermentations; it is formed by microbial degradation of sugars in products such as sauerkraut and pickles. The acid produced in such fermentations decreases the pH to levels unfavourable for growth of spoilage organisms such as putrefactive anaerobes and butyric-acid-producing bacteria. Yeasts and moulds that can grow at such pH levels can be controlled by the inclusion of other preservatives such as sorbate and benzoate.

Acetic Acid: Acetic acid is a general preservative inhibiting many species of bacteria, yeasts and to a lesser extent moulds. It is also a product of the lactic-acid fermentation, and its preservative action even at identical pH levels is greater than that of lactic acid. The main applications of vinegar (acetic acid) includes products such as pickles, sauces and ketchup.

Other acidulants:

- Malic and tartaric (tartric) acids is used in some countries mainly to acidify and preserve fruit sugar preserves, jams, jellies, etc.
- Citric acid is the main acid found naturally in citrus fruits; it is widely used (in carbonated beverages) and as an acidifying agent of foods because of its unique flavour properties. It has an unlimited acceptable daily intake and is highly soluble in water. It is a less effective antimicrobial agent than other acids.
- Ascorbic acid or vitamin C, its isomer isoascorbic or erythorbic acid and their salts are highly soluble in water and safe to use in foods.

Commonly Used Lipophilic Acid Food Preservatives

Benzoic acid in the form of its sodium salt, constitutes one of the most common chemical food preservative. Sodium benzoate is a common preservative in acid or acidified foods such as fruit juices, syrups, jams and jellies, sauerkraut, pickles, preserves, fruit cocktails, etc. Yeasts are inhibited by benzoate to a greater extent than are moulds and bacteria.

Sorbic acid is generally considered non toxic and is metabolised; among other common food preservatives the WHO has set the highest acceptable daily intake (25 mg/kg body weight) for sorbic acid. Sorbic acid and its salts are practically tasteless and odourless in foods, when used at reasonable levels (< 0.3 %) and their antimicrobial activity is generally adequate.

Sorbates are used for mould and yeast inhibition in a variety of foods including fruits and vegetables, fruit juices, pickles, sauerkraut, syrups, jellies, jams, preserves, high moisture dehydrated fruits, etc.

Potassium sorbate, a white, fluffy powder, is very soluble in water (over 50%) and when added to acid foods it is hydrolysed to the acid form. Sodium and calcium sorbates also have preservative activities but their application is limited compared to that for the potassium salt, which is employed because of its stability, general ease of preparation and water solubility.

Gaseous Chemical Food Preservatives

Sulphur Dioxide and Sulphites: Sulphur dioxide (SO_2) has been used for many centuries as a fumigant and especially as a wine preservative. It is a colourless, suffocating, pungent-smelling, non-flammable gas and is very soluble in cold water (85 g in 100 ml at 25°C).

Sulphur dioxide and its various sulphites dissolve in water, and at low pH levels yield sulphurous acid, bisulphite and sulphite ions. The various sulphite salts contain 50-68% active sulphur dioxide. A pH dependent equilibrium is formed in water and the proportion of SO_2 ions increases with decreasing pH values. At pH values less than 4.0 the antimicrobial activity reaches its maximum.

Sulphur dioxide is used as a gas or in the form of its sulphite, bisulphite and metabisulphite salts which are powders. The gaseous form is produced either by burning Sulphur or by its release from the compressed liquefied form.

Metabisulphite are more stable to oxidation than bisulphites, which in turn show greater stability than sulphites.

The antimicrobial action of sulphur dioxide against yeasts, moulds and bacteria is selective, with some species being more resistant than others.

Sulphur dioxide and sulphites are used in the preservation of a variety of food products. In addition to wines these include dehydrated/dried fruits and vegetables, fruit juices, acid pickles, syrups, semi-processed fruit products, etc. In addition to its antimicrobial effects, sulphur dioxide is added to foods for its antioxidant and reducing properties, and to prevent enzymatic and non-enzymatic browning reactions.

Carbon dioxide (CO_2) is a colourless, odourless, non-combustible gas, acidic in odour and flavour. In commercial practice it is sold as a liquid under pressure (58 kg per cm^3) or solidified as dry ice.

Carbon dioxide is used as a solid (dry ice) in many countries as a means of low-temperature storage and transportation of food products. Beside keeping the temperature low, as it sublimes, the gaseous CO_2 inhibits growth of psychrotrophic microorganisms and prevents spoilage of the food (fruits and vegetables, etc.).

Carbon dioxide is used as a direct additive in the storage of fruits and vegetables. In the controlled/ modified environment storage of fruit and vegetables, the correct combination of O_2 and CO_2 delays respiration and ripening as well as retarding mould and yeast growth.

The final result is an extended storage of the products for transportation and for consumption during the off-season. The amount of CO2 (5-10%) is determined by factors such as nature of product, variety, climate and extent of storage.

Chlorine: The various forms of chlorine constitute the most widely used chemical sanitiser in the food industry. These chlorine forms include chlorine (Cl_2), sodium hypochlorite (NaOCl), calcium hypochlorite ($Ca(OCl)_2$) and chlorine dioxide gas (ClO_2).

These compounds are used as water adjuncts in processes such as product washing, transport, and cooling of heat-sterilised cans; in sanitising solutions for equipment surfaces, etc.

Important applications of chlorine and its compounds include disinfection of drinking water and sanitation of food processing equipment.

General Rules for Chemical Preservation

1. Chemical food preservatives have to be used only at a dosage level which is needed for a normal preservation and not more.
2. "Reconditioning" of chemical preserved food, *e.g.* a new addition of preservative in order to stop a microbiological deterioration already occurred is not recommended.
3. The use of chemical preservatives MUST be strictly limited to those substances which are recognised as being without harmful effects on human beings' health and are accepted by national and international standards and legislation.

Factors which determine/ influence the action of chemical food preservatives:

1. Factors related to the chemical preservatives:
   a. chemical composition;
   b. concentration.
2. Factors related to microorganisms:
   (a) microorganism species; as a general rule it is possible to take the following facts as a basis:
   - sulphur dioxide and its derivatives can be considered as an "universal" preservative; they have an antiseptic action on bacteria as well as on yeasts and moulds;
   - benzoic acid and its derivatives have a preservative action which is stronger against bacteria than on yeasts and moulds;
   - sorbic acid acts on moulds and certain yeast species; in higher dosage levels it acts also on bacteria, except lactic and acetic ones;
   - formic acid is more active against yeasts and moulds and less on bacteria.

   (b) the initial number of microorganisms in the treated product determines the efficiency of the chemical preservative.

The efficiency is less if the product has been contaminated because of preliminary careless hygienic treatment or an incipient alteration. Therefore, with a low initial number of microorganisms in the product, the preservative dosage level could be reduced.

3. Specific factors related to the product to be preserved:
   a. product chemical composition;
   b. influence of the pH value of the product: the efficiency of the majority of chemical preservatives is higher at lower pH values, *i.e.* when the medium is more acidic.
   c. physical presentation and size which the product is sliced to: the chemical preservative's dispersion in food has an impact on its absorption and diffusion through cell membranes on microorganisms and this determines the preservation effect.

Therefore, the smaller the slicing of the product, the higher the preservative action. Preservative dispersion is slowed down by viscous foods (concentrated fruit juices, etc.)

4. Miscellaneous factors
   a. *Temperature:* Chemical preservative dosage level will be established as a function of product temperature and characteristics of the micro-flora;
   b. *Time:* At preservative dosage levels in employed in industrial practice, the time period needed in order to obtain a "chemical sterilisation" is a few weeks for benzoic acid and shorter for sulphurous acid.

Usual accepted chemical food preservatives:

Chemical Food Preservatives

Agent	Acceptable Daily intake (mg/Kg body weight)	Commonly used levels (%)
Lactic acid	No limit	No limit
Citric acid	No limit	No limit
Acetic acid	No limit	No limit
Sodium Diacetate	15	0.3-0.5
Sodium benzoate	5	0.03-0.2
Sodium propionate	10	0.1-0.3
Potassium sorbate	25	0.05-0.2
Methyl paraben	10	0.05-0.1
Sodium nitrite	0.2	0.01-0.02
Sulphur dioxide	0.7	0.005-0.2

Source: FDA, 1991

For the purpose of this document, some food products in common usage are summarised as follows:

Citric Acid: fruit juices; jams; other sugar preserves;

Acetic Acid: vegetable pickles; other vegetable products;

Sodium Benzoate: vegetable pickles; preserves; jams; jellies; semi-processed products;

Sodium Propionate: fruits; vegetables;

Potassium Sorbate: fruits; vegetables; pickled products; jams, jellies;

Methyl Paraben: fruit products; pickles; preserves;

Sulphur Dioxide: fruit juices; dried / dehydrated fruits and vegetables; semi-processed products.

Preservation of Vegetables by Acidification

Food acidification is a means of preventing their deterioration in so far as a non-favourable medium for microorganisms development is created. This acidification can be obtained by two ways: natural acidification and artificial acidification.

Natural Acidification

This is achieved by a predominant lactic fermentation which assures the preservation based on acidoceno-anabiosys principle; preservation by lactic fermentation is called also biochemical preservation. Throughout recorded history food has been preserved by fermentation. In spite of the introduction of modern preservation methods, lactic acid fermented vegetables still enjoy a great popularity, mainly because of their nutritional and gastronomic qualities.

The various preservation methods discussed thus far, based on the application of heat, removal of water, cold and other principles, all have the common objective of decreasing the number of living organisms in foods or at least holding them in check against further multiplication. Fermentation processes for preservation purposes, in contrast, encourage the multiplication of microorganisms and their metabolic activities in foods. But the organisms that are encouraged are from a select group and their metabolic activities and end products are highly desirable. The extent of this desirability is emphasised by a partial list of fermented fruits and vegetable products from various parts of the world.

There are some characteristic features in the production of fermented vegetables which will be pointed out below using cucumbers as an example. In the production of lactic acid fermented cucumbers, the raw material is put into a brine without previous heating. Through the effect of salt and oxygen deficiency the cucumber tissues gradually die. At the same time, the semi-permeability of the cell membranes is lost, whereby soluble cell components diffuse into the brine and serve as food substrate for the microorganisms.

Under such specific conditions of the brine the lactic acid bacteria succeed in overcoming the accompanying microorganisms and lactic acid as the main metabolic products is formed. Under favourable conditions (for example moderate salt in the brine, use of starter cultures) it takes at least 3 days until the critical pH value of 4.1 or less-desired for microbiological reasons - is reached.

Beside the typical taste, for the consumer a crisp texture is the most important quality criterion for fermented vegetables. Because there is no heating step before the fermentation, the indigenous plant enzymes in the fermenting materials are still present during the very first phase. After the destruction of the cell membranes they easily get to their active sites and under favourable conditions they can easily cause softening.

The environmental conditions act in a different manner on single enzymes or enzymes systems: some enzymes are strongly inhibited by salt, others are activated, and in the acid pH region many enzymes are irreversibly inactivated. Beside indigenous enzymes also enzymes produced by microorganisms can be responsible for the undesired soft products.

In technically advanced societies the major importance of fermented foods has come to be variety they add to the diet. However, in many less developed areas of the world, fermentation and natural drying are the major food preservation methods and as such are vital to survival of a large proportion of the world's current population.

Artificial acidification is carried out by adding acetic acid which is the only organic acid harmless for human health and stable in specific working conditions; in this case biological principles of the preservation are acidoanabiosys and, to a lesser extent, acidoabiosys.

Combined acidification is a preservation technology which involves as a preliminary processing step a weak lactic fermentation followed by acidification (vinegar addition).

The two main classes of vegetables preserved by acidification are sauerkraut and pickles; the definitions of these products adapted from US Code of Federal Register (7 CFR 52, 1991) are as follows. Bulk sauerkraut. Bulk or barrelled sauerkraut is the product of characteristic acid flavour, obtained by the full fermentation, chiefly lactic, of properly prepared and shredded cabbage in the presence of 2-3% salt. On completion of fermentation, it contains not more than 1.5% of acid, expressed as lactic acid.

Canned sauerkraut. Canned (or packaged) sauerkraut, is prepared from clean, sound, well-matured heads of the cabbage plant (Brassica oleracea var. capitata L.) which have been properly trimmed and cut; to which salt is added and which is cured by natural fermentation.

The product may or may not be packed with pickled peppers, pimientos, or tomatoes or contain other flavouring ingredients to give the product specific flavour characteristics.

The product:

a) may be canned by processing sufficiently by heat to assure preservation in hermetically sealed containers; or

b) may be packaged in sealed containers and preserved with or without the addition of benzoate of soda or any other ingredient permissible under the provisions of Food and Drug Administration (FDA).

"Pickles" means the product prepared entirely or predominantly from cucumbers (Cucumis sativus L.). Clean, sound ingredients are used which may or may not have been previously subjected to fermentation and curing in a salt brine (solution of sodium chloride, NaCl).

The prepared pickles are packed in a vinegar solution to which may be added salt and other vegetables, nutritive sweeteners, seasonings, flavourings, spices, and other ingredients permissible under FDA regulations. The product is packed in suitable containers and heat treated, or otherwise processed to assure preservation. Sauerkraut and pickle products can be preserved under the effect of natural or added acidity, followed by pasteurization when this acidification is not sufficient.

Sauerkraut is a very good source of vitamin C; the importance of this product should be emphasised in developing countries as a simple technology which can be applied mainly for consumption of the finished products in remote, isolated areas during the cold season. It is also a excellent technology to be learned to schools which have their own source of cabbage and cucumbers through school agricultural farms.

Sauerkraut and pickles are manufactured on an industrial scale in significant quantities worldwide. However, the basic technology is simple and could be applied at home, farm and community level after some explanation and training. The natural acidification preservation could be considered similar to sun/solar drying in terms of training and development.

Some Industrial Fermentation Processes in Food Industries

I. Lactic acid bacteria	
- cucumbers	dill pickles, sour pickles
- cabbage	sauerkraut
- turnips	sauerruben
- lettuce	lettuce kraut
- mixed vegetables, turnips, radish, cabbage	
- mixed Chinese vegetables,	
cabbage	Kimchi
- vegetables and milk	Tarhana
- vegetables and rice	Sajur asin
II. Lactic acid bacteria with other micro-organisms	
- with yeasts	Nukamiso pickles
- with moulds	tempeh, soy sauce
III. Acetic acid bacteria - wine, cider or any alcoholic and sugary or starchy products may be converted to vinegar	
IV. Yeasts	
- fruit	wine, vermouth

Source: Pederson (1)

Preservation with Sugar

The principle of this technology is to add sugar in a quantity that is necessary to augment the osmotic pressure of the product's liquid phase at a level which will prevent microorganism development. From a practical point of view, however, it is usual to partially remove water (by boiling) from the product to be preserved, with the objective of obtaining a higher sugar concentration. In concentrations of 60% in the finished products, the sugar generally assures food preservation.

It is important to know the ratio between the total sugar quantity in the finished product and the total sugar concentration in the liquid phase because this determines, in practice, the sugar preserving action. The percent composition of a product preserved with sugar, for example marmalade, can be expressed as follows: $[i + S + s + n + w] = 100$;

i = insoluble substance;

s = sugar from fruits;

S = added sucrose;

n = soluble "non sugar"

w = water.

In this case, total sugar concentration, in the liquid phase, of the finished product is:

$$X = 100 (S + s) / 100 - (n + i) [\%]$$

Therefore, in the case of a standard marmalade with 55 % sugar added (calculated on the finished product basis), the real concentration in the liquid phase is for example:

$$X = 100 (55 + 8) / 100 - (5 + 3) = 68.5\%$$

In the food preservation with sugar, the water activity cannot be reduced below 0.845; this value is sufficient for bacteria and neosmophile yeast inhibition but does not prevent mould attack. For this reason, various means are used to avoid mould development:

- finished product pasteurization (jams, jellies, etc.);
- use of chemical preservatives in order to obtain the antiseptisation of the product surface.

It is very important from a practical point of view to avoid any product contamination after boiling and to assure an hygienic operation of the whole technological process (this will contribute to the prevention of product moulding or fermentation). Storage of the finished products in good conditions can only be achieved by ensuring the above level of water activity.

Heat Preservation/heat Processing

Various Degrees of Preservation

There are various degrees of preservation by heating; a few terms have to be identified and understood.

a. *Sterilisation:* By sterilisation we mean complete destruction of microorganisms. Because of the resistance of certain bacterial spores to heat, this frequently means a treatment of at least 121° C (250° F) of wet heat for 15 minutes or its equivalent. It also means that every particle of the food must receive this heat treatment. If a can of food is to be sterilised, then immersing it into a 121° C pressure cooker or retort for the 15 minutes will not be sufficient because of relatively slow rate of heat transfer through the food in the can to the most distant point.

b. "*Commercially Sterile*": Term describes the condition that exists in most of canned or bottled products manufactured under Good Manufacturing Practices procedures and methods; these products generally have a shelf-life of two years or more.

c. Pasteurized means a comparatively low order of heat treatment, generally at a temperature below the boiling point of water. The more general objective of pasteurization is to extend product shelf-life from a microbial and enzymatic point of view; this is the objective when fruit or vegetable juices and certain other foods are pasteurized. Pasteurization is frequently combined with another means of preservation - concentration, chemical, acidification, etc.

d. Blanching is a type of pasteurization usually applied to vegetables mainly to inactivate natural food enzymes. Depending on its severity, blanching will also destroy some microorganisms.

Determining Heat Treatment/thermal Processing Steps

Since heat sufficient to destroy microorganisms and food enzymes also usually has adverse effects on other properties of foods, in practice the minimum possible heat treatment should be used which can guarantee freedom from pathogens and toxins and give the desired storage life; these aims will determine the choice of heat treatment. In order to safely preserve foods using heat treatment, the following must be known:

(a) what time-temperature combination is required to inactivate the most heat resistant pathogens and spoilage organisms in one particular food?

(b) what are the heat penetration characteristics in one particular food, including the can or container of choice if it is packaged?

Preservation processes must provide the heat treatment which will ensure that the remotest particle of food in a batch or within a container will reach a sufficient temperature, for a sufficient time, to inactivate both the most resistant pathogen and the most resistant spoilage organisms if it is to achieve sterility or "commercial sterility", and to inactivate the most heat resistant pathogen if pasteurization for public health purposes is the goal. Different foods will support growth of different pathogens and different spoilage organisms so the target will vary depending upon the food to be heated.

Food acidity/pH value has a tremendous impact on the target in heat preservation/ processing.

Heat Processing Requirements—Dependence on Product Acidity

FRUITS	Drying Conditions			Finished Product	
	Load kg/m²	Temperature °C	Time	Moisture %	Yield %
Plums	15	I 40-50	6 H	18-20	25-35
		II 75-80	14 H		
Apples (Rings)	10	75-55	5-6 H	20	10-12
Apricots (Halves)	10	70-60	10- 15	15-20	10-15
Cherries (w. stones)	10	55-70	6-8	12-15	25
Pears (Halves and quarters)	15	70-65	15-22	18-20	10-15
	15	70-60	10-15	15-20	10-15

Source: Desrosier and Desrosier (1977)

Sequence of operations employed in heat preservation of foods (fruit and vegetables, etc.) In a simplified manner, the main operations employed in heat preservation can be described as follows:

Food Preparation: Preparation procedures will vary with the type of food. For fruit, washing, sorting, grading, peeling, cutting to size, pre-cooking and pulping operations may be employed.

Can/receptacle: This may be carried out manually or by using sophisticated filling machinery. The ratio of liquid to solid in the can must be carefully controlled and the can must not be overfilled. A headspace of 6-9 mm depth (6-8% of the container volume) above the level of food in the can is usual.

Vacuum Production: This can be achieved by filling the heated product into the can, by heating the can and contents after filling, by evacuating the headspace gas in a vacuum chamber, or by injecting superheated steam into the headspace. In each case the can end is seamed on immediately afterwards.

Thermal Processing: The filled sealed can must be heated to a high temperature for a sufficient length of time to ensure the destruction of spoilage microorganisms. This is usually carried out in an autoclave or retort, in an environment of steam under pressure.

Cooling: The processed cans must be cooled in chlorinated water to a temperature of 37°C. At this temperature the heat remaining is sufficient to allow the water droplets on the can to evaporate before labelling and packing.

Labelling and Packing: Labels are applied to the can body, and the cans are then packed into cases.

In principle, all these operations can also be carried out at the farm/community level using the appropriate small scale equipment, preferably only glass jars (*e.g.* no metal cans).

Technological Principles of Pasteurization

Physical and chemical factors which influence pasteurization process are the following:

a. temperature and time;

b. acidity of the products;

c. air remaining in containers.

Pasteurization Processes: In pasteurising certain acid juices for example, there are two categories of processes:

(a) Low pasteurization where pasteurization time is in the order of minutes and related to the temperature used; two typical temperature/time combinations are as following:

63° C to 65° C over 30 minutes or

75° C over 8 to 10 minutes.

Pasteurization temperature and time will vary according to:

- nature of product; initial degree of contamination;
- pasteurized product storage conditions and shelf life required.

In this first category of pasteurization processes it is possible to define three phases:

- heating to a fixed temperature;
- maintaining this temperature over the established time period (= pasteurization time);
- cooling the pasteurized products: natural (slow) or forced cooling.

(b) Rapid, high or flash pasteurization is characterized by a pasteurization time in the order of seconds and temperatures of about 85° to 90° C or more, depending on holding time. Typical temperature/time combinations are as follows:

88° C (190° F) for 1 minute;

100° C for 12 seconds;

121°C for 2 seconds.

While bacterial destruction is very nearly equivalent in low and in high pasteurization processes, the 121° C/2 seconds treatment give the best quality products in respect of flavour and vitamin retention. Such short holding times, however, require special equipment which is more difficult to design and generally is more expensive than the 63-65 ° C/30 minutes type of processing equipment.

In flash pasteurization the product is heated up rapidly to pasteurization temperature, maintained at this temperature for the required time, then rapidly cooled down to the temperature for filling, which will be performed in aseptic conditions in sterile receptacles. Taking into account the short time and rapid performance of this operation, flash pasteurization can only be achieved in continuous process, using heat exchangers.

Industrial applications of pasteurization process are mainly used as a means of preservation for fruits and vegetable juices and specially for tomato juice.

During pasteurization it is necessary that a sufficient heat quantity is transferred through the receptacle walls; this is in order that the product temperature rises sufficiently to be lethal to microorganisms throughout the product mass.

The most suitable and practical method to speed up thermopenetration is the movement of receptacles during the pasteurization process. Rapid rotation of receptacles around their axis is an efficient means to accelerate heat transfer, because this has the effect, among others of rapidly mixing the contents.

The critical speed of for this movement is generally about 70 rotations per minute (RPM). This enables a more uniform heating of products, reducing heating time and organoleptic degradation.

The food processor will employ no less than that heat treatment which gives the necessary degree of microorganism destruction. This is further ensured by periodic inspection by local sanitary authorities or by the importing countries sanitary services. However, the food processor also will want to use the mildest effective heat treatment to ensure highest food quality.

It is convenient to separate heat preservation practices into two broad categories: one involves heating of foods in their final containers, the other employs heat prior to packaging.

The latter category includes methods that are inherently less damaging to food quality, where the food can be readily subdivided (such as liquids) for rapid heat exchange. However, these methods then require packaging under aseptic or nearly aseptic conditions to prevent or at least minimise recontamination.

On the other hand, heating within the package frequently is less costly and produces quite acceptable quality with the majority of foods and most of our present canned food supply is heated in the package.

In practice, therefore, most of the canned food produced locally in developing countries should be heated within the package.

Osmotic Dehydration Processing

Osmotic dehydration is a useful technique for the concentration of fruit and vegetables, realised by placing the solid food, whole or in pieces, in sugars or salts aqueous solutions of high osmotic pressure. It gives rise to at least two major simultaneous counter-current flows: a significant water flow out of the food into the solution and a transfer of solute from the solution into the food.

Process Variables: Main process variables are:

a. pre-treatments;

b. temperature;

c. nature and concentration of the dehydration solutions;

d. agitation;

e. additives.

In the light of the published literature, some general rules can be noted:

- water loss and solid gain are mainly controlled by the raw material characteristics and are certainly influenced by the possible pre-treatments;
- it is usually not worthwhile to use osmotic dehydration for more than a 50% weight reduction because of the decrease in the osmosis rate over time. Water loss mainly occurs during the first 2 hr and the maximum solid gain within 30 min.
- the rate of mass exchanges increases with temperature but above 45°C enzymatic browning and flavour

deterioration begin to take place. High temperatures, *i.e.* over 60° C, modify the tissue characteristics so favouring impregnation phenomena and thus the solid gain;

- the best processing temperature depends on the food; mass exchanges are favoured by using high concentration solutions;
- phenomena which modify the tissue permeability, such as over-ripeness, pre-treatments with chemicals (SO_2), blanching or freezing, favour the solid gain compared to water loss because impregnation phenomena are enhanced;
- the kind of sugars utilised as osmotic substances strongly affects the kinetics of water removal, the solid gain and the equilibrium water content. Low molar mass saccharides (glucose, fructose, sorbitol, etc.) favour the sugar uptake;
- addition of NaCl to osmotic solutions increases the driving force for drying.

Synergistic effects between sugar and salt have also been observed. A pilot plant equipment used for detailed study of process parameters in osmotic concentration of fruits. (Source: Garrote, R.L. et al., 1992).

Applications: The effects of osmotic dehydration as a pre-treatment are mainly related to the improvement of some nutritional, organoleptic and functional properties of the product.

As osmotic dehydration is effective at ambient temperature, heat damage to colour and flavour is minimised and the high concentration of the sugar surrounding fruit and vegetable pieces prevents discoloration.

Furthermore, through the selective enrichment in soluble solids high quality fruit and vegetables are obtained with functional properties "compatible" with different food systems. These effects are obtained with a reduced energy input over traditional drying process. The main energy-consuming step is the reconstitution of the diluted osmotic solution that could be obtained by concentration or by addition of sugar.

Various applications of the technique as a unit operation in the food area are summarised in Pig. 8.4.1 together with the process parameters regarded as optimal in the light of the published literature.

Drying: Air drying following osmotic dipping is commonly used in tropical countries for the production of so-called "semi-candied" dried fruits. The sugar uptake, owing to the protective action of the saccharides, limits or avoids the use of SO2 and increases the stability of pigments during processing and subsequent storage period.

The organoleptic qualities of the end product could also be improved because some of the acids are removed from the fruit during the osmotic bath, so a blander and sweeter product than ordinary dried fruits is obtained. Owing to weight and volume reduction, loading of the dryer can be increased 2-3 times.

The combination of osmosis with solar drying has been put forward, mainly for tropical fruit. A 24 hour cycle has been suggested combining osmodehydration, performed during the night, with solar drying during the day. Two-three-fold increase in the throughput of typical solar dryers is feasible, while enhancing the nutritional and organoleptic quality of the fruits.

A two-step drying process, OSMOVAC, for producing low moisture fruit products was described. The osmotic step is performed with sucrose syrup 65-75 Brix until the weight reduction reaches 30-50%.

By osmotic dehydration followed by vacuum drying puffy products with a crisp, honeycomb-like texture can be obtained at a cost comparatively lower than freeze-drying.

Commercial feasibility of the process on bananas has been studied, based on the results of a semi-pilot scale operation; the process scheme is reported. Osmotically dried bananas retained more puffiness and a crisper texture than simple vacuum dried ones, and the flavour lasted longer at ambient temperature.

The combination of osmotic dehydration with freeze-drying has been proposed only at laboratory scale.

Appertisation: A combination of osmotic dehydration with appertisation has been proposed to improve canned fruit preserves. The feasibility of a process, called osmo-appertisation, to obtain high quality fruit in syrup, has been assessed on a pilot scale.

The key point of this technique is the pre-concentration of the fruit to about 20-40 Brix, that causes, together with the enhancement of the natural flavour, an increase of the resistance of the fruit to the following heat treatment, especially for colour and texture stability.

The products obtained are stable up to 12 months at ambient temperature and show a higher organoleptic quality than canned preserved alternatives. Furthermore, because of their higher specific weight and diminished volume, the filling capacity of jars or pouches is increased.

Freezing: The frozen fruit and vegetable industry uses much energy in order to freeze the large quantity of water present in fresh products. A reduction in moisture content of the material reduces refrigeration load during freezing.

Other advantages of partially concentrating fruits and vegetables prior to freezing include savings in packaging and distribution costs and achieving higher product quality because of the marked reduction of structural collapse and dripping during thawing.

The products obtained are termed "dehydro-frozen" and the concentration step is generally carried out through conventional air drying, the additional cost of which has to be taken into account.

Osmotic dehydration could be used instead of air drying to obtain an energy saving or a quality improvement especially for fruit and vegetable sensitive to air drying.

Extraction of Juices

An osmotic pre-step before juice extraction was reported to give highly aromatic fruit or vegetable juice concentrates.

Further developments: So far only applications on a pilot plant scale are reported in the literature. For further developments on a larger scale, theoretical and practical problems should be solved.

The industrial application of the process faces engineering problems related to the movement of great volumes of concentrated sugar solutions and to equipment for continuous operations. The use of highly concentrated sugar solutions creates two major problems.

The syrup's viscosity is so great that agitation is necessary in order to decrease the resistance to the mass transfer on the solution side.

The difference in density between the solution (about 1.3 kg/litre) and fruit and vegetables (about 0.8 kg/litre), makes the product float. Another important aspect, so far not investigated, is the microbiological safety of the process, which should be studied thoroughly before further industrial development.

Osmoappertisation in the Processing of Apricots

In order to obtain an alternative to the canned fruit preserves and to maintain a high quality of the fruits, a research has been carried out on the osmoappertisation of apricots, a "combined" technique that consists in the appertisation of the osmodehydrated apricots.

This technique could contributes also to the reduction of energy consumption, limits the cost of production and combines "convenience" (ready-to-eat, medium shelf-life) with many market outlets (retail, catering, bakery, confectionery, semi-finished products).

Osmoappertisation combines two unit operations: dehydration by osmosis and appertisation (packaging + pasteurization).

Apricot Processing

a) Fresh apricot puree.

After washing, cutting and removal of stones, apricot halves are dipped in 2% solution of sodium or potassium metabisulphite for 10 minutes. After draining, the resulting material is passed through a 0.045-in. screen pulper - finisher to produce a fresh apricot puree.

The fresh apricot puree obtained in this way could be further processed in different semiprocessed (*i.e.* chemically or otherwise preserved products) or finished fruit products (fruit leathers, fruit bars, jams, etc.).

b) Concentrated apricot pulp.

Fresh apricot halves could also be steam blanched for 5 min., passed through a 0.045-in pulper - finisher and transformed in a puree with about 14 Brix depending on the fruit quality. This puree may be concentrated in steam jacketed kettles up to 20 Brix or in other adequate equipment (*e.g.* a stirred vacuum evaporator) up to 28°Brix.

As for fresh apricot puree, the concentrate may be further processed in various semiprocessed or finished fruit products as mentioned above and as will be described below.

c. Dried apricot leather.

- From fresh fruit puree by drum drying. The fresh apricot puree at about 14 Brix could be dried to 12% moisture apricot sheets, using a double-drum dryer operating at 132 degrees C with a drum clearance of 0.008 in and speed of 45 sec per revolution.
- From fruit concentrate by drum drying. The concentrate could also be dried to 12% moisture fruit sheets by the same process as described above.
- From fresh fruit puree or from apricot concentrate by sun/solar drying or by dehydration.

a) *Trays:* For sun / solar drying or dehydration of fruit pulp, the trays must have a solid base in order to retain the liquid contents. They may be made of metal, timber or plastic. Stainless steel or plastic trays are most suitable because they are unaffected by acid fruit pulp; they are, however, expensive.

A metal tray could be 75 x 50 cm in size and with side 5 cm high. The trays must keep level during drying; if the tray is not level the pulp will run to the lowest point, giving a layer of irregular depth which will dry unevenly. Any tray which is not made of stainless steel or plastic must be covered inside with a sheet of heavy gauge plastic film to protect the pulp from chemical or bacteriological contamination.

Standard sun/solar trays as described can be used by covering them inside with a sheet of plastic film to create a solid base.

b) *Preparation before Drying/dehydration:* Fresh apricot puree can be directly used for next processing steps. Fruit concentrate needs to be added to potassium metabisulphite to obtain a 0.3% concentration of SO_2 in the material.

c) *Drying/dehydration:* The apricot/fruit puree or concentrate is poured into the trays to a depth of about 1.5 cm. When stainless steel or plastic trays are used they should be coated with a thin layer of glycerine to prevent sticking.

The pulp is then sun/solar dried or tunnel/cabinet dehydrated; moisture content in the dried product should not exceed 14% and the SO_2 content should not be less than 1500 pp.

The dried product is wrapped in cellophane to prevent sticking, then put inside polythene bags and stored at best in tight fitting tins and sealed to prevent moisture transfer.

From fresh fruit puree or from apricot concentrate, with sugar addition, and then processed by sun / solar drying or by dehydration.

In some countries preference is for finished products with added sugar; and this is also interesting from a point of view of energy consumption (concentration is partially achieved by sugar dry matter) and of shelf life. The overall content in SO_2 could also be reduced as sugar is a preservation agent, the product will be close to a fruit "paste".

Dried/dehydrated Products

In reconstitution water is added to the product which is restored to a condition similar to that when it was fresh. This enables the food product to be cooked as if the person was using fresh fruit or vegetable.

All vegetables are cooked but many of the dried fruits can be used for eating after they have been soaked in water. The following reconstitution test is used to find out the quality of the dried product.

Reconstitution Test

1. Weigh out a sample of 35 grams from the bulked and packed final product of the previous day's production.
2. Put the sample into a small container (beaker) and add 275 ml of cold water (and 3.5 g salt).
3. Cover the container (with a watch-glass) and bring the water to the boil.
4. Boil GENTLY for 30 minutes.
5. Turn out the sample onto a white dish.
6. At least two people should then examine the sample for palatability, toughness, flavour and presence or absence of bad flavours. The testers should record their results independently.
7. The liquid left in the container should be examined for traces of sand/soil and other foreign matter.

This test can be used also to examine dried products after they have been stored for some time. Evaluation of rehydration ratio may be performed according to the following calculations.

Rehydration ratio. If weight of the dried sample is 10 g (Wd) and the weight of the sample after rehydration is 60 g (Wr), rehydration ratio is:

Rehydration coefficient. The weight of rehydrated sample is 60 g (Wr); the weight of dried sample is 10 g (Wd) and its moisture is 5% (Wu); raw material before drying had 87% water (A); rehydration coefficient is:

A simpler test for eating quality can be carried out without weighing and measuring. The material is placed in a cooking pot with water (and a little salt). The pot is then covered and boiled as described above.

Except for a few products which are eaten in the dry state, most dried fruit and all dried vegetables are prepared by soaking and cooking. Often this preparation is carried out incorrectly and dried products get a bad reputation.

Good quality dried products, after cooking and if properly treated should be similar to cooked fresh produce. In order to get good results, the following methods are recommended:

Quick Method: Cold water, ten times the weight of the dry product, is added to the dried product. The container is covered, brought to the boil and simmered GENTLY until the product is tender. The cooking time may be 15 to 45 minutes after the boiling point has been reached.

Slow Method: This gives better results than the quick method. Cold water is added to the dry food and is left to soak for 1 to 2 hours before cooking. The product is then cooked in the same water as that in which it was soaked. The actual cooking time will probably be shorter than that for the quick method.

Other points to remember are:

- if too much water is added the cooked product will have little flavour. However, if too little water is added the product may dry and burn. This can be avoided by adding small quantities of water during cooking;
- always cook with a lid on the container;
- salt, if required, should be added when the cooking is almost complete;
- partly used packages of dry products should be reclosed tightly or kept in containers with good fitting lids.

Composition of Some Dried Fruits

Fruit	Moisture content %	Sugar as monosaccharides, %	Other carbohydrates	Vitamin C mg/100 g
Raisins	21.5	64	7	0.0
Dates	12.5	55	7	0.0
Prunes	20.0	45	15	0.0
Apricots	15.0	45	24	4.8
Peaches	15.5	55	14	8.0

Handling, sorting, packing and storage of dried and dehydrated fruit and vegetables.

It is not easy to assess when drying has been completed. In absence of instrumentation, the characteristics of the various products after drying/dehydration can only be assessed by experience. Although this cannot be conveyed adequately on paper, some general indications can be given.

Fruit Products: When a handful of fruit is squeezed tightly together in the hand and then released, the individual pieces should drop apart readily and no moisture be left behind on the hand. It should not be possible to separate the skin by rubbing unpeeled fruit and the fruit centre should no longer reveal any moist area. Banana should be leathery and not too tough to eat in their dry state.

Vegetable Products: Onions should be dried until they are crisp whereas tomatoes should be leathery.

In general, the lower the moisture content, the better the keeping quality will be, but overdried products generally have an inferior quality. Also the loss in weight from excessive drying cannot be tolerated in a commercial operation designed to run profitably.

It is, however, essential to dry up to an optimum / safe moisture level, related to the type of the product and his designed shelf life, and to avoid running the risk of the products becoming spoiled due to excess water content. When drying is completed, the material should be sorted either on trays or on a table in order to remove pieces of poor quality and colour and any foreign matter.

Very fine material should be separated from the bulk of the material by using a sieve (12 or 16 mesh per inch). Bad quality products which show poor colour need to be removed from the bulk of finished product. After selection and grading, dried products should be packed immediately, preferably in polythene bags which must be folded and closed / tied tightly. However, plastic bags are easily damaged and therefore they must be packed into cartons or jute sacks before they are transported.

Deterioration of Dried Fruit during Storage

Dried fruit must be considered as a relatively perishable commodity in the same category as cereals, pulses and similar stored products. It is subject to deterioration resulting from mould growth, insect and mite infestation and physical and chemical changes.

Mould Growth

When the moisture content of dried fruit is allowed to exceed the maximum permissible level for safe storage then mould growth may occur. The moisture levels applicable to various types of fruit; and it can be seen that the safe moisture levels for dried fruit are much higher than those for other similar commodities.

At the present time, suitable field moisture meters for use with dried fruit are not readily available, and moisture determinations can only be satisfactorily carried out where laboratory facilities are available.

Various species of drought resisting fungi may develop on dried fruit when the moisture content is just above the safe level, and a number of osmophilic yeasts are quite commonly associated with spoilage in dried fruit.

Many of the yeasts bring about fermentation with the production of lactic acid or alcohol, and yeasts are frequently present in wart-like crystalline growths which occur in fruit which has become "sugared". In very moist fruit mucoraceous fungi may predominate and are visible as white fluffy growths on and within the fruit.

Mite Infestation: Severe mite infestations are often associated with the growth of osmophilic yeasts in fermenting dried fruit products. Many of these mites are unable to complete their development in the absence of yeast. They have been reported as occurring on dried fruit, and particularly figs and prunes in Mediterranean countries. Such infestations are difficult to eradicate and affect consumer acceptance of the contaminated products.

Insect Infestation: Insect infestations may begin in the field before harvest, may continue during bulk storage after drying, and unless measures are taken to prevent it, may occur in the finished packaged product during storage prior to distribution and consumption.

Regular treatments of the stack of dried fruit with a suitable insecticide will be necessary as a routine to combat light insect infestations. Pyrethrins synergised with piperonyl butoxide are commonly used as a surface spray or as an aerosol fog for this purpose. Heavy infestations will require that the fruit be fumigated.

Technology of Semi-processed Fruit Products

The semi-processed fruit products are manufactured in order to be delivered to industry processing centres (in the fruit producing country itself or in importing countries) where they will be further manufactured in consumer oriented finished products: jams, jellies, syrups, fruits in syrup, etc.

In the practice of semi-processed fruit products and for the purpose of this document the following categories are defined:

a. *Fruit "Pulps":* semi-processed products, not refined, obtained by mechanical treatment (or, less often, by thermal treatment) of fruit followed by their preservation. Either whole fruit, halves or big pieces are used which enables easy identification of the species. "Pulps" can be classified in boiled or non boiled (raw).

b. *Fruit "Purees-marks":* semi-processed products obtained by thermal and mechanical treatment or, very rare, raw and then refined, operations by which all nonedible parts (cores, peels, etc.) are removed. "Purees-marks" are classified in boiled (the more usual case) and non boiled (raw).

c. *Semi-processed Juices:* products obtained by cold pressure or very rare by other treatments (diffusion, extraction, etc.) followed by the preservation.

Technical Processes for Preservation of Semi-processed Fruit Products

Preservation can be achieved by chemical means, by freezing or by pasteurization. The choice of preservation process for each individual case is a function of the semi-processed product type and the shelf life needed.

Chemical Preservation: In many countries, in practice, this is carried out with sulphur dioxide, sodium benzoate, formic acid and, on a small scale, with sorbic acid and sorbates.

Preservation with sulphur dioxide is a widespread process because of its advantages: universal antiseptic action and very economic application. The drawbacks of SO_2 are: SO_2 turn firms the texture of some fruit species (pomaces), desulphiting is not always complete and recolouring of red fruits is not always complete after desulphitation.

Practical preservation dosage levels with SO_2 for about 12 months is 0.18-0.20% SO_2 (with respect to the product to be preserved).

This level could be reduced to 0.09% SO_2 for 3 months and to 0.12% SO_2 for 6 months preservation.

The preservation with sulphur dioxide is in use mainly for "pulps" and for "purees-marks".

Chemical preservation can be performed from a practical point of view by the utilisation of 6% SO_2 water solutions or by direct introduction of sulphur dioxide gas in the product (for "purees-marks"). The preparation of 6% SO_2 solutions is done by bubbling the gas from cylinders in cold water; from a 50 kg SO2 compressed gas cylinder results 830 l of 6% SO_2 solution.

These SO_2 solutions have to be stored in cool places, in closed receptacles and with periodic concentration control/check by titration or by density measurements approximate results.

Preservation with sodium benzoate has the following advantages: it does not firm up the texture and does not modify fruit colour. The disadvantages are: it is not a universal antiseptic, its action needs an acid medium and the removal is partial. Sodium benzoate is in use for "pulps" and for "purees-marks" but less for fruit juices.

Practical dosage level for 12 months preservation is 0.18-0.20 % sodium benzoate, depending on the product to be preserved. Sodium benzoate is used as a solution in warm water; the dissolution water level has to be at maximum 10% reported to semi-processed product weight.

Formic acid preservation is performed mainly for semi-processed fruit juices at a dosage level of 0.2 % pure formic acid (100%). Formic acid is an antiseptic effective against yeasts, does not influence colour of products and is easily removed by boiling.

Formic acid could be diluted with water in order to insure a homogeneous distribution in the product to be preserved; water has to be at maximum 5 % of the product weight. Because of a potential effect of pectic substance degradation, formic acid is less in use for "pulps" and "purees-marks" preservation.

Sorbic acid used as potassium sorbate (easily water soluble) can be used for preservation of fruit semi-processed products at a dosage level of 0. 1% maximum. Advantages of sorbates are: they are completely harmless and without any influence on the organoleptic properties of semi-processed fruit products.

Preservation by Pasteurization: As fruit has a low pH, preservation of semiprocessed fruit products could also be performed by pasteurization (heat treatment step at maximum temperature of 100° C), the length of this step varying with the size of the receptacles.

The advantages of this type of treatment are: hygienic process, which assure a long term preservation; the disadvantages are: need for air tight receptacles, and pectic substances could begin to deteriorate if the thermal treatment is too long.

Thermal preservation of fruit semi-processed products could also be done by a "selfpasteurization": very hot semi-processed products are filled into receptacles (*e.g.* metal cans) which are sealed and then inverted in order to sterilise the air which goes through the hot fruit mass.

Preservation by Freezing: This is done on an industrial scale in some countries and can be done with or without sugar addition.

The advantages of this process are: absence of added substances; very good preservation of quality of fruit constituents (pectic substances, vitamins, etc.) and good preservation of organoleptic properties (flavour, taste, colour). Freezing is done at about -20 to -30° C and storage at -10 to -18° C.

Freezing is applied mainly to semi-processed fruit products aimed at very high quality and high cost finished products.

Technological flow-sheet for semi-processed fruit "pulps": chemical preservation.

Sorting is needed in order to remove sub-standard fruit (with moulds, with diseases, etc.) and all foreign bodies.

Washing is obligatory in order to remove all impurities which cannot be eliminated at the processing step in finished products.

Coring and Cutting, mainly for pomace fruits, has as main objective a better utilisation of preservation "space" in receptacles and is not mandatory; this will be defined by customer/ supplier agreements / standards. This operation is preferably performed by mechanical means.

Preservation is carried out with the 6% SO_2 solution which is added to the prepared fruits (placed in bulk in receptacles) in the quantity needed to obtain the preservation dosage level. For a better / homogeneous preservative distribution, the initial 6% SO2 solution could be diluted with water; however, the diluted solution (which will be filled in receptacles) has to be at a dosage level of less than 10% of the semi-processed product weight.

For some soft fruit, especially strawberries, preservation is done with a mix of 6% SO2 solution and calcium bisulphite solution (containing also 6% SO_2).

Preparation of calcium bisulphite solution is done by the introduction of 30 kg of CaO in 1 m^3 SO_2 solution and mixing up to clarification. The resulting solution is mixed with the initial 6% SO_2 solution, generally in a 1:1 ratio, but the ratio can be adapted to the fresh fruit texture. Firming of soft fruit texture by this treatment is based on the formation of calcium pectate with pectic substances from fruit tissues.

In the case of sodium benzoate, formic acid or potassium sorbate, the dosage levels to be used are as indicated above with the rule that it is not allowed to add more that 10% liquid in receptacles on the prepared fruits.

Preservation by pasteurization or "self-pasteurization" will need as additional steps: (a) boiling with a minimum water addition (maximum 10%); (b) filling of receptacles; (c) hermetic closing followed by (d) pasteurization or "self-pasteurization".

Some general technical data for the preparation of chemical preserved semi-processed fruit "pulps".

General Technical Processing Data for Semi-processed Fruit "Pulps"

Fruit species	Preliminary operations	Preservation means
Apples, pears, quinces	Sorting, washing, coring, cutting	Sulphur dioxide
Prunes, peaches, wax cherries, apricots	Sorting, washing, stone removal (pitting)	Sulphur dioxide or sodium benzoate
Cherries	Sorting, washing	Sulphur dioxide or sodium benzoate, sometimes with calcium bisulphite addition
Strawberries	Sorting, washing	Sulphur dioxide in mix with calcium bisulphite
Wild berries	Sorting, washing	Sulphur dioxide or sodium benzoate; in some cases with calcium bisulphite addition

Technological Flow-sheet for Semi-processed "Puree-Marks"

The general technological flow-sheet includes the following operations: Sorting and Washing are obligatory and are carried out in a similar manner as for "pulps".

Heat Treatment/Boiling is needed in order to soften the fruit tissues before refining. For some fruits as strawberry and wild berries, this step is not done and fruits are refined "raw" in order to preserve their flavour.

Pulping is performed with specific equipment - refiners or pulpers - which eliminate seeds, pits and other non edible parts (peels, cores, etc.). Preservation is carried out by chemical means, by freezing or by pasteurization.

The general technical / processing data for the manufacture of "puree-marks" are seen in Table.

General Technical Processing Data for Semi-processed "Pureesper—Marks"

Fruit species	Preliminary operations	Preservation means
Apples, pears, quinces	Sorting, washing, boiling, refining	Sulphur dioxide; in less frequent cases formic acid or sodium benzoate
Prunes, peaches, wax cherries, apricots,	Sorting, washing, boiling, refining	Sulphur dioxide, formic acid or sodium benzoate; cherries freezing with or without sugar; self-pasteurization
Strawberry, wild berries	Sorting, washing, refining	Chemical preservation, freezing or self pasteurization

Correlation between Density and Concentration for SO_2 Solutions

Density at 15°C	Concentration of solutions % SO_2
1.0168	3.00
1.0181	3.25
1.0194	3.50
1.0206	3.75
1.0221	4.00
1.0234	4.25
1.0248	4.50
1.0261	4.75
1.0275	5.00
1.0289	5.25
1.0302	5.50
1.0315	5.75
1.0340	6.25
1.0353	6.50
1.0365	6.75
1.0377	7.00

Fruit Sugar Preserves Technology

As a overall rule of thumb, a sugar concentration of about 60% in finished or processed fruit products generally insures their preservation. Preservation is not only determined by the osmotic pressure of sugar solutions but also by the water activity values in the liquid phase, which can be lowered by sugar addition; and by evapcration down to 0.848 aw; this value however does not protect products from mould and osmophile yeast attack. Maximum saccharose concentration that can be achieved in the liquid phase of the product is 67.89%; however higher total sugar quantities (up to 70-72%) found in products are explained by an increased reducing sugar solubility resulting from saccharose inversion.

Jams

The preservation of fruit by jam making is a familiar process carried out on a small scale by housewives in many parts of the world. Factory jam making has become a highly complex operation, where strict quality control procedures are employed to ensure a uniform product, but the manufacturing operations employed are in essence the same as those employed in the house.

Fresh or pre-cooked fruit is boiled with a solution of cane or beet sugar until sufficient water has been evaporated to give a mixture which will set to a gel on cooling and which contains 32-34% water.

Gel formation is dependent on the presence in the fruit of the carbohydrate pectin, which at a pH of 3.2 - 3.4 and in the presence of a high concentration of sugar, has the property of forming a viscous semi-solid.

During jam boiling, all microorganisms are destroyed within the product, and if it is filled hot into clean receptacles which are subsequently sealed, and then inverted so that the hot jam contacts the lid surface, spoilage by microorganisms will not take place during storage.

The composition of jam made from stone fruit and berry fruit. About 30% of the vitamin C present in fresh fruit is destroyed during the jam-making process, but that which remains in the finished product is stable during storage.

The high moisture content of jam (equivalent to an equilibrium relative humidity of about 82%) makes it susceptible to mould damage once the receptacle has been opened and exposed from some time to the air. No problems of microbiological spoilage are likely to arise in the canned product during storage.

Composition of Some Fruit Jams

Type of Jam	Moisture content %	Sugar (as invert sugar, %)	Vitamin C mg/100 g
Jam made from berry fruits: strawberry, raspberry, etc.	29.8	69.0	10 - 25
Jam made from stone fruits: apricot, peach, etc.	29.6	69.3	10 - 35

Source: FAD/WFP, 1970

Marmalade

This sugar preserve is defined as "semisolid or gel-like product prepared from fruit ingredients together with one or more sweetening ingredients and may contains suitable food acids and food pectins; the ingredients are concentrated by cooking to such a point that the TSS—Total Soluble Solids- of the finished marmalade is not below 65%".

Fruit Paste

Fruit paste is a product obtained in the same way as special non-gelified fruit marmalade but with a lower water content- about 25% TSS in fruit paste.

Lowering water content could be achieved by continuing boiling of the product or by drying the product by natural or artificial drying. An example of paste without sugar is the sun dried apricot or prune paste.

General procedure for the preparation of jams, jellies and marmalade

a. Boil the pulp or the juice (with water when necessary)
b. Add the pectin
   * to the batch while stirring very vigorously
   * Pectin which has previously been mixed with 5 times its weight in sugar taken from the recipe)
c. Boil for about 2 minutes to assure a complete dissolution
d. Add the sugar while keeping the batch boiling
e. Boil down quickly to desired Brix
f. Add the acid (usually citric acid) and remove the froth
g. Fill hot into the (previously cleaned) jars and close
h. Invert the jars for three minutes to pasteurize the cover

* ***Note:*** the pectin in solution can also be added at the end of the step (e) and has to be prepared as follows: use a strong blender. For one litre of water add slowly into the blender 25 g of pectin mixed with 100 g of sugar taken from the recipe.

Basic Recipes: The following recipes must be considered only as guidelines because the composition of the fruit can vary; also the taste of the consumers varies concerning the consistency, the sweetness and acidity.

Before starting to make jam it is important to know the yield to settle the question on containers. The calculation is made as follows:

$$\frac{\text{Total Soluble Solids content of all the raw ingredients}}{\text{Percentage of Total Soluble Solids in the final product}} \times 100$$

In these basic recipes it is assumed that the fruits are poor in pectin content.

Recipe 1. Fruit: sugar = 50:50; desired Brix in the finished product is 68.

	Soluble Solids
10 kg of fruit at 10% TSS	1.000 kg
10 kg of sugar	10.000 kg
60 g of pectin (grade 200)	0.060 kg
55 g of citric acid	0.055 kg
	11.115 kg

Recipe 2. Fruit: sugar = 45:55; desired Brix in the finished product is 68.

	Soluble Solids
d10 kg of fruit at 10% TSS	1.000 kg
2.5 litre of water	—
12.2 kg of sugar	12.200 kg
65 g of pectin grade 200	0.060 kg
60 g of citric acid	0.060 kg
	13.325 kg

Recipe 3. Fruit: sugar = 40:60; desired Brix in the finished product is 68.

	Soluble Solids
d10 kg of fruit at 10% TSS	1.000 kg
3.3 litre of water	—
15 kg of sugar	15.000 kg
85 g of pectin grade 200	0.085 kg
80 g of citric acid	0.080 kg
	16.165 kg

Processing of Pineapple-papaya Jam

The fruit should be prepared as per previous instructions.

For pineapples, the ends are removed and discarded; the cores and outer parts of the fruits are also removed. The fruit cylinders obtained are pulped through a special extractor (Fitzpatrick communiting machine) equipped with a 0.40-in screen sieve; the pulp thus obtained is used for making jam.

The papaya are prepared by hand-peeling the fruit; the fruit is then halved and the seeds removed. It is then pulped in the communiting machine using a 0.40-in screen sieve.

When ginger root is used as flavouring, it is peeled and macerated in a Kenwood blender to a very fine consistency.

A typical formula for a pineapple-papaya jam (50:50 ratio) with ginger flavouring is given as follows:

Pineapple pulp	25.0
Papaya pulp	25.0 pounds
Cane sugar	50.0
Apple pectin (150 grade)	6.0 ounces
Citric acid	6.4
Fresh ground ginger	7.5

Processing is carried out in the following way:

The weighed fruit pulp is placed in a stainless steel steam-jacketed kettle and heated to about 110°F under constant stirring.

When the product reaches this temperature, the heat is turned off. The pectin (mixed in about ten times its weight with some of the weighed sugar), is then mixed into the fruit pulp, stirring constantly in order to prevent the pectin from clotting.

When the pectin has dissolved, the remainder of the sugar is added and dissolved completely in the mixture. The heat is then turned on and the jam mixture is stirred constantly until it starts boiling vigorously. During the remainder of the cooking, the product is stirred occasionally. Near the finishing point (approximately 221° F), the citric acid and the ginger (if it is used) are also added.

Determination of the finishing point is done by removing samples at intervals, cooling, and reading the soluble solids by means of a refractometer equipped with a Brix scale. After the jam reaches the proper Total Soluble Solids content, the heat is turned off and the surface scum/foam is removed.

The jam then is quickly put into receptacles which have been cleaned and sterilised with boiling in water for 30 minutes. The filling operation is done rapidly in order to prevent the temperature of the jam from falling below 190° F.

After filling, sterilised lids (boiled for 30 minutes in water) are placed on the receptacles and they are then sealed.

After this operation the receptacles are inverted for about 3 minutes to insure that the lids are sterilised. The receptacles are then placed upright. At this stage it is not necessary to do any further processing, therefore the receptacles are cooled in running cold water until they reach a temperature slightly above room temperature. They are then dried in air and labelled.

Evaluation of Finished Products

During production at medium/large scale, it is recommended that quality controls be performed during manufacturing. After ten weeks of storage at room temperature it is recommended that an examination of finished products be performed. The receptacles are opened and contents carefully emptied on to enamel trays without disturbing the formation of the jam.

The empty cans (if metal cans were used) are then inspected for signs of corrosion. Factors other than flavour include colour, appearance, syrup separation, firmness and spreading quality. For flavour, jam is tested on pieces of bread. Samples are taken for measurement of pH (with a glass electrode pH meter) and Total Soluble Solids (with a refractometer equipped with a Brix scale).

This evaluation enables to have a quality check during product shelf life and to obtain data needed for necessary improvements of future productions.

For pineapple-papaya jam, products made with 30% pineapple and 70% papaya with added ginger has the highest score for flavour. The use of plain tin cans causes corrosion problems which is not the case when acid resistant lacquer cans are used.

Pineapple Jam Making

1. Boil 40 lb. of pulp and 12 lb. of water.
2. Add 225 g of pectin to the batch while stirring rapidly.
3. Boil for about 90 sec to assure complete dissolution.
4. Add 60 lb. of sugar gradually if possible in several portions, while keeping the batch boiling.
5. Boil down quickly to 69 deg. Brix (223 deg. F).
6. Take off steam, remove foam.
7. Add 300 cc citric acid solution 50%.
8. Fill hot (180 deg. F).
9. Invert receptacles for 3 minutes.

Check each batch for Brix 68-70 deg.; acidity/pH = 3.2 +/- 0.2.

Evaluate: absence of defects; colour; flavour; consistency.

Various fruit:sugar ratios can be manufactured; some basic recipes are as follows:

	Ratio 50:50	Ratio 45:55	Ratio 40:60
Fruit	55 lb	49.5 lb	44 lb
Water	-	11 lb	13.2 lb
Sugar	55 lb	60.5 lb	66 lb
Pectin (150 grade)	225 g	237.5 g	250 g
Citric acid (50% sol.)	300 cc	320 cc	335 cc

6

Gelified Sugar Fruit Preserves

Technology of Fruit Jellies

Jellies are gelified products obtained by boiling fruit juices with sugar, with or without the addition of pectin and food acids. Jellies are usually manufactured from juices obtained from a single fruit species only, obtained by boiling in order to extract as much soluble pectin as possible.

Jellies have to be clear, shiny, transparent and with a colour specific to the fruit from which they are obtained. Once the product is removed from the glass receptacles where it was packed, jellies must keep their shape and gelification and not flow, without being sticky or of a too hard consistency.

Technological flow-sheet for jellies manufacturing covers two categories of operations: those to obtain gelifying juices and those related to the manufacturing of jelly itself.

(a) Production of gelifying juices:

Washing & Sorting are carried out in usual conditions;

Cutting is applied eventually only to pomaces (apples, quinces) and are limited to cutting in halves or quarters;

Boiling Is Performed With Water Addition With 50-100% water, needed for pectin extraction. The boiling time is 30-60 min., it should not be longer so as to avoid pectin degradation; at the same time boiling must not be too violent.

Juice Separation is carried out by a simple drain through metallic sieve or cloths; in these cases the yield is lower and the residue can be used for marmalade production. In bigger productions it juice separation by hydraulic press is preferred, yield being in these cases greater.

Juice Clarification is strictly necessary in order to obtain clear jellies. This step can be achieved by sedimentation during 24 hours or by filtration.

(b) Manufacturing of jellies

Basic Recipe Setting is done starting with equal parts in weight of sugar and juice (for example 1000 g juice and 1000 g sugar). As final jelly has to contain about 60% added sugar, weight of finished product must be of about 1600 g, by evaporation of about 400 g water.

Boiling is carried out as following: juice is boiled up to removal of about half of the water that has to be evaporated, then the calculated sugar quantity is added gradually; the remainder of the water is evaporated until a concentration in soluble substances (refractometric extract) of 65-67% is reached, in which is incorporated also the sugar from juice.

During boiling it is necessary to remove foam / scum formed. Product acidity must be brought to about 1% (malic acid) corresponding to pH > 3. Any acid addition is performed always at the end of boiling.

For juices rich in pectin, gelification will occur without pectin addition. If at the trial boiling test the gelification has not occurred, because of pectin absence, in this case 1-2% powder pectin will be added by operating as indicated: pectin is mixed with 10-20 fold sugar quantity and is introduced directly in the partially evaporated juice and then boiling is conducted rapidly up to final point. Evaluation of final point is done not only by refractometry but also by gelifying test.

A rapid test for evaluation of juice pectin content is possible by mixing a small sample of juice with an equal volume of 96% alcohol; the apparition of a compact gelatinous precipitate indicates a sufficient pectin content for gelification.

Boiling of jellies is performed in small batches (25-75 kg) in order to avoid excessively long boiling time which brings about pectin degradation.

Cooling is optional and is carried out up to 85 °C, in double wall baths with water circulation.

Filling is performed at a temperature not below 85 °C in receptacles (glass jars, etc.), which must be maintained still about 24 hours to allow cooling and product gelification.

Receptacle Closing is done after product gelification.

Usual jelly types are: quinces, strawberries, cherries, wild berries, alone or in mixes with apples.

Grading of Marmalades

Three categories can be defined:

- fine marmalade, manufactured from one fruit;
- superior marmalade, obtained from a mix of fruit in which 30% are "noble" species (cherries, strawberries, apricots, etc.) and 70% from other species;
- marmalade from fruit mixes; apples, pears, plums, quinces, ungrafted apricots and wax cherries may be used, with the optional addition of "superior" fruit which was rejected at sorting but which was sound.

The content in total soluble substances (refractometric extract) of marmalades must be 64% minimum; the acidity must be between 0.5% and 1.8% expressed as malic acid.

Basic Recipe Setting for a normal composition - marmalade without pectin addition the following is a basic recipe:

100 kg semi-processed fruit product:

(10% refractometric extract) ...10 kg soluble substances

55 kg sugar ....55 kg soluble substances

155 kg 65 kg soluble substances

55 kg water to be evaporated

100 kg marmalade with 65% refractometric extract

This marmalade satisfies many standards and at same time has a good shelf-life since it contains less than 35% water. Semi-processed fruit products must have a minimum 8% refractometric extract; in this case the recipe should use 125 kg of raw material, with 80 kg water to be evaporate.

The use of semi-processed fruit products with a low refractometric extract presents the following drawbacks:

a. higher water quantity to be evaporated;

b. longer boiling times with negative impact of pectin degradation;

c. loss of flavour and

d. lower equipment efficiency.

Pectin addition in marmalade manufacture produces the following advantages:

a. improvement of gelification,

b. economy in fruit;

c. shorter boiling time; this maintains taste and flavour and produces higher equipment efficiency.

Pectin addition makes it possible to obtain the "fine" type of marmalade from "noble" fruits which do not contain enough pectin (cherries, peaches, apricots, etc.). In marmalades from fruit mixes, low pectin content can be compensated by addition of semi-processed fruit products which are rich in this component (for example apples).

When pectin is to be added, the above recipe should be modified as follows:

80 kg semi-processed fruit product

(10% refractometric extract) ...8 kg soluble substances

55 kg sugar ....55 kg soluble substances

10 kg pectic extract (10 % R.E.) ......1 kg soluble substances

145 kg 64 kg soluble substances

45 kg water to be evaporated

100 kg marmalade with 64% refractometric extract

Pectin can be added as pectic extract with about 10% refractometric extract (R.E.) in the recommended proportion or in the form of a powder considered with 100% dry matter (*e.g.* 100% soluble substances) in a quantity of about 1%.

In some countries it is usual to add 5-12% of corn syrup (calculated to finished product weight) to replace the corresponding quantity of sugar (100 parts corn syrup can replace 80 parts sugar). Corn syrup has to be liquefied by heating before use. Corn syrup addition reduces the too sweet taste of marmalade, avoids sugar crystallisation and gives a special shine to finished products.

Marmalade manufacturing follows the technological line drawing and covers the following steps:

"Mark" Preparation can be achieved from fresh fruits (as indicated in the section related to semi-processed fruit products) or starting from chemical preserved semiprocessed fruit products: "marks" or "pulps". In the latter case, pulps will be processed in marks which then will be desulphitated.

Desulphitation is carried out by boiling at atmospheric pressure, under vacuum in specialised equipment or under pressure in special retorts built in acid-resistant material. In any case, the desulphitation must be carried out before sugar addition because sugar will bound to the sulphur dioxide. The desulphitation operation must be conducted so as to be, if possible, fully completed; the finished product must contain less than 0.005% free SO_2.

The technological flow for marmalade production according to the following: fresh fruit after sorting on control belt (1) is washed in a washing machine (2) is brought to the continuous boiling equipment (3) then to the pulper (4); the semi-finished mark is passed on to the storage tank (6). Pulps are boiled and desulphited in continuous boiling equipment (3), then are brought to pulper (4) and to the storage tank (6).

Boiling aims at evaporating the required water quantity, to facilitate the formation of pectin-sugar-acid gel and to partially invert sugar (about 40% from total sugar). The boiling operation can be carried out in open kettles or in evaporators under vacuum. In the latter case, the warm "mark" from storage tank (6) is aspirated in a concentrator (7) in a vacuum and submitted to a partial boiling up to removal of half of water quantity which needs to be evaporated; the calculated sugar quantity is then added by aspiration, keeping the boiling on.

After this the pectin extract or powder pectin which has previously been dissolved in warm water, is added; when the final concentration is reached, as indicated by refractometric control, the required quantity of acid is added. Sugar is added in proportion of 55% in finished product, pectic extract (10% refractometric extract) at a level of about 10-15% and the acid (citric, tartaric, lactic) in a quantity needed to obtain a finished product acidity of about 1%.

Boiling at atmospheric pressure affects not only the appearance but also the nutritional value of the products, mainly if these contain proteins, as some albuminoids coagulate even at 60°C.

Food products for which flavour is an essential property as for example fruit juices, etc., are also affected by the action of heat. Heat treatment has an impact on vitamin losses, mainly of vitamin C, in the presence of oxygen as is the case at concentration in open vessels.

Technological Line for Marmalade Production

Sugars are generally less damaged by heat at temperatures below 100° C; as the boiling point is increasing above 100° C, a risk of partial sugar caramelization exists.

In the study of heat effects on products submitted to concentration operation, it is necessary to take into account not only the evaporation surface temperature but also the distribution of the temperature in the whole liquid mass.

The length of the heating period also has a major influence because in many cases it is preferable to concentrate the liquid at a relatively high temperature in a short time avoiding the drawbacks of lower temperatures acting during a long time.

In order to maintain the food value and organoleptic properties, it is necessary that concentration take place at a low temperature which can be achieved by concentration under partial vacuum, taking into account that boiling point decreases when the residual pressure decreases, respectively with the increase of vacuum degree.

Advantage of concentration under partial vacuum are the following:

- lowering of boiling point;

- the total time needed for concentration of food products under a residual pressure of about 200 mm Hg is about half as compared to the that of concentration by boiling at atmospheric pressure;
- by lowering the concentration temperature and time, organoleptic properties and of nutritional value are maintained better particularly as far as the vitamins are concerned;
- when products are concentrated in a vacuum, it is possible to recover volatile aromatic substances by using adequate installations.

Technical procedures of concentration by vaporisation can be classified in:

a) concentration at atmospheric pressure: continuous or discontinuous;

b) concentration under partial vacuum: discontinuous (in vacuum equipment with simple or multiple effect) or continuous (in vacuum installations with continuous action or in thin film vaporisation installations).

Even if open kettle equipment is less expensive than evaporators in a vacuum, it is necessary to take into account that boiling under vacuum has the following advantages:

a) low boiling temperature (60-70° C), depending the degree of the vacuum; this give the fruit better taste and flavour-keeping qualities;

b) easy feeding with raw and auxiliary materials;

c) shorter boiling time;

d) better working conditions (vapour elimination in condensed water and not in open air).

There are small size evaporators under vacuum which can be well suited to the needs of medium size operations in developing countries.

Cooling of marmalade to about 50-60°C can even be done in a vacuum evaporator by closing the heating steam and maintaining vacuum degree or by discharge in storage tanks.

Filling in receptacles (boxes, jars, glasses, etc.) is done preferably with filling machines followed by labelling. Small packages can be closed warm or after complete cooling; big packages (boxes, etc.) must be closed only after cooling, *e.g.* 24 hours after processing.

Storage of marmalade must be done in dry rooms (air relative humidity at about 75%), well ventilated, medium cool places (temperature 10-20 degrees C), disinfected and away from direct sunlight and heat. These measures are necessary because marmalade is a hydroscopic product and, by water absorption, favourable conditions for mould development are created.

Technology of Special Fruit Jams

Special fruit jams are products similar to marmalades but in which fruit partially keep their shape (whole, halves, etc.). Special fruit jams are obtained by boiling fruit with sugar, with or without pectin addition, with acid addition followed by n concentration by evaporation.

Special fruit jams present a pronounced gelification at their cooling and can be considered as fruits included in a pectin-sugar-acid gel.

High quality special fruit jams are obtained only from fresh fruit or possibly frozen and from only one fruit species.

Special fruit jams are classified in:

a) non-pasteurized (min. 68% refractometric extract) and

b) pasteurized (min. 65% refractometric extract); minimum acidity, expressed in malic acid, is 0.5%.

Basic Recipe Setting is done taking into account following elements: - maintenance, as much as possible, of fruit shape, because this is specific to these finished products;

- added sugar, in relation to finished product, must be at 60-65%; this high percentage is needed in order to obtain a rapid and strong gelification and to facilitate conservation of fruit shape;
- a satisfactory gelification cannot generally be obtained without pectin addition, at a level of 1-2%, or respectively 10-20% pectic extract (10% refractometric extract);
- partial (about 40%) inversion of added sugar is necessary.

The basic recipe is: 80 kg fruit + 100 kg sugar + 1.6-2.0 kg pectin powder and 1 kg citric or tartric acid; this will yield about 165 kg of special fruit jam with 60% added sugar; water quantity to be evaporated is about 18 kg. If the finished product has to contain 65 % added sugar, the boiling has to be continued up to evaporation of about 30 kg water, the resulting finished product quantity will be about 153 kg.

Technological flow sheet for manufacturing of special fruit jams is as follows:

Fruit Preparation: sorting, washing, peeling and coring (for apples, pears, quinces), or removal of quetches and stones/ pits (for plums, peaches, apricots, cherries) or of short tails (for strawberries and wild berries). Pomace fruits are then cut in slices or quarters.

Boiling With Sugar is the most important operation in production of special fruit jams and has as objective to evaporate water until gel formation and partial inversion of sugar. Boiling has to be conducted in such a way as to avoid fruit disintegration, but fruits must be well penetrated with sugar.

By boiling an equilibrium is reached, by osmosis, between sugar solution and cellular juice. The initial concentration gap between sugar syrup and cellular juice is very high and if the equilibration process is forced, juice comes out of cells and fruit loses its shape and may even disintegrate.

The boiling process accelerates the equilibration, intensity of which increases with temperature and boiling time.

Pectin addition shortens boiling and thus delays the equilibration; for this reason there are different methods for special fruit jams preparation:

- maintaining fruit in sugar, at ambient temperature, over 8-24 hours; fruit to sugar ratio is that indicated in recipe. After this sugar impregnation, fruits and resulting sugar syrup are brought together to boil; toward the end of boiling, pectin, dissolved in warm water, and necessary acid quantity are added.
- preparation of a very concentrated sugar syrup (at least 75%), in which fruit is introduced in order to be boiled.

The rest of the operations are similar to those described in the previous method

Soft fruit (strawberries, wild berries) can be mixed with sugar directly in evaporating open kettles, without added water and then heated gradually up to boiling, which is continued as in previous method.

Boiling is preferably carried out in small open kettles (50-100 kg) in order to avoid too long a boiling and fruit disintegration.

Gelification corresponds generally to the reaching of a concentration of 65% soluble extract, respectively 68% refractometric extract. Practical test for gelification is done as for jellies and marmalades.

Cooling is a technological step strictly necessary in order to avoid fruit rising to the surface in preservation receptacles. Cooling is done in a double bottomed water bath in which water circulates at about 80° C.

Filling of receptacles (jars, boxes, glasses, etc.) is carried out and it is necessary to assure at this stage that the finished product is homogeneous (equal quantities of fruits and gel).

Pasteurization is only applied to special fruit jams with 65 % refractometric extract packed in jars or boxes and is performed at about 100° C for about 20 min.

Gelification is carried out during product cooling and intensifies during storage.

Storage must follow the same conditions as for marmalades.

Non-gelified Sugar Fruit Preserves

Technology of Fruit Jams

Fruit jams are products obtained by boiling of fruits (whole, halves, etc.) with sugar syrup until the reaching a viscous consistency. Jams can be defined as fruits included in a concentrated syrup.

Jams are only prepared from fresh fruits; the usual product range covers the following species: strawberries, cherries, apricots, wild berries, peaches, plums, raisins, quinces, rose petals, etc.; manufacturing of jams from fruit mixes is not an accepted practice. Standards indicate that fruit jams have to conform with following conditions:

- fruit or fruit pieces content 45-55%;
- soluble substances, refractometric degree min. 72%;
- acidity (expressed in malic acid) min. 0.7%.

Basic Recipe Setting is presented below as an indication:

100 kg fruits

150 kg sugar

------

250 kg

35 kg water to be evaporated

------

215 kg jam at about 72% refractometric extract

Fruit Preparation is similar to special fruit jams with the difference that stone / seed removal is mandatory for all species.

Boiling fruits with sugar can be achieved by the same three methods described above for special fruit jams. By boiling a sugar content equilibration is foreseen and the operation must be conducted in such a way that texture, flavour and colour of fruits be preserved. Foam/ scum has to be removed during boiling.

Generally boiling in concentrated sugar syrup (at least 75%) after a previous diffusion during 2-4 hours is in practical use. Boiling must to be carried out in small portions (about 15 kg) in order to avoid fruit disintegration.

For some fruit species, boiling has to be conducted in many phases / steps with "stops" to enable sugar diffusion. At the end of boiling, vanillin at a ratio of 125 g for 100 kg jam may be added for some fruits (white cherries, raisins, etc.). At the same time, it is also possible to add citric or tartric acid in order to avoid the "sugaring" defect.

Cooling of jams, necessary in order to avoid fruit rising to the surface, is carried out as for special fruit jams. Filling of receptacles and Storage for jams are performed in same way as for special fruit jams.

Special Non-gelified Fruit Marmalades

Special non gelified fruit marmalades are products resulting from fruit without stones or seeds, sieved or squashed, concentrated by boiling, without sugar added and non gelified. Their consistency results from a low water content (about 35%) and a high percentage of insoluble substances (5-10%). Sugar the from fruits acts as a preserving agent.

Plum special non gelified fruit marmalade is the product representative of this category. Other fruit is used very rarely, because they have a reduced sugar content as compared to plums; though there are some countries producing special non gelified fruit marmalade from pears or sweet apples.

For plums, the finished product in this category must contain minimum 55-60% soluble substances (refractometric extract), rising up to 70% for a high quality product.

Basic Recipe Setting for plums is done in relation with sugar content of fruits; when this is higher, the product quality is better and the yield is higher. Thus, if plums with 18% refractometric extract are used, 300-350 kg of fruits are needed for 100 kg finished product.

At the same time, concentration of sugar by boiling, also increases by three fold acidity and astringency of this special plum non gelified marmalade; for this reason it is recommended to use only sweet and completely mature plums.

Washing is Performed in Usual Conditions

Pre-boiling of fruits can be carried out in water or vapour, preferably with continuous running and has as its objective the softening of tissues.

De-stoning is Performed in a Pulper

Boiling of the sieved mass is done in double bottom open kettles with a big evaporation surface or in vacuum evaporators.

Boiling in open kettles enables production of a more tasty slightly caramelized, product; boiling in vacuum evaporators has the technological advantages indicated in marmalade production.

At the end of boiling and once of necessary concentration is reached, the product is poured directly into receptacles (drums, etc.) and left to cool in order to form a hard surface layer (crust).

Storage of well closed receptacles is carried out as for marmalade. Special non gelified fruit marmalade can also be prepared from chemically preserved semi-processed fruit products, but the quality is lower than that obtained from fresh fruits.

Sometimes dried prunes in a mix with preserved semi-processed products can be used for plum special non gelified marmalade preparation. In some countries plum finished products in this category are sweetened by the addition of maximum 30% sugar, calculated in relation to the finished product.

Technology of Fruit Pastes

These products are obtained in a similar way to marmalades and special non gelified marmalades, but have a lower water content (about 25%). Reduction of water content can be achieved by continuing the boiling of the product or by natural or artificial drying. A typical example of fruit paste without sugar added is the apricot paste - "pistil", etc. which is a concentrated special non gelified fruit marmalade poured in thin layers and sun dried.

An example of fruit paste with sugar added is quince paste which is a marmalade concentrated by evaporation. Sugar content must be 65%; soluble substances content, 7075% refractometric extract and acidity at least 0.5% expressed as malic acid. Packing is done usually in polyethylene sheets and then in boxes or tins; storage conditions are similar to those for marmalade.

Technology of Fruit Syrups

Fruit syrups are products obtained by dissolving sugar in juices obtained from direct pressing of fruits. Sugar dissolving can be done at room temperature or by heating.

Syrups have to contain 68% soluble substances (refractometric extract) and minimum 1 g/100 ml malic acid.

Up to a maximum 10% of sugar can be replaced by corn syrup. Syrups must be manufactured from the juice of only one fruit species.

Juice Preparation is carried out at room temperature as described above.

Sugar Dissolving In Fruit Juice can be performed by one of the following methods:

(a) `Boiling in open kettles, done by using the basic recipe:

350-400 kg fruit juice;

650-660 kg sugar;

max. 10 kg citric or tartric acid.

The juice is brought to boiling and the sugar is dissolved; the total time has to be as short as possible in order to avoid flavour loss and a too high sugar inversion degree (optimum inversion degree is 40%). Acid is added preferably towards the end of boiling.

During all boiling processes it is necessary to remove foam / scum. In order to avoid caramelization, the syrup has to be cooled rapidly, and this can be carried out in baths with double bottoms through which are circulated water.

One alternative to this method is to boil syrup in closed vessels to avoid flavour losses.

(b) *Boiling in a Vacuum:* The basic recipe is the same as above. Sugar and fruit juice are mixed previously in a pre-heating kettle and then transported to vacuum equipment.

Boiling is performed at 50° C and at the end the temperature is raised slowly up to 65-70° C. The syrup can be cooled directly in vacuum equipment by closing the steam inlet and by increasing the vacuum. In this boiling method it is possible to incorporate a flavour recuperation device.

(c) A continuous process for syrup preparation can be carried out by dissolving components with heat while passing them through a horizontal cylinder with a screw inside.

In the methods where sugar is dissolved by heat, it is also possible to use chemically preserved juices.

In this case it is necessary to first perform the desulphitation of juices preserved with SO2. This can be performed by boiling juice with optional water addition (and before any sugar addition). High quality syrups are obtained however from fresh juices.

(d) Sugar can also be dissolved at room temperature by using continuous flow percolators. These are similar to those used for salt solution preparation in vegetable canning processes. The juice goes over a sugar layer and is concentrated progressively until saturation (about 65 %). The syrup is then passed through a filtration section in the bottom of the percolator.

Syrup Filtration is needed in order to clarify crystals; the filtration of syrup is done in warm conditions through cloth.

Filling of syrup in bottles is done in aseptic conditions as much as possible in order to avoid syrup infection with osmophile yeasts. Syrup preservation is assured by the high sugar content with respect to a low water activity unclear.

Storage takes place in well ventilated storage rooms; avoiding sunlight at 10-15° C. The usual product range is: strawberries, cherries, wild berries, citrus fruits.

Technology of "Fruit in Syrup" Products

This type of product is represented by fruit (whole, halves or pieces) covered by a sugar solution and preserved by pasteurization. In these products sugar does not have a preservation effect but only a sweetening role.

Technological general flow-sheet covers the following steps:

Sorting is necessary in order to choose mature fruit, whole, unblemished, undamaged and with a specific texture. Fruit of good quality but with a texture not compatible with this type of finished product are used for the production of semi-processed products, marmalades, etc. This step is done manually on a sorting belt.

Washing is performed in equipment with fan and sprays. Cleaning covers removal of leaves, etc., skin removal by one of the described methods for some fruits and coring / pitting for others.

Cutting is applied only to pomace fruits and is performed preferably by mechanical means.

Preliminary Heat Treatment depends very much upon the fruit types and can be as different as a light blanching up to a real boiling; this step is aimed toward softening of tissues for hard fruits, elimination of waxy pruin layer (plums, etc.) or enzyme inactivation for pomace fruits.

This treatment has to be reduced to the minimum necessary in order to avoid sugar losses. Some fruit (for example apricots, black cherries, grapes, etc.) does not undergo preliminary heat treatment.

Cooling is carried out in water (as cold as possible) and should not be too long to avoid soluble substances loss.

Receptacle Filling Fruit is introduced manually or sometimes mechanically (filling tables, etc.) in receptacles and then sugar syrup is added.

Sugar Syrup Preparation is performed by dissolving crystal sugar in hot water (90100° C). After the sugar has been dissolved, the syrup is boiled briefly; removal of impurities and coagulated substances is then performed by foaming/scumming. An addition of about 0.3% citric acid helps syrup clarification, followed by filtration through cloth.

Preheating/Emptying of open receptacles has as its objective to eliminate air from fruit tissues and this step is largely replaced today by receptacle closing under vacuum. Preheating may be done in steam or in water in such a way that syrup temperature reaches 8090° C and is maintained for 10-15 min.

Hermetic Sealing is performed according to the type of receptacle and of cover/cap.

Pasteurization is carried out at 100° C and may be performed in continuous installations or in normal water baths. In some countries fruits in syrup are pasteurized at 80-90° C; this contributes to a greater flavour retention.

Cooling should be intensive in order to avoid discolorations and colour modifications and to reduce corrosion in metal cans.

Refractometric extract is to be measured 30 days after manufacturing. In order to establish exact concentration of sugar syrup to be added in each individual case it is possible to use the following formula:

$C = 1.5 \times [(2 \times E) - F]$

C = added syrup concentration;

E = refractometric extract needed in finished products, according to Table.

F = fruit refractometric extract.

Hot syrup is put into receptacles with a minimum temperature of 80° C in order to enable a corresponding vacuum after cooling.

It is possible to add in the syrup 0.3-0.5% citric or tartaric acid for fruit with a low acidity. Adding ascorbic acid in proportion of 0.8% assures colour maintenance and taste improvement based on its reducing action.

Avoidance of excessively soft texture in finished products may be achieved for soft fruit (strawberries, apricots, etc.) by dipping fruit in a solution of calcium chloride (5% $CaCl_2$ in water).

Finished product defects and production "accidents" for "Fruit in syrup" and the means of preventing them

- Change of colour in red fruit (red cherries, plums, etc.) to violet; to avoid this it is necessary to use glass jars or varnished/lacquered tinplate;
- Colour change to red for pears and quinces (pink shade) could arise in products which are over-pasteurized and insufficiently cooled;
- Change of colour of whole peaches peeled by chemical means is due to an incomplete inactivation of oxidative enzymes; to avoid this it is necessary to boil fruit after chemical peeling followed by washing with 0.5% citric or tartric acid solutions or by adding ascorbic acid in syrup;
- Softening of strawberries tissues can be prevented by a pre-treatment with warm syrup and also by addition of calcium salt.

Some technical data for processing are seen in Table.

Technical Data for "Fruits in Syrup"

Type of fruit/product		Minimum fruit	Soluble substances (in syrup), RE** at 20°C,
name	Size/shape	content, %	minimum
Apricots	Halves	55	25
Strawberries	Whole	47	25
Cherries	Whole	53	22
Quinces	Slices, 30 mm	50	26
Apples	Halves or quarters	47	25
Pears	Halves or quarters	57	22
Plums	Whole	45	25
Melon	Pieces	55	26
Wax cherries	Whole	53	27
Wild cherries	Whole	47	27
Pineapple	Slices	55	26
Mangos	Pieces	50	25
Papaya	Pieces	55	27
Fruit mixes	According to the	50	25
	conditions indicated		
	above, for each fruit		

** RE = Refract metric extract

Fruit Juice Technologies

Fruit juices are products for direct consumption and are obtained by the extraction of cellular juice from fruit, this operation can be done by pressing or by diffusion.

For the purpose of this document, the technology of fruit juice processing will cover two finished product categories:

- juices without pulp ("clarified" or "not clarified");
- juices with pulp ("nectars").

We will also define as :

a. "natural juices" products obtained from one fruit; and
b. "mixed juices" products obtained from the mix of two or three juices from different fruit species or by adding sugar.

Juices obtained by removal of a major part of their water content by vacuum evaporation or fractional freezing will be defined as "concentrated juices".

Technological Steps for Processing of Fruit Juices without Pulp

Fruit juices must be prepared from sound, mature fruit only. Soft fruit varieties such as grapes, tomatoes and peaches should only be transported in clean boxes which are free from mould and bits of rotten fruit.

Washing: Fruit must be thoroughly washed. Generally, fruit will be submitted to a pre-washing before sorting and a washing step just after sorting.

Sorting: Removal of partially or completely decayed fruit is the most important operation in the preparation of fruit for production of first quality fruit juices; sorting is carried out on moving inspection belts or sorting tables.

Crushing/Grinding/Disintegration Step is applied in different ways and depends on fruit types:

Crushing for grapes and berries;

Grinding for apples, pears;

Disintegration for tomatoes, peaches, mangoes, apricots.

This processing step will need specific equipment which differs from one type of operation to another.

Enzyme Treatment of crushed fruit mass is applied to some fruits by adding 2-8% pectolitic enzymes at about 50° C for 30 minutes.

This optional step has the following advantages: extraction yield will be improved, the juice colour is better fixed and finished product taste is improved.

However, for fruit which is naturally rich in pectic substances, this treatment makes the resulting "exhausted" material useless for industrial pectin production.

Heating of crushed fruit mass before juice extraction is an optional step used for some fruit in order to facilitate pressing and colour fixing; at same time, protein coagulation takes place.

Pressing to extract juice.

Diffusion is an alternative step for juice extraction and can be carried out discontinuously or in batteries at water temperature of about 80-85°C.

Juice Clarifying can be performed by centrifugation or by enzyme treatment. Centrifugation achieves a separation of particles in suspension in the juice and can be considered as a pre-clarifying step. This operation is carried out in centrifugal separators with a speed of 6000 to 6500 RPM.

Enzyme clarifying is based on pectic substance hydrolysis; this will decrease the juices' viscosity and facilitate their filtration. The treatment is the addition of pectolitic enzyme preparations in a quantity of 0.5 to 2 g/l and will last 2 to 6 hours at room temperature, or less than 2 hours at 50° C, a temperature that must not be exceeded.

The control of this operation is done by checking the decrease in juice viscosity. Sometimes, the enzyme clarifying is completed with the step called "sticking" by the addition of 5-8 g/hl of food grade gelatine which generates a flocculation of particles in suspension by the action of tannins.

Filtration of clarified juice can be carried out with kieselgur and bentonite as filtration additive in press-filters (equipment).

De-tartarisation is applied only to raisin juice and is aimed to eliminate potassium bitartrate from solution. This step can be performed by the addition of 1% calcium lactate or 0.4% calcium carbonate.

Pasteurization of juice can be done for temporary preservation (pre-pasteurization) and in this case this operation is carried out with continuous equipment (heat exchangers, etc.); warm juice is stored in drums or large size receptacles (20-30 kg). Pasteurization conditions are at 75°C in continuous stream.

Pasteurization of bottled juice is then carried out just before delivery to the market; this is performed in water baths at 75° C until the point where the juice reaches 68° C. In cases when the final pasteurization is done without pre-pasteurization and temporary storage, modern methods use a rapid pasteurization followed by aseptic filling in receptacles.

Rapid pasteurization conditions are as follows: temperature about 80° C, over 10-60 sec., followed by cooling; all operations are carried out in continuous stream.

Preservation under CO_2 pressure may be done at a concentration of 1.5% CO_2 under a pressure of 7 kg/cm². At the distribution step, proceed at CO_2 decompression and the juice is then submitted to a sterilising filtration and aseptic filling in receptacles.

Preservation by freezing is carried out at about -30° C, after a preliminary de-aeration; storage is at -15 to -20° C.

Production of concentrated juices by evaporation is performed under vacuum (less than 100 mm Hg residual pressure) up to a concentration of 65-70% total sugar which assures preservation without further pasteurization. Modern evaporation installations recover flavours from juices which are then reincorporated in concentrated juices.

Additional operations for juice manufacturing are the vacuum de-aeration and mixing with other fruit juices or with sugar.

For the production of non clarified juices the centrifugation is the only specific step, enzyme clarifying and subsequent filtration being eliminated.

The optimum sugar/acid ratio for the majority of fruit, mainly for pomaces, is 10/1 to 15/1.

Fruit which is rich in carotenoids (apricots, peaches, etc.) is only processed as juices with pulp ("nectars").

Technological flow-sheet for fruit juices with pulp ("Nectars")

This process is divided at industrial scale in two categories of operations:

a. processing for obtaining juices;
b. juice conditioning for preservation.

(a) Operations in the first category are differ according to the type of fruit which to be processed.

Pomaces (apples, pears) are washed and sorted and then crushed in a colloid mill, fruit puree is then passed through a screw type heating equipment where direct steam is used as a source of heat. Warm fruit mass is treated in a pulper with a 2 mm screen and then through an extractor similar with the equipment used for tomato juice.

Stone fruits (apricots, peaches, cherries, etc.) after washing and sorting are submitted to steam in a continuous heater, then the warm fruit mass is passed through a pulper and then an extractor (as mentioned above). Berries (strawberry, wild berries, etc.) are washed, sorted and then crushed, preheated and then introduced in extractor.

In order to avoid browning and undesirable taste modifications it is usual to add about 0.05% ascorbic acid.

(b) Second category type of operations are similar for all fruit species. Partial elimination of cellulose is achieved with a continuous centrifugal separator; the resulting juice is then processed in order to adjust sugar and acid content for viscosity.

Sugar (about 8-10%) is added as a syrup (in water or in the juice of same fruit obtained by pressure). Acidity is adjusted with citric or tartaric acid. The adjusted juice is then deaerated under vacuum at about 40° C. This step aims at avoiding oxidative reactions and vitamin C loss reduction.

An important subsequent step is an intensive homogenisation (under pressure at 150-180 A) in order to obtain particles with dimensions below 100. The homogenised juice obtained is then continuously pasteurized in plate heat exchanger equipment at a temperature of about 130° C, cooled down to about 90° C and aseptically packed in receptacles.

The principal characteristics of fruit "nectars" are uniformity and stability of the content provided by the advanced disintegration of fruits. Stability can be obtained by increasing product viscosity by adding pectin for fruit which is deficient in this component. In order to avoid "separation", intensive homogenisation is carried out as described above.

Fruit "nectars" contain all the important components of the original fruit and to a large extent maintain their taste and flavour. The sugar/acidity (as citric acid) ratio is to a large extent determined by the type of fruit and the correction applied; for example, this ratio is 30 for apricots, 40 for peaches, 160 for pears, etc.

A general technological flow-sheet for fruit juice processing is presented below.

Banana and Plantain Processing Technologies

Traditional Processing

1 *Products:* Uses and Dietary Significance

Most of the world's bananas are eaten either raw, in the ripe state, or as a cooked vegetable, and only a very small proportion are processed in order to obtain a storable product. This is true both at a traditional village level with both dessert and cooking bananas and when considering the international trade in dessert bananas.

In general, preserved products do not contribute significantly to the diet; however, in some localised areas the products are important in periods when food are scarce.

Probably the most widespread and important product is flour preparation from unripe banana and plantains by sun-drying. In Uganda, dried slices known as "mutere" are prepared for storage from green bananas, the dried slices being either used directly for cooking or after grinding into a flour. "Mutere" is used chiefly as a famine reserve and does not feature largely in the diet under normal conditions.

In Gabon, plantains are sometimes made into dried slices which can be stored and used on long journeys, and plantains are used in Cameroon to prepare dried pieces which are stored and ground as needed into flour for use in cooking a paste known as "fufu". Dried green banana slices are also used in parts of South and Central America and West Indies for preparing flour.

The other nutritionally important product is beer which is a major product in Uganda, Rwanda and Burundi where green banana utilisation is particularly high.

2. Preservation Methods and Processes.

Drying. Both ripe and unripe bananas and plantains are normally peeled and sliced before drying, although banana figs are sometimes prepared from whole ripe fruit. Sun drying is the most widespread technique where the climate is suitable but drying in ovens or over fires is also practiced. In west Africa, plantains are often soaked and sometimes parboiled before drying.

The slices of unripe fruit are normally spread out on bamboo frameworks; or a cemented area; or on a mat; or on a swept-bare patch of earth; or on a roof; or sometimes on stones outcrops or sheets of corrugated iron.

Oven-drying of ripe bananas is practiced in Polynesia as a mean of preserving the fruits, which are then wrapped in leaves and bound tightly to store until needed. In East Africa a method has been reported that involves drying the peeled bananas on a frame placed over a fire for 24 hr before drying in the sun, to accelerate the process.

3. Product stability and storage problems

There is little experimental data on the storage life of the traditionally made banana and plantain products.

4. Potential for scaling up of traditional processes to industrial level

Many banana products are now produced on an industrial scale, including the traditional banana figs and flour, and the processing techniques are described below. One of the main problems encountered has been the susceptibility of banana products to flavour deterioration and discoloration and in the past many products reaching the market have been of poor quality.

A great deal of research has been directed to overcoming these problems, although however good the resultant products are they cannot compare in flavour and other characteristics with the fresh banana fruit. Indeed, an important constraint on the large-scale development of banana processing is the lack of demand for banana products since the fresh fruit is available throughout the year in most parts of the tropical world.

The production of beer from banana and plantains has not been scaled up to an industrial level, and while an important product in localised areas of tropical Africa, the market is rapidly declining in favour of European-type brews produced locally.

Industrial Processing

Products and Uses: The main commercial products made from bananas are canned or frozen puree, dried figs, banana powder, flour, flakes, chips (crisps), canned slices and jams.

Banana products can be divided roughly into two types - those for direct consumption, such as figs, and those for use in food manufacturing industry, for example purees and powder.

Banana figs, or fingers as they are sometimes known, are usually whole, peeled fruit carefully dried so as to retain their shape, although sometimes the fruit is sliced or halved to facilitate drying. Banana and plantain chips (crisps) are thinly sliced pieces of fruit fried in oil and eaten as a snack like potato chips (crisps).

The main use of canned slices is in tropical fruit salads. Banana flakes are used as a flavouring or in breakfast cereals. Banana purée find use mainly in the production of baby foods. Banana flour is said to be highly digestible and is used in baby and invalid foods, but can also be used in the preparation of bread and beverages.

Banana powder is used chiefly in the baking industry for the preparation and fillings for cakes and biscuits and is also used for invalid and baby foods.

Processing Technology: In general, to obtain a good-quality product from ripe-bananas the fruit is harvested green and ripened artificially under controlled conditions at the processing factory. After ripening, the banana hands are washed to remove dirt and any spray residues, and peeled. Peeling is almost always done by hand using stainless steel knives, although a mechanical peeler for ripe bananas has been developed, capable of peeling 450 Kg of fruit per hour (Banana Bulletin, 1974).

The peeling of unripe bananas and plantains is facilitated by immersing the fruit in hot water. For example, immersion in water at 70-75 ° C for 5 min. has been suggested as an aid for peeling green bananas for flour production, while the peeling of green bananas for freezing has been facilitated by immersion in water at 93° C for 30 min.

Banana Figs: Fully ripe fruits with a sugar content of about 19.5% are used and are treated with sulphurous acid after peeling, then dried as soon as possible afterwards. Various drying systems have been described using temperatures between 50 and 82° C for 10 to 24 hr to give a moisture content ranging from 8 to 18% and a yield of dried figs of 12 to 17% of the fresh banana on the stem.

One factory in Australia uses a solar heat collector on the roof to augment the heat used for drying bananas. Bananas can also be dried by osmotic dehydration, using a technique which involves drawing water from 1/4-in. thick banana by placing them in a sugar solution of 67 to 70 deg. Brix for 8 to 10 hr. followed by vacuum-drying at 65 to 70° C, at a vacuum of 10 mm Hg for 5 hr. The moisture content of the final products is 2.5% or less, much lower than that achieved by other methods.

Banana Puree: Banana puree is obtained by pulping peeled, ripe bananas and then preserving the pulp by one of three methods: canning aseptically, acidification followed by normal canning, or quick-freezing.

The bulk of the world's puree is processed by the aseptic canning technique. Peeled, ripe fruits are conveyed to a pump which forces them through a plate with 1/4-in. holes, then onto a homogeniser, followed by a centrifugal de-aerator, and into a receiving tank with 29in. vacuum, where the removal of air helps prevent discoloration by oxidation.

The puree is then passed through a series of scraped surface heat exchangers where it is sterilised by steam, partially cooled, and finally brought to filling temperature. The sterilised puree is then packed aseptically into steam-sterilised cans which are closed in a steam atmosphere.

Banana Slices: Several methods for canning of banana slices in syrup are used. Best-quality slices are obtained from fruit at an early stage of ripeness. The slices are processed in a syrup of 25 deg. Brix with pH about 4.2, and in some processes calcium chloride (0.2%) or calcium lactate (0.5%) are added as firming agents.

A method for producing an intermediate-moisture banana product for sale in flexible laminate pouches has been developed.

Banana slices are blanched and equilibrated in a solution containing glycerol (42.5%), sucrose (14.85%), potassium sorbate (0.45%), and potassium metabisulphite (0.2%) at 90 deg. C for 3 min. to give a moisture content of 30.2%.

Banana Powder: In the manufacture of banana powder, fully ripe banana pulp is converted into a paste by passing through a chopper followed by a colloid mill. A 1 or 2 % sodium metabisulphite solution is added to improve the colour of the final product. Spray- or drum-drying may be used, the latter being favoured as all the solids are recovered.

A typical spray dryer can produce 70 kg powder per hour to give yields of 8 to 11% of the fresh fruit, while drum-drying gives a final yield of about 13% of the fresh fruit. In the latter method the moisture content is reduced to 8 to 12 % and then further decreased to 2 % by drying in a tunnel or cabinet dryer at 60° C.

Banana Flour: Production of flour has been carried out by peeling and slicing green fruit, exposure to sulphur dioxide gas, then drying in a counter-current tunnel dryer for 7 to 8 hr. with an inlet temperature of 75° C and outlet temperature of 45° C, to a moisture content of 8%, and finally milling.

Banana Chips (crisps): Typically, unripe peeled bananas are thinly sliced, immersed in a sodium or potassium metabisulphite solution, fried in hydrogenated oil at 180 to 200° C, and dusted with salt and an antioxidant.

Alternatively, slices may be dried before frying and the antioxidant and salt added with the oil. Similar processes for producing plantain chips have been developed.

Banana Beverages: In a typical process, peeled ripe fruit is cut into pieces, blanched for 2 min. in steam, pulped and pectolytic enzyme added at a concentration of 2 g enzyme per 1 kg pulp, then held at 60 to 65° C and 2.7 to 5.5 pH for 30 min.

In a simpler method, lime is used to eliminate the pectin. Calcium oxide (0.5%) is added to the pulp and after standing for 15 min. this is neutralised giving a yield of up to 88% of a clear, attractive juice. In another process banana pulp is acidified, and steam-blanched in a 28-in Hg vacuum which ensures disintegration and enzyme inactivation. The pulp is then conveyed to a screw press, the resulting puree diluted in the ratio 1:3 with water, and the pH adjusted by further addition of citric acid to 4.2 to 4.3, which yields an attractive drink when this is centrifuged and sweetened.

Jam: A small amount of jam is made commercially by boiling equal quantities of fruit and sugar together with water and lemon juice, lime juice or citric acid, until setting point is reached.

Product Stability and Spoilage Problems

All dried banana products are very hydroscopic and susceptible to flavour deterioration and discoloration, but this can be overcome to some extent by storing in moisture-proof containers and sulphiting the fruit before drying to inactivate the oxidases. The dried products are also liable to attack by insects and moulds if not stored in dry conditions, although disinfestation after drying by heating for 1 hr to 80° C or by fumigation with methyl bromide ensures protection against attack. Banana powder is said to be stored for up to a year commercially and flakes have been stored in vacuum-sealed cans with no deterioration in moisture, colour or flavour for 12 months.

Banana chips tend to have a poor storage life and to become soft and rancid. However, chips treated with an antioxidant have been stored satisfactorily at room temperature in hermetically sealed containers up to 6 months with no development of rancidity.

Quality Control Methods

In general a good quality product is obtained if fruit is harvested at the correct stage of maturity and, where appropriate, ripened under controlled conditions. For example, in the case of banana figs, the fruit should be fully mature (sugar content of 19.5% or above) or the final product is liable to be tough and lacking in flavour. However, if over-ripe fruit is used, the figs tend to be sticky and dark in colour, so the fruit must be fully yellow but still firm.

For banana flour, which is prepared from unripe bananas, the fruit is harvested at three-quarters the full-ripe stage and is processed within 24 hr. prior to the onset of ripening. If less mature fruit is used, the flour tastes slightly astringent and bitter due to the tannin content. Bananas harvested between 85 and 95 days after the emergence of the inflorescence, with a pulp-to-peel ratio of about 1.7, were considered to be most suitable for the deep-fat frying.

Other criteria suggested for assessing maturity were beta-carotene and reducing sugar content, both of which increase with increasing maturity and pH which decreases as the fruit ripens, and these should be, respectively, about 2000 µg/100 g, less than 1.5% and 5.8 or above. Browning was found to occur if the sugar content was higher than 1.5%. The determination of crude fat in processed chips is also considered to be a necessary quality control measure.

It is important to remove all impurities prior to processing of products, and this is done by washing to remove dirt and spray residues and control on the processing line so that substandard fruit can be removed.

Preparation Methods for Fresh Bananas and Plantains

The main ways of preparing fresh bananas for consumption are boiling or steaming, roasting or baking and frying. Boiling followed by pounding into "fufu" is also widely adopted in certain areas of the tropics.

Boiling or Steaming: Plantains and bananas are often prepared simply by boiling in water, either in their peel or after peeling, and either ripe or unripe; if unripe, the fruit is scraped thoroughly after peeling to remove all traces of fibrous material. The boiled fruit is eaten alone or more usually accompanied by a sauce. This preparation technique is widely used in West Africa.

Roasting or Baking: Unpeeled or peeled fruit, either ripe or unripe, is roasted simply by placing in the ashes of a fire or in an oven. This method is widely used in West Africa, East Africa and the South Pacific islands. For example, ripe plantains are placed unpeeled in an oven and when partly brown and tender, removed and peeled, then replaced in the oven and roasted evenly.

Frying: Ripe or unripe plantains or bananas are often peeled, sliced and cooked in oil, particularly in West Africa and in parts of South America and the West Indies. Similar products are also made in East Africa. Typically, ripe plantains are peeled, cut into slices or split lengthways, and fried in palm oil or with groundnut oil, the pieces being served either hot with a sauce or with fried eggs, or cold as a snack.

Pounding: Pounding is a process, used particularly in West Africa, for preparing most perishable staple food crops including plantains, cassava, yams and cocoyams to obtain a paste or dough known as "fufu" (also spelled "foofoo", "foutou", "foufou"). The plantains are peeled or boiled and peeled after boiling and pounded in a wooden mortar, the resulting paste normally being eaten with soup or a spiced sauce of meat and vegetables, but sometimes after wrapping in leaves and steaming.

Mango and Guava Processing Technologies

Mangoes are processed at two stages of maturity. Green fruit is used to make chutney, pickles, curries and dehydrated products. The green fruit should be freshly picked from the tree. Fruit that is bruised, damaged, or that has prematurely fallen to the ground should not be used. Ripe mangoes are processed as canned and frozen slices, puree, juices, nectar and various dried products. Mangoes are processed into many other products for home use and by cottage industry.

The mango processing presents many problems as far as industrialization and market expansion is concerned. The trees are alternate bearing and the fruit has a short storage life; these factors make it difficult to process the crop in a continuous and regular way. The large number of varieties with their various attributes and deficiencies affects the quality and uniformity of processed products.

The lack of simple, reliable methods for determining the stage of maturity of varieties for processing also affects the quality of the finished products. Many of the processed products require peeled or peeled and sliced fruit. The lack of mechanised equipment for the peeling of ripe mangoes is a serious bottleneck for increasing the production of these products.

A significant problem in developing mechanised equipment is the large number of varieties available and their different sizes and shapes. The cost of processed mango products is also too expensive for the general population in the areas where most mangoes are grown. There is, however, a considerable export potential to developed countries but in these countries the processed mango products must compete with established processed fruits of high quality and relatively low cost.

Green Mango Processing

Pickles: The optimum stage of maturity should be determined for each variety used to make pickles.

There are two classifications of pickles - salt pickles and oil pickles. They are processed from whole and sliced fruit with and without stones. Salt is used in most pickles.

The many kinds of pickles vary mainly in the proportions and kinds of spices used in their preparation. One basic recipe for the study of the preparation and storage of pickles in oil is as follows:

Mango pieces	250 g	Tumeric powder	2 to 4 g
Salt	60 g	Fenugreek seeds	2 to 4 g
Mustard powder	30 g	Bengal gram seeds	2 to 4 g
Chili powder	20 g	Gingelly oil	20 to 30 g

The ingredients are mixed together and filled into wide-mouthed bottles of 0.5 kg capacity. Three days later the contents are thoroughly mixed and refilled into the bottles. Extra oil is added to form a 1-2 cm layer over the pickles.

Chutney: The product is prepared from peeled, sliced or grated unripe or semi-ripe fruit by cooking the shredded fruit with salt over medium heat for 5 to 7 minutes, mixed and then sugar, spices and vinegar are added. Cook over moderate heat until the product resembles a thick puree, add remaining ingredients and simmer another 5 min. Cool and preserve in sterilised jars. Spices usually include cumin seeds, ground cloves, cinnamon, chili powder, ginger and nutmeg. Other ingredients such as dried fruits, onions, garlic and nuts may be added.

Drying/dehydration: Immature fruit is peeled and sliced for sun-drying. The dried mango slices can be powdered to make a product called amchoo. The use of blanching, sulphuring and mechanical dehydration gives a product with better colour, nutrition, storability and fewer microbiological problems.

Ripe Mango Processing

Puree: Mangoes are processed into puree for re-manufacturing into products such as nectar, juice, squash, jam, jelly and dehydrated products. The puree can be preserved by chemical means, or frozen, or canned and stored in barrels. This allows a supply of raw materials during the remainder of the year when fresh mangoes are not available.

It also provides a more economical means of storage compared with the cost of storing the finished products, except for those which are dehydrated, and provides for more orderly processing during peak availability of fresh mangoes.

Mangoes can be processed into puree from whole or peeled fruit. Because of the time and cost of peeling, this step is best avoided but with some varieties it may be necessary to avoid off-flavours which may be present in the skin. The most common way of removing the skin is hand-peeling with knives but this is time-consuming and expensive. Steam and lyepeeling have been accomplished for some varieties.

Several methods have been devised to remove the pulp from the fresh ripe mangoes without hand-peeling. A simplified method is as follows: the whole mangoes were exposed to atmospheric steam for 2 to 2 1/2 min in a loosely covered chamber, then transferred to a stainless steel tank.

The steam-softened skins allowed the fruit to be pulped by a power stirrer fitted with a saw-toothed propeller blade mounted 12.7 to 15.2 cm below a regular propeller blade. The pulp is removed from the seeds by a continuous centrifuge designed for use in passion fruit extraction. The pulp material is then passed through a paddle pulper fitted with a 0.084 cm screen to remove fibre and small pieces of pulp.

Mango puree can be frozen, canned or stored in barrels for later processing. In all these cases, heating is necessary to preserve the quality of the mango puree. In one process, puree is pumped through a plate heat exchanger and heated to 90°C for 1 min and cooled to 35° C before being filled into 30 lb tins with polyethylene liners and frozen at -23.50 C.

In an other process, pulp is acidified to pH 3.5, pasteurized at 90°C, and hot-filled into 6 kg high-density bulk polyethylene containers that have been previously sterilised with boiling water. The containers are then sealed and cooled in water. This makes it possible to avoid the high cost of cans.

Wooden barrels may be used to store mango pulp in the manufacture of jams and squashes. The pulp is acidified with 0.5 to 1.0% citric acid, heated to boiling, cooled, and SO_2 is added at a level of 1000 to 1500 ppm in the pulp. The pulp is then filled into barrels for future use.

Slices: Mango slices can be preserved by canning or freezing, and recent studies have shown the feasibility of pasteurized-refrigerated and dehydro-canned slices. The quality of the processed product in all of these procedures will be dependent upon selection of a suitable variety along with good processing procedures. Thermal process canning of mango slices in syrup is the most widely used preservation method.

Beverages: The commercial beverages are juice, nectar and squash. Mango nectar and juice contain mango puree, sugar, water and citric acid in various proportions depending on local taste, government standards of identity, pH control, and fruit composition of the variety used. Mango squash in addition to the above may contain SO2 or sodium benzoate as a preservative. Other food grade additives such as ascorbic acid, food colouring, or thickeners may be used in mango beverages.A short description of finished products found in literature is as follows:

- *Mango Juice:* prepared by mixing equal quantities of pulp (puree) and water together and adjusting the total soluble solids (TSS) and acidity to taste (12 to 15% TSS and 0.4 to 0.5% acidity as citric acid);

- Mango nectar containing 25% puree can be prepared using the following procedure.

	Brix of purée		
Nectar components	15°	17°	20°
Purée	100	100	100
Sugar	45	43	40
Water	255	257	260

Commercial processing conditions may require the use of a preservative.

The pH is adjusted to approximately 3.5 by adding citric acid as a 50% solution.

The time of heat processing will vary with filling temperature, can size and viscosity of the juice or nectar.

Mango squash may be prepared according to flow-sheet described below; the finished product may contain 25% juice, 45% TSS and 1.2 to 1.5% acidity and may be preserved with sulphur dioxide (350ppm) or sodium benzoate (1000 ppm) in glass bottles.

Mango squash simplified flow-sheet.

Ingredients

Mango pulp	900	900
Sugar	900	1100
Citric acid	18	15
Water	900	900

Mangoes are washed, stored, peeled with stainless steel knives. The pulp is prepared by using a pulper with fine sieve (0.025-in); Sugar is mixed with water and citric acid = syrup; The pulp is added to the syrup and mixed well; The mixture is strained trough cloth; The squash is heated at 85° C and bottles are filled and closed. For additional heat treatment bottles may need to be maintained at a product temperature of 80°C for 30 minutes if the product is to be processed without preservatives. The bottles are then left to cool in water and stored at room temperature.

Two negative points must be avoided: presence of air bubbles (which is a source of quick deterioration) and separation of squash solids (giving an undesirable appearance). The means to avoid these two phenomena are described in the fruit juices section. A type of "squash type" beverage may also be manufactured with '/a pulp + '/a water + i/a sugar and pH adjusted to 3.7 by addition of citric acid. Using different sieve sizes affects the quality and reduces air bubbles to a certain extent but homogenisation and de-aeration of puree or squash seem to be important in order to avoid separation and air bubbles.

The squash quality is evaluated on the basis of the following characteristics: pH, titrable acidity, soluble solids, ascorbic acid (by 2,6 dichlorophenol indophenol method), specific gravity.

Dried/dehydrated: Ripe mangoes are dried in the form of pieces, powders, and flakes. Drying procedures such as sun-drying, tunnel dehydration, vacuum-drying, osmotic dehydration may be used. Packaged and stored properly, dried mango products are stable and nutritious.

One described process involves as pre-treatment dipping mango slices for 18 hr (ratio 1:1) in a solution containing 40° Brix sugar, 3000 ppm SO_2, 0.2% ascorbic acid and 1% citric acid; this method is described as producing the best dehydrated product. Drying is described using an electric cabinet through flow dryer operated at 60° C. The product showed no browning after 1 year of storage.

Drum-drying of mango puree is described as an efficient, economical process for producing dried mango powder and flakes. Its major drawback is that the severity of heat preprocessing can produce undesirable cooked flavours and aromas in the dried product. The drum-dried products are also extremely hydroscopic and the use of in-package desiccant is recommended during storage.

Canning: This preservation technology is described in various technological flow-sheets in this bulletin. Mango bar or "fruit leather" is presented in various flow-sheets.

Guava Processing Technologies

Guava Puree: Guava puree is used in the manufacture of guava nectar, various juice drink blends and in the preparation of guava jam. The washed sound fruit is first passed through a chopper or slicer to break up the fruit and this material is fed into a pulper. The pulper will remove the seeds and fibrous pieces of tissue and force the reminder of the product through a perforated stainless steel screen. The holes in the screen should be between 0.033 and 0.045 in. The machine should be fed at a constant rate to ensure efficient operation.

The pureed material coming from the pulper is next passed through a finisher. The finisher is equipped with a screen containing holes of approximately 0.020 in. The finisher will remove the stone cells from the fruit and provide the optimum consistency to the product. Perhaps the best way to preserve the guava puree is by freezing and the material passing through the finisher can be packaged and frozen with no further treatment. It is not necessary to heat the product to inactivate enzymes or for other purposes. The material can be frozen in a number of types of cartons and cans; however, a fibre box with a plastic bag inside is commonly used and is probably the less expensive.

It is also possible to can and heat process the guava puree and this can be accomplished by heating the puree to 195° F in an open double bottom kettle, filling into cans, closing the cans, inverting the cans for a few seconds, followed by cooling. Cans should be cooled rapidly to approximately 100° F before they are cased and stacked into warehouses.

Guava Juice and Concentrate: Guava juice can be used in the manufacture of a clear guava jelly or in various drinks. A clear juice may be prepared from guava puree that is depectinised enzymatically. About 0.1% pectin-degrading enzyme is mixed into the puree at room temperature; heating of the product at approximately 120° F will greatly speed the action of the enzyme. After 1 hr. clear juice is separated from the red pulp by centrifuging or by pressing in a hydraulic juice press. A batch-type or continuous-flow centrifuge can be used on the depectinised puree with no further treatment.

The clear juice after centrifuge or after press (and subsequent filtration) can be preserved by freezing or by pasteurization in hermetically sealed cans. For shipment to overseas markets it may be advantageous to concentrate either the puree or the juice.

Recent Trends in Fruit and Vegetable Processing

New Products: The number and variety of fruit and vegetable products available to the consumer has increased substantially in recent years. The fruit and vegetable industry has undoubtedly benefited from the increased recognition and emphasis on the importance of these products in a healthy diet. Traditional processing and preservation technologies such as heating, freezing and drying together with the more recent commercial introduction of chilling continue to provide the consumer with increased choice. This has been achieved by new heating (*e.g.* UHT, microwave, ohmic) and freezing (*e.g.* cryogenic) techniques combined with new packaging materials and technologies (*e.g.* aseptic, modified atmosphere packaging).

The overall trend in new fruit and vegetable products is "added value", thus providing increased convenience to the consumer by having much greater variety of ready prepared fruit and vegetable products. These may comprise complete meals or individual components. The suitability of products and packages for microwave re-heating has been an important factor with respect to added convenience. The major trends in the development of fruit and vegetable heat processed products in recent years are shown in table; the number of new fruit and vegetable products.

Numbers of New Fruit and Vegetable Products

	1990	1991	1992 Jan-June
Frozen vegetable products	66	95	21
Chilled vegetable products	76	81	78
Heat processed vegetables	51	60	38
Heat processed fruits	13	14	5
Fruit juices and drinks	73	83	46
Potato crisps	32	33	16

Source: C. Dennis (1993)

New product development in the fruit and vegetable sector is most important in meeting the continued challenge of providing the consumer with choice and high quality products.

A Fresh Look at Dried Fruit

New fruit varieties and advance in drying technologies are putting a fresh twist on dried fruit applications. Fruits that have been introduced to the drying process include cranberries, blueberries, cherries, apples, raspberries and strawberries - not to mention the traditional mainstays of raisins, dates, apricots, peaches, prunes and figs.

Perceived as a "value-added" ingredient, dried fruit adds flavour, colour, texture and diversity with little alteration to an existing formula. The growing interest in ethnic cuisines in U.S.A. and the change to a more healthy way of eating, has also moved dried fruit considerably closer to the mainstream.

Found primarily in the baking industry, dried fruit is coming into its own in various food products, including entrees, side dishes and condiments. Compotes, chutneys, rice and grain dishes, stuffings, sauces, breads, muffins, cookies, deserts, cereals and snacks are all food categories encompassing dried fruit.

Since some dried fruit is sugar infused (osmotic drying), food processors can decrease the amount of sugar in formula-this is especially the case in baked products. Processors are making adjustments in moisture content of the dried fruit so that a varied range is available for different applications. An added bonus is dried fruits' shelf stability (a shelf life of at least 12 months). Dried fruit is more widely available in different forms, including whole dried, cut, diced and powders.

Citric acid and its use in fruit and vegetable processing.

Citric acid may be considered as "Nature's acidulant".

It is found in the tissues of almost all plants and animals, as well as many yeasts and moulds.

Commercially citric acid is manufactured under controlled fermentation conditions that produce citric acid as a metabolic intermediate from naturally-occurring yeasts, moulds and nutrients. The recovery process of citric acid is through crystallization from aqueous solutions.

Citric acid is widely used in carbonated and still beverages, to impart a fresh-fruit "tanginess". Citric acid provides uniform acidity, and its light fruity character blends well and enhances fruit juices, resulting in improved palatability. The amount of citric acid used depends on the particular desired flavour (*e.g.*, High-acid: lemonade; Medium-acid: orange, punch, cherry; Low-acid: strawberry, black cherry, grape).

Sodium citrate is often added to beverages to mellow the tart taste of high acid concentrations. It provides a cool, distinctive smooth taste and masks any bitter aftertaste of artificial sweeteners. In addition, it serves as a buffer to stabilise the pH at the desired level. The high water solubility of citric acid (181 g/100 ml) makes it an ideal additive for fountain fruit syrups and beverages concentrates as a flavour enhancer and microbial growth inhibitor (preferably at pH < 4.6).

In processed fruits and vegetables, citric acid performs the following functions:

a. It reduces heat-processing requirements by lowering pH: inhibition of microbial growth is a function of pH and heat treatment. Higher heat exposure and lower pH result in greater inhibition. Thus the use of citric acid to bring pH below 4.6 can reduce the heating requirements. In canned vegetables, citric acid usage is greatest in tomatoes, onions and pimentos. For tomato packs, the National Canners Association recommends a pH of 4.1 to 4.3. In general, 0. 1% citric acid will reduce the pH of canned tomatoes by 0.2 pH units.

b. Optimise flavour: citric acid is added to canned fruits to provide for adequate tartness. Recommended usage level is generally less then 0.15%.

c. Supplement antioxidant potential: citric acid is used in conjunction with antioxidants such as ascorbic and erythorbic acids, to inhibit colour and flavour deterioration caused by metal-catalysed enzymatic oxidation. Recommended usage levels are generally 0.1% to 0.3% with the antioxidant at 100 to 200 ppm.

d. Inactivate undesirable enzymes: oxidative browning in most fruits and vegetables is catalysed by the naturally present polyphenol oxidase. The enzymatic activity is strongly dependent on pH.

Addition of citric acid to reduce pH below 3 will result in inactivation of this enzyme and prevention of browning reactions.

Cherry and Apricot Oils are Safe for Food Use

The oils obtained by cold pressing the kernels of the cherry (Prunus cerasus) and the apricot (Prunus armeniaca) have been declared acceptable for food use by the UK Ministry of Agriculture, Fisheries & Food's Advisory Committee on Novel Foods and Processes (ACNFP) by June 1993.

In its assessments of the safety in use of the cherry and apricot kernel oils, the ACNFP consider specifications that included data on fatty acid composition, the presence of natural antioxidants and the content of cyanide, mycotoxins and heavy metals.

The Committee says that it gave particular consideration to the possible presence in the oils of the cyanogenic glucoside amygdalin, from which cyanide is released by enzymic action when the kernels of cherry and apricot are crushed. Amygdalin was found to be absent from the cherry and apricot kernel oils.

The ACNFP is satisfied that there are no food safety reasons why the use of cherry and apricot kernel should not be acceptable provided there is compliance with the specifications.

Specification of Purity for Cherry and Apricot Kernel Oils as Determined by UK ACNFP

	Cherry	Apricot
Contaminants limits:		
Heavy metals (total)	0.5 mg/kg	0.5 mg/kg
Aflatoxins (total)	4.0 g/kg	4.0 g/kg
Cyanide	0.15 mg/kg	0.15 mg/kg
Pesticide residues	0.01 mg/kg	0.01 mg/kg
Tocopherols:		
alpha/delta/gamma (mg/kg)	356-886	569-899

Source: Anon. (1993)

The oils are obtained by the mechanical mincing and cold pressing of kernels extracted from cleaned cherry or apricot stones. After filtering, the oils are stored and are to be sold in a raw, unrefined state. The cherry and apricot kernel oils are high unsaturated and are expected to be used as speciality oils for salad dressings, baking and shallow frying applications.

The Use of Fruit Juices in Confectionery Products

During the last decade, the concept of fruit juices has gained immensely on consumer popularity. The majority of new non-alcoholic and alcoholic fruit drink products were a combination of syrups, fruit juices and flavours.

The confectionery industry followed suit and new products incorporated fruit juices as part of their confectionery formulations and processes. Fruit juice concentrates of high solids are often used instead of normal or single-fold juices.

Juice concentrates are made of pure fruit juices. The process starts with pressing fruits and obtaining pure fruit juice; this is stabilised by heat treatment which inactivates enzymes and microorganisms. The next processing step is concentration under vacuum up to 40-65° Brix or 4-7 fold. The concentrates are then blended for standardisation and stored.

These fruit juice concentrates are often further stabilised by the addition of sodium benzoate and potassium sorbate and are usually stored away from light and are refrigerated or frozen.

Depectinised fruit juices are also used to prevent foaming in confectionery processes and are essential for use in clear beverage products. Fruit juice concentrates which are depectinised, and have added preservatives are called stabilised, clarified, fruit juice concentrates.

Fruit juices are used in confectionery products in conjunction with natural and artificial flavours which provides intense flavour impact and are cost-effective for a confectionery product.

The traditional concern in using fruit juice concentrates in confectionery applications has been the effect of the natural acids on the finished product, particularly the formation of invert sugar during processing.

This is a logical concern since concentrates contain differing amounts and types of acids. For example: apple, cherry, strawberry and other berries contain primarily malic acid. Grapes mainly contain tartaric acid. Cranberry is high in quinic acid. Citrus fruits and pineapple contain differing amounts of citric acid. The concentrates, when used, are normally buffered to a pH of 5-7 with sodium hydroxide.

In formulating products with fruit juice concentrates, the solids of the concentrate are considered as mostly reducing sugars and a reduction in corn syrup is made to compensate for equivalent amount of reducing sugar being added in the concentrate.

The exact replacement can be determined by measuring the D.E. of the concentrate to be added. In formulations when small amounts of concentrate are used (less than 1%), no adjustment is made since the reducing sugar contribution of the concentrate is not significant.

Fruit juice concentrates can also be used to provide a source of natural colour, in particular red colour. Grape, raspberry, cherry, strawberry and cranberry concentrates in small amounts are very effective in colouring cream centres.

The inclusion of fruit juices in confectionery products is now left up to the imagination of the manufacturer. These products must, of course, hold up to the standards of flavour integrity, and product excellence, during the shelf-life of these products.

Vegetable Specific Processing Technologies

Vegetables Varieties

Vegetable processors must appreciate the substantial differences that varieties of a given vegetable will possess. In addition to variety and genetic strain differences with respect to weather, insect and disease resistance, varieties of a given vegetable will differ in size, shape, time of maturity, and resistance to physical damage.

Varietal differences then further extend into warehouse storage stability, and suitability for such processing methods as canning, freezing, pickling or drying. A variety of peas that is suitable for canning may be quite unsatisfactory for freezing and varieties of potatoes that are preferred for freezing may be less satisfactory for drying or potato chip manufacture.

This should be expected since different varieties of a given vegetable will vary somewhat in chemical composition, cellular structure and biological activity of their enzyme system.

Harvesting and Pre-processing

When vegetables are maturing in the field they are changing from day to day. There is a time when the vegetable will be at peak quality from the stand-point of colour, texture and flavour.

This peak quality is quick in passing and may last only a day. Harvesting and processing of several vegetables, including tomatoes, corn and peas are rigidly scheduled to capture this peak quality. After the vegetable is harvested it may quickly pass beyond the peak quality condition. This is independent of microbiological spoilage; these main deteriorations are related to:

(a) loss of sugars due to their consumption during respiration or their conversion to starch; losses are slower under refrigeration but there is still a great change in vegetable sweetness and freshness of flavour within 2 or 3 days;

(b) production of heat when large stockpiles of vegetables are transported or held prior to processing.

At room temperature some vegetables will liberate heat at a rate of 127,000 kJ/ton/day; this is enough for each ton of vegetables to melt 363 kg of ice per day. Since the heat further deteriorates the vegetables and speeds microorganisms growth, the harvested vegetables must be cooled if not processed immediately.

But cooling only slows down the rate of deterioration, it does not prevent it, and vegetables differ in their resistance to cold storage. Each vegetable has its optimum cold storage temperature which may be between about 0-100 C (32-500 F).

(c) the continual loss of water by harvested vegetables due to transpiration, respiration and physical drying of cut surfaces results in wilting of leafy vegetables, loss of plumpness of fleshy vegetables and loss of weight of both.

Moisture loss cannot be completely and effectively prevented by hermetic packaging. This was tried with plastic bags for fresh vegetables in supermarkets but the bags became moisture fogged, and deterioration of certain vegetables was accelerated because of buildup of CO2 and decrease of oxygen in the package.

It therefore is common to perforate such bags to prevent these defects as well as to minimise high humidity in the package which would encourage microbial growth.

Shippers of fresh vegetables and vegetable processors, whether they can, freeze, dehydrate, or manufacture soups or ketchup, appreciate the instability and perishability of vegetables and so do everything they can to minimise delays in processing of the fresh product. In many processing plants it is common practice to process vegetables immediately as they are received from the field.

To ensure a steady supply of top quality produce during the harvesting period the large food processors will employ trained field men; they will advise on growing practices and on spacing of plantings so that vegetables will mature and can be harvested in rhythm with the processing plant capabilities. This minimises stockpiling and need for storage.

Cooling of vegetables in the field is common practice in some areas. Liquid nitrogen-cooled trucks may next provide transportation of fresh produce to the processing plant or directly to market. Upon arrival of vegetables at the processing centre the usual operations of cleaning, grading, peeling, cutting and the like are performed using a moderate amount of equipment but a good deal of hand labour also still remains.

Reception: This covers qualitative and quantitative control of delivered vegetables. The organoleptic control and the evaluation of the sanitary state, even if they are very important steps in vegetables' characteristics assessment, cannot establish their technological value.

On the other hand, laboratory controls do not precisely establish their technological properties because of the difficulty in putting into showing some deterioration when using rapid control methods.

One correct method of vegetable quality appraisal is their overall evaluation based on the whole complex of data that can be obtained by combining an extensive organoleptic evaluation with simple analysis that can be performed rapidly in plant laboratory. These analysis can be:

a. refractometric extract (tomatoes, fruit, etc.);
b. specific weight (potatoes, peas, etc.);
c. consistency (measured with tenderometers, penetrometers, etc.);
d. boiling tests, etc.

Temporary storage.

This step should be as short as possible and better completely eliminated. Vegetables can be stored in:

a. simple stores, without artificial cooling;
b. in refrigerated stores; or, in some cases,
c. in silos (potatoes, etc.).

Simple stores should be covered, fairly cool, dry and well ventilated but without forced air circulation which can induce significant losses in weight through intensive water evaporation; air relative humidity should be at about 70-80%.

Refrigerated storage is always preferable and in all cases a processing centre needs a cold room for this purpose, adapted in volume I capacity to the types and quantities of vegetables (and fruits) that are further processed.

Washing: Washing is used not only to remove field soil and surface microorganisms but also to remove fungicides, insecticides and other pesticides, since there are laws specifying maximum levels of these materials that may be retained on the vegetable; and in most cases the allowable residual level is virtually zero. Washing water contains detergents or other sanitisers that can essentially completely remove these residues.

The washing equipment, like all equipment subsequently used, will depend upon the size, shape and fragility of the particular kind of vegetable:

- flotation cleaner for peas and other small vegetables;
- rotary washer in which vegetables are tumbled while they are sprayed with jets of water; this type of washer should not be used to clean fragile vegetables;

Sorting: This step covers two separate operations:

(a) Removal of non-standard vegetables (and fruit) and possible foreign bodies remaining after washing;

(b) Quality grading based on variety, dimensional, organoleptical and maturity stage criterion.

Skin Removal/peeling: Some vegetables require skin removal. This can be done in various ways.

(a) Mechanical

This type of operation is performed with various types of equipment which depend upon the result expected and the characteristics of the fruit and vegetables, for example:

i. a machine with abrasion device (potatoes, root vegetables);

ii. equipment with knives (apples, pears, potatoes, etc.);

iii. equipment with rotating sieve drums (root vegetables). Sometimes this operation is simultaneous with washing (potatoes, carrots, etc.) or preceded by blanching (carrots).

(b) Chemical

Skins can be softened from the underlying tissues by submerging vegetables in hot alkali solution. Lye may be used at a concentration of about 0.5-3%, at about 93° C (2000 F) for a short time period (0.5-3 min). The vegetables with loosened skins are then conveyed under high velocity jets of water which wash away the skins and residual lye.

In order to avoid enzymatic browning, this chemical peeling is followed by a short boiling in water or an immersion in diluted citric acid solutions.

It is more difficult to peel potatoes with this method because it is necessary to dissolve the cutin and this requires more concentrated lye solutions, up to 10%.

(c) Thermal

Wet Heat (steam): Other vegetables with thick skins such as beets, potatoes, carrots and sweet potatoes may be peeled with steam under pressure (about 10 at) as they pass through cylindrical vessels. This softens the skin and the underlying tissue. When the pressure is suddenly released, steam under the skin expands and causes the skin to puff and crack. The skins are then washed away with jets of water at high pressure (up to 12 at).

Dry Heat (flame): Other vegetables such as onions and peppers are best skinned by exposing them to direct flame (about 1 min at 1000° C) or to hot gases in rotary tube flame peelers. Here too, heat causes steam to develop under skins and puff them so that they can be washed away with water.

Manual peeling only use when the other methods are impossible or sometimes as a completion of the other three ways. Average losses at this step are given in Table.

Losses at Vegetable Peeling, in %

	Peeling methods		
Vegetables	Manual	Mechanical	Chemical
Potatoes	15-19	18-28	-
Carrots	13-15	16-18	8-10
Beets	1416	13-15	9-10

Size Reduction: This step is applied according to specific vegetable and processing technology requirements.

Blanching: The special heat treatment to inactivate enzymes is known as blanching. Blanching is not indiscriminate heating. Too little is ineffective, and too much damages the vegetables by excessive cooking, especially where the fresh character of the vegetable is subsequently to be preserved by processing.

This heat treatment is applied according to and depends upon the specificity of vegetables, the objectives that are followed and the subsequent processing / preservation methods.

Two of the more heat resistant enzymes important in vegetables are catalase and peroxidase. If these are destroyed then the other significant enzymes in vegetables also will have been inactivated. The heat treatment to destroy catalase and peroxidase in different vegetables are known, and sensitive chemical tests have been developed to detect the amounts of these enzymes that might survive a blanching treatment. Catalase and peroxidase inactivation tests are presented in section 9.2.9.

Because various types of vegetables differ in size, shape, heat conductivity, and the natural levels of their enzymes, blanching treatments have to be established on an experimental basis. As with sterilisation of foods in cans, the larger the food item the longer it takes for heat to reach the centre. Small vegetables may be adequately blanched in boiling water in a minute or two, large vegetables may require several minutes.

Blanching as a unit operation is a short time heating in water at temperatures of 100° C or below. Water blanching may be performed in double bottom kettles, in special baths with conveyor belts or in modern continuous blanching equipment.

In order to reduce losses of hydrosoluble substances (mineral salts, vitamins, sugars, etc.) occurring during water blanching, several methods have been developed:

- temperature setting at 85-95° C instead of 100° C;
- blanching time has to be just sufficient to inactivate enzymes catalase and peroxidase;
- assure elimination of air from tissues.

An illustration of blanching parameters is seen in Table.

Blanching Parameters for Some Vegetables

Vegetables	*Temperature, °C*	*Time, min.*
Peas	85-90	2-7
Green beans	90-95	2-5
Cauliflower	Boiling	2
Carrots	90	3-5
Peppers	90	3

Steam heat treatment can also be applied instead of water blanching as a preliminary step before freezing or drying, as long as the preservation method is only used for enzyme inactivation and not to modify consistency.

For drying, the vegetables are conveyed directly from steaming equipment to drying installations without cooling.

Vegetable steaming is carried out in continuous installations with conveyer belts made from metallic sieves.

Cooling of vegetables after water blanching or steaming is performed in order to avoid excessive softening of the tissues and has to follow immediately after these operations; one exception is the case of vegetables for drying which can be transferred directly to drying equipment without cooling.

Natural cooling is not recommended because is too long and generates significant losses in vitamin C content. Cooling in pre-cooled air (from special installations) is sometimes used for vegetables that will be frozen

Cooling in water can be achieved by sprays or by immersion; in any case the vegetables have to reach a temperature value under 37° C as soon as possible. Too long a cooling time generates supplementary losses in valuable hydrosoluble substances; in order to avoid this, the temperature of the cooling water has to be as low as possible.

Canning: Large quantities of vegetable products are canned. A typical flow sheet for a vegetable canning operation (which also applies to fruit for the most part) covers some food process unit operations performed in sequence: harvesting; receiving; washing; grading; heat blanching; peeling and coring; can filling; removal of air under vacuum; sealing/closing, retorting/heat treatment; cooling; labelling and packing. The vegetable may be canned whole, diced, puréed, as juice and so on.

On-line Simplified Methods for Enzyme Activity Check

Peroxidase Test:

(a) *Solutions:* In order to check the peroxidase activity two solutions have to be prepared:

- 1% guaiacol in alcohol solution (1 g guaiacol is dissolved in about 50 cm^3 of 96% ethylic alcohol and then this preparation is brought to 100 cm° with the same solvent);
- peroxide solution 0.3% (1 cm^3 perhydrol is brought to 100 cm^3 with distilled water.

(b) *Sampling:* From various parts of the material samples are taken (about 20-30 pieces, etc.) the material is then crushed in a laboratory bowl in order to obtain an average sample.

(c) *Check:* Prom the average sample, 10-20 g of material is taken in a medium capacity test tube; on this sample are poured: 20 cm^3 distilled water; 1 cm^3 of 1% guaiacol solution; 1.6 cm^3 of peroxide solution.

The contents of the test tube is shaken well. The gradual appearance of a weak pink colour indicates an incomplete peroxidase inactivation - reaction slightly positive. If there are no tissue colour modifications after 5 minutes, the reaction is negative and the enzymes have been inactivated. As an orientative check it is also possible to simply pour a few drops of 1% guaiacol solution and 0.3% peroxide solution directly on blanched and crushed vegetables. A rapid and intensive brown-reddish tissue colouring indicates a high peroxidase activity (positive reaction).

Catalase Test: In order to identify the catalase enzyme activity, 2 g of dehydrated vegetables are well crushed and mixed with about 20 cm^3 of distilled water. After 15 min softening, 0.5 cm^3 of a 0.5% or 1% peroxide solution is poured on prepared vegetables. In the presence of catalase, a strong oxygen generation is observed for about 2-3 minutes.

These tests are of a paramount importance in order to determine the vegetable blanching treatments (temperature and time); incomplete enzyme inactivation has a negative effect on finished product quality. For cabbage catalase inactivation by blanching is sufficient; blanching further to peroxidase inactivation would have negative effects on product quality and even complete browning. For all other vegetables and for potatoes, both tests MUST be negative, for catalase and for peroxidase.

Fresh Vegetable Storage

The vegetables can be stored, in some specific natural conditions, in fresh state, that is without significant modifications of their initial organoleptic properties. Fresh vegetable storage can be short term; this was briefly covered under temporary storage before processing. Also fresh vegetable storage can be long term during the cold season in some countries and in this case it is an important method for vegetable preservation in the natural state. In order to assure preservation in long term storage, it is necessary to reduce respiration and transpiration intensity to a minimum possible; this can be achieved by:

a. maintenance of as low a temperature as possible (down to 0° C),

b. air relative humidity increased up to 85-95 % and

c. CO_2 percentage in air related to the vegetable species.

Vegetables for storage must conform to following conditions: they must be of one of the autumn or winter type variety; be at edible maturity without going past this stage; be harvested during dry days; be protected from rain, sun heat or wind; be in a sound state and clean from soil; be undamaged. From the time of harvest and during all the period of their storage vegetables are subject to respiration and transpiration and this is on account of their reserve substances and water content. The more the intensity of these two natural processes are reduced, the longer sound storage time will be and the more losses will be reduced.

For this reason, vegetables have to be handled and transported as soon as possible in the storage conditions (optimal temperature and air relative humidity for the given species). Even in these optimal conditions storage will generate losses in weight which are variable and depend upon the species.

Some optimal storage conditions are shown in table.

Optimal Conditions for Fresh Vegetable Storage

Vegetables	Storage conditions	
	Temperature, °C	Relative humidity, %
Potatoes	+1...+3	85-90
Carrots	0 ... +1	90-95
Onions	0 ... +1	75-85
Leeks	0 ... +0.5	85-90
Cabbage	-1 ... 0	90-97
Garlic	0 ... +1	85-90
Beets	0 ... +1	90-95

7

Vegetable Drying/dehydration

Technology for Vegetable Powder Processing

This technology has been developed in recent years with applications mainly for potatoes (flour, flakes, granulated), carrots (powder) and red tomatoes (powder). In order to obtain these finished products there are two processes:

(a) Drying of vegetables down to a final water content below 4% followed by grinding, sieving and packing of products;

(b) Vegetables are transformed by boiling and sieving into purees which are then dried on heated surfaces (under vacuum preferably) or by spraying in hot air.

Industrial installations that can be used for these products and technological data are summarised below:

- Dryers with plates under vacuum are equipped with plates heated with hot water. Stainless steel plates containing the puree to be dried are placed on them. Process conditions are at low residual pressure (about 10-20 mm Hg) and a product temperature of 50-70° C. This equipment is discontinuous but easy to operate.
- Drum dryers have one or two drums heated with hot water or steam as heating elements. Feeding is continuous between the two drums which are rotating in reverse direction (about 2-6 rotations per minute) and the distance of which is adjustable and determines the thickness of layer to be dried, he product is dried and removed by mechanical means during rotation.

- Drying installations by spraying in hot air; the product is introduced in equipment and sprayed by a special device in hot air. Drying is instantaneous (1/50 s) and therefore can be carried out at 130-15O° C.

Packing and Storage of Dried and Powdered Vegetables

Dried vegetables can suffer significant modifications that bring about their deterioration during storage. The factors that determine these degradations impose at same time the type of packaging materials and storage conditions for packaged products.

The main factor in maintaining the quality of dried products is to follow the maximum moisture contents that have to be as close as possible to the limits indicated.

The moisture content of dried vegetables is not constant because of their hygroscopicity and is always in equilibrium with relative humidity of air in storage rooms. Technical solutions for maintaining a low dehydrated products moisture are:

(a) storage in stores with air relative humidity below 78%;

(b) use packages that are water vapour proof. The most efficient packages are tin boxes or drums (mainly for long term storage periods); combined packages (boxes, bags, etc.) from complexes (carton with metallic sheets, plastic materials, etc.) mainly for small packages. One solution for some dried vegetables may be the use of waterproof plywood drums.

Modern solutions are oriented not only to the maintaining product moisture during storage but also reducing this parameter by the use of desiccants (substances which absorb moisture) introduced in packages, hermetically closed.

A desiccant in current use is calcium oxide. Granulated calcium oxide is introduced in small bags from a material which is permeable to water vapour but which does not permit the desiccant to escape into products.

With desiccants, product moisture can be reduced to even below 4%, and this inhibits or reduces the biochemical and microbiological processes during storage.

Another factor that can deteriorate dried/dehydrated vegetables is atmospheric oxygen through the oxidative phenomena that it produces. In order to eliminate the action of this agent some packing methods under vacuum or in inert gases (carbon dioxide or nitrogen) are in use, applied mainly for packing dried carrots in order to avoid beta-carotene oxidation in beta-ionone (foreign smell, discoloration, etc.). In order to avoid the action of oxygen it is also possible to add ascorbic acid as antioxidant (for example in carrot powder).

Sun or artificial light action on dehydrated vegetables generally causes discoloration which can be avoided by using opaque packaging materials.

Dehydrated vegetable compression (especially for roots) to form blocks with a weight of 50-600 g, is practiced sometimes; it has as advantages the reduction of evaporation surface and contact with atmospheric oxygen and volume reduction. Dehydrated vegetables are compressed at about 300 at. Compressed blocks are packaged in heat sealed plastic materials.

Storage temperature has an important role because this reduces or inhibits the speed of all physico-chemical, biochemical and microbiological processes, and thus prolongs storage period. The storage temperature should be below 25° C (and preferably 15° C); lower temperatures (0-10° C) help maintain taste, colour and water rehydration ratio and also, to some extent, vitamin C.

Potato Crisp/chip Processing

The most important steps involved in potato crisps processing are:

1. Selecting, procuring and receiving potatoes
2. Storage of potato stock under optimum conditions
3. Peeling and trimming the tubers
4. Slicing
5. Frying in oil
6. Salting or applying flavoured powders
7. Packaging

Moisture and Shipping Factors for Some Dehydrated Vegetables

Product	Form/cut	Moisture %	Weight kg/m³
Bean (green)	20 nun cut	5	1.6
Bean (lima)	5	3.3	
Beet	6 mm strips	5	1.6-1.9
Cabbage	6-12 mm shreds	4	0.7-0.9
Carrots	5-8 mm strips	5	3-5
Celery	Cut	4	
Garlic	Cloves	4	
Okra	6 mm slices	8	
Onion	Slices	4	0.4- 0.6
Pea (fresh)	Whole	5	3.4
Pepper (hot)	Ground	5	
Pepper (sweet)	5 mm strips	7	
Potato (Irish)	5-8 mm strips	6	2.9-3.2
	Diced	5	3.3-3.6
Tomato	7-10 mm slices	35	

Selection and Storage

It is important to select potatoes of high specific gravity since this characteristic indicates superior yield and lower oil absorption. It is even more important to select potatoes with low reducing sugar contents or to store them at temperatures conducive to the minimising of these substances.

Sprouting and fungal damage must also be minimised by the storage conditions.

Peeling

The ideal peeling operation should only remove a very thin outer layer of the potato, leaving no eyes, blemishes, or other material for later removal by hand trimming. It should not significantly change the physical or chemical characteristics of the remaining tissue.

Preferably peeling should use small amount of water and result in minimal effluent; compromises will have to be made in all of these aspects of peeling. First, the potatoes are thoroughly washed, not only for sanitary reasons, but also to prevent dirt of grit from abrading the equipment the tubers will dater contact. Washing may take place in streams, as the potatoes are being conveyed by water streams, or in equipment provided with means for scrubbing the potato with brushes or rubber rolls.

In barrel-type washers, potatoes are cleaned by being tumbled and rubbed against each other and against the sides of the barrel while they are immersed in, or sprayed with, water. After washing, the potatoes are allowed to drain, usually on mesh conveyors, and they travel over an inspection belt where foreign material and defective tubers are removed. The more common peeling methods are abrasion, lye immersion, and steam.

Abrasion peelers which may be either batch or continuous, use disks or rollers coated with grit to grind away the potato surface. An important design feature is to ensure that all surfaces of the tuber are equally exposed to the rasping action. The peel fragments are flushed out of the unit by water sprays.

Such systems work best with uniform, round, undamaged potatoes. Some of the advantages of abrasion peelers are their simplicity, compactness, low cost, and convenience. They are particularly suitable for peeling potatoes intended for chipping, since they do not chemically alter the surface layers. About 10% of the original tuber weight is lost through abrasion peeling prior to chipping.

Slicing

The peeled potatoes are cut into slices from 1/15 to 1/25 in. by rotary slicers. Centrifugal force presses the tuber against stationary gauging shoes and knives. Thickness is varied, not only to meet consumer preferences, but also to fit the condition of the tubers and the frying temperature and time.

Slices produced at any one time must be very uniform in thickness, however, in order to obtain uniformly coloured chips. Slices with rough or torn surfaces lose excess solubles from ruptured cells and absorb larger amounts of fat.

It is necessary to remove the starch and other material released from the cut cells from the surface of slices so that the slices will separate readily and completely during frying. The slices are washed in stainless steel wire-mesh cylinders or drums rotating in a rectangular stainless steel tank. After washing and an additional rinse in similar equipment, the potatoes may or may not be dried.

Frying

The capacity of the fryer is generally the limiting factor in the process line. Most manufacturers currently use continuous fryers but some batch equipment is still employed.

Modern continuous fryers have the following essential elements: (1) a tank of hot oil in which the chips are cooked; (2) a means for heating and circulating the oil; (3) a filter for removing particles from oil; (4) a conveyor to carry chips out of the tank; (5) a reservoir in which oil is heated for adding to the circulating frying oil and (6) vapour-collecting hoods above the tank. Temperatures normally used are from 350 to 375° F at the receiving end and 320 to 345° F at the exit end.

The oil used for deep-fat frying of potato chips has two functions:

(i) it serves as a medium for transferring heat from a thermal source to the tuber slices;

(ii) it becomes an ingredient of the finished product.

Use of highly refined oil is of great importance in flavour and stability of the crisps. Flavour, texture, and appearance are affected both by the amount of oil absorbed and its characteristics as it exists in the crisp (*i.e.* not necessarily its initial chemical and physical parameters).

Oils change continuously during the frying process but the heat abuse resulting from the crisp cooking is relatively mild. Temperatures rarely rise above 385° F at any point.

Better control over crisp colour could be obtained if the final stage of moisture removal could be achieved without the browning reaction that always accompanies it in the frying process.

Crisps may be sorted for size after frying, with the larger crisps being diverted to the bulk packs and larger pouches and the smaller pieces used for vending machine packs and other individual service containers. Potato crisp sizing is also accomplished by separating the peeled potatoes into large and small sizes, which are then sliced and fried separately.

The crisps are salted immediately after they leave the fryer. It is important that the fat be liquid at this point to cause maximum adherence of the granules. Powders containing barbecue spices, cheese, or other speciality materials may be added at this point. The salt may contain added enrichment materials or antioxidants.

After salting, the crisps pass on to a conveyor belt where they are visually inspected and off-colour material is removed. I the crisps are allowed to cool before packaging, better adherence of salt and flavour powders is obtained.

Some consumers prefer the hard, curled-up crisp that is characteristic of the hand-kettle type of operation. The special flavour of the hand-kettle crisp is said to be due, at least partly, to the starch retained on the cut surfaces of the potato slices as a result of the omission of a washing process after slicing. Starch-covered slices tend to stick together in the fryer so it is necessary to use devices to prevent clumping.

The principal factors affecting potato crisp acceptability are piece size, colour, and of course, flavour. These factors are controllable primarily by selection of the raw material, adjustment of processing conditions, and packaging.

Storage Stability

If the frying oil is stabilised and has not deteriorated through use, and if the packaging is opaque and has a low moisture vapour transmittance rate (MVTR), a shelf-life of 4-6 weeks should be achieved when crisps are stored at temperatures of about 70° F.

Once potato crisps are in the bag, the three forms of quality loss which have the greatest effect on consumer acceptance are breakage, absorption of moisture with loss of crispiness, and fat oxidation leading to development of rancid odours.

The mechanical abuse causing breaking of the crisps can be partially prevented by using stiff packaging material, making the package "plump" with contained air, and avoiding crushing in the shipping case.

Absorption of moisture is prevented largely by proper choice of packaging material. Cellophane coated with various moisture barriers has proved to be a satisfactory pouch films for the relatively short shelf-life expected (generally stated to be 4-6 weeks).

Light (especially fluorescent light) accelerates oxidation, so that opaque packaging material must be used to obtain maximum shelf-life.

Potato crisps are considered commercially unacceptable when they have a moisture content above 3%, which is in equilibrium with a relative humidity of about 32%. The containers should have a high degree of resistance to moisture-vapour transfer.

If pouches are used, foil-containing films are preferable, since they not only resist moisture-vapour transfer but reflect light.

Vegetable Juices and Concentrated Products

Vegetable Juices

Vegetable juices are natural products constituted from cellular juice and a part of crushed pulp, from the tissues of some vegetables. These juices contain all valuable substances from the vegetables: vitamins, sugars, acids, mineral salts and pectic substances. The most important of these products is tomato juice; in a lower proportion there are also other juices (carrots, beet, sauerkraut, etc.).

Tomato Juice: This product is characterised not only by its organoleptical properties (taste, colour, flavour) but also by its vitamin content close to those of fresh tomatoes. Modem technology is oriented to a maximum maintenance of organoleptic properties and of vitamin content.

At same time, it is important to assure juice uniformity by avoiding cellulosic particle sedimentation. Juice stability is assured by a flash pasteurization which assures the destruction of natural micro-flora, while keeping the initial properties.

The modern technological flow-sheet covers the following main operations: Pre-washing is carried out by immersion in water, cold or heated up to 50° C (possibly with detergents to eliminate traces of pesticides). This operation is facilitated by bubbling compressed air in the immersion vessel/equipment.

Washing is performed with water sprays, which in modern installations have a pressure of 15 at or more.

Sorting/Control on rolling sorting tables enables the removal of non-standard tomatoes - with green parts, yellow coloured, etc.

Crushing in Special Equipment

Preheating at 55-60° C facilitates the extraction, dissolves pectic substances and contributes to the maintaining of vitamins and natural pigments. In some modern installations, this step is carried out under vacuum at 630-680 mm Hg and in very short time.

Extraction of juice and part of pulp (maximum 80%) is performed in special equipment / tomato extractors with the care to avoid excessive air incorporation. In some installations, as an additional special care, a part of pulp is removed with continuous centrifugal separators.

De-aeration under high vacuum of the juice brings about its boiling at 35-40° C.

Homogenisation is done for mincing of pulp particles and is mandatory in order to avoid future potential product "separation" in two layers.

Flash Pasteurization at 130-150° C, time = 8-12 see, is followed by cooling at 90° C, which is also the filling temperature in receptacles (cans or bottles).

Aseptic Filling

Closing Of Receptacles is followed by their inversion for about 5 to 7 minutes. Cooling has to be carried out intensely.

Full cans do not need further pasteurization because the bacteria that have potentially contaminated the tomato juice during filling are easily destroyed at 90° C due to natural juice acidity.

For bottles, it may be possible to avoid further sterilisation if the following conditions can be respected: washing and sterilising of receptacles, cap sterilisation (with formic acid), filling and capping under aseptic conditions, in a space with UV lamps. In so far as this is quite difficult to achieve it may be necessary to submit bottles to a pasteurization in water baths. The main characteristics of high quality tomato juice are:

- natural red colour;
- taste and flavour of fresh tomatoes;
- uniformity (without pulp sedimentation);
- total soluble solids: 6% minimum;
- total soluble substances (by refractometer): 5% minimum;
- vitamin C: 15 mg/100ml minimum.

In traditional processes it is recommended to:

- thoroughly wash and rinse the empty receptacles (including jar caps / covers and bottle crown corks) and then "sterilize" by keeping in boiling water for 30 mini
- add salt and lemon juice to the prepared receptacles just before filling;
- pasteurize closed glass receptacles (bottles or jars) according to conditions recommended in technological flow-sheets and which is summarised as follows:

Receptacle size	Pre-heating	Time of pasteurization
0.33 1	60° C	40 minutes
0.501	60° C	45 minutes
0.66 1	60° C	55 minutes
0.751	60° C	60 minutes
1.0 litre	60° C	70 minutes

Carrot Juice: This product represents an important dietetic product due to its high soluble pectin content. Technological flow-sheet is oriented to the maintaining of as high as possible a pectin content and covers the following steps:

Pre-washing

Cleaning

Washing

Blanching in steam for 20 minutes

Grating

Pressing

Juice in the pressed juice will then be incorporated 25% of grated carrot (non pressed)

Homogenisation in colloidal mills

Acidification with 0.25% citric or tartric acid

De-aeration

Filling in receptacles (bottles or tinplate cans)

Airtight sealing

Pasteurization at 100° C for 30 minutes.

The main characteristics of a good quality carrot juice:

- uniformity (no separation in layers occurs during storage);
- good orange colour;
- pleasant taste, close to fresh carrot taste;
- total soluble solids: 12 %;
- total sugar content: 8%;
- beta-carotene: 1.3 mg/100 ml;
- soluble pectin: 0.4 %.

Red Beet Juice: The product is obtained following this technological flow-sheet: washing, cleaning, steam treatment/ steaming (30-35 min at 1050 C), pressing, strain through small hole sieve, filling in receptacles, tight sealing/closing, sterilisation (25 min at 1 15° C). In order to improve taste, the juice is acidified with 0.3% citric or tartric acid.

Sauerkraut Juice: Sauerkraut juice is produced in some countries for its dietetic value (lactic acid and vitamin C content) and its refreshing taste. The juice which is the result of the fermentation of lactic acid from cabbage, mainly from sliced sauerkraut, is used.

The juice must be the result of a normal lactic fermentation, *i.e.* without butyric fermentation or other deterioration.

A good quality juice must have an acidity of 1.4% lactic acid and a content of maximum 2.5% salt; this is obtained by the mixing of various sauerkraut qualities.

The collected juice (from sauerkraut production) is heated slightly in order to eliminate CO2 gas and to obtain protein coagulation. Filtration of juice is the next technological step, followed by filling in receptacles, closing of receptacles and pasteurization at 75-80° C for 4-5 minutes.

Concentrated Tomato Products

Tomato Paste: The product with highest production volumes among concentrated products is tomato paste which is manufactured in a various range of concentrations, up to 44% refractometric extract. Tomato paste is the product obtained by removal of peel and seeds from tomatoes, followed by concentration of juice by evaporation under vacuum.

In some cases, in order to prolong production period, it may be advisable or possible to preserve crushed tomatoes with sulphur dioxide as described under semi-processed fruit "pulps".

Technological flow-sheets run according to equipment/ installation lay-outs, which are especially designed for this finished product. Manufacturing steps fall into three successive categories:

a. obtaining juice from raw materials;
b. juice concentration and
c. tomato paste pasteurization.

(a) Obtaining juice from raw material. Preliminary operations (pre-washing, washing and sorting / control) are carried out in the same conditions as for manufacturing of "drinking" tomato juice described above. Next operation is removal of seeds from raw tomatoes: tomato crushing and seed separation with a centrifugal separator.

Tomato pulp is pre-heated at 55-60° C and then passed to the equipment group for sieving: pulper, refiner and superrefiner with sieves of 1.5 mm, 0.8 mm and 0.4-0.5

mm respectively in order to give the smoothest possible consistency to the tomato paste.

(b) Juice is concentrated by vacuum evaporation, a technological step which in modern installations runs continuously, tomato paste from the last evaporation step being at the specified concentration.

In continuous installations with three evaporation steps (evaporating bodies), the juice is submitted in step / body I to pasteurization at 85-900 C for 15 min and this will determine the microbiological stability of finished product. Vacuum degree corresponding to this temperature is 330 mm Hg.

In evaporating bodies II and III, temperatures are around 42-46° C and vacuum at 680700 mm Hg.

Juice concentration occurs gradually and continuously in the three evaporating bodies.

The advantages of continuous concentration are as follows: the taste, colour, flavour, "shine" and consistency of tomato paste are improved because:

(i) the real concentration is performed in evaporating bodies II and III at low temperatures (42-46° C) and

(ii) the whole concentration process time from the input of juice in body I until the output of paste from body III is of about 1 hour (for paste with 30-35% refractometric extract).

- production capacity is raised by about 30% as compared to discontinuous installations with the same evaporation surface;
- the steam consumption is reduced by 60% because heating of bodies II and II is done with vapours resulting from juice evaporation in body I (double effect); water and electricity consumptions are also reduced by 30-40%.

(c) Tomato paste pasteurization assures the microbiological stability of the product. For this purpose, the paste coming out from concentration equipment is passed continuously and in a "forced" mode through a tubular pasteurizer from which it emerge at a temperature of 90-92° C.

Usual commercial tomato paste types are at concentrations of 24%, 28% and 32% refractometric extract. Sometimes it is possible to obtain a tomato paste with a concentration of 44% refractometric extract; for this purpose it is necessary to eliminate a part of cellulose from tomatoes, an operation performed in a separating turbine.

Tomato paste storage and preservation is carried out after packing which is done usually in drums, metallic cans or glass jars; some modern equipment has been developed for packing in aluminium bags. As far as the concentration of tomato paste is concerned it is not possible to reduce water content down to 30% which corresponds to a water activity aw of 0.700.75 (minimum limit of mould growing), it is necessary to take special measures (e. g pasteurization, cold storage or salt addition).

Salt is not a preservative in itself but contributes to the lowering of water activity.

In drums, the preservation of tomato paste with minimum 30% refractometric extract is carried out in two ways:

- the hot paste (about 90° C) flows directly from pasteurization equipment into drums that have been previously steamed;
- the paste is cooled down to 30° C through a heat exchanger and is introduced into drums that have been previously steamed.

For preservation purposes, it is possible to add 3-8% salt.

Preservation with 3% salt must be carried out respecting the following criteria:

(a) processing of a healthy raw material;

(b) thorough washing and control;

(c) pasteurization of concentrated paste and use of well prepared drums. Paste in drums has to be stored in cold storage rooms during the hot season.

Preservation in big metal cans of 5 and 10 kg capacity of tomato paste with a minimum of 30% refractometric extract can be achieved without sterilisation if the following conditions are respected:

(a) sterilisation by steam of cans and covers;

(b) filling of paste at 92-94° C;

(c) airtight sealing/ closing of cans;

(d) invert cans and then

(e) air cooling.

For small packages (tinplate cans of 1/10-1/1 or glass jars of same capacity) it is usual to use pasteurized paste, as hot as possible (92-94° C). The receptacles are first sterilised by steam. After airtight sealing, the receptacles are kept in boiling water for a short time in order to sterilize their inner surface and the paste in contact with inner receptacle surface. In some countries small receptacles are not further sterilised if the manufacturing is carried out in perfect hygienic and sanitary conditions.

Packing in small tinned aluminium tubes is carried out with concentrated paste, pasteurized and hot.

Good quality tomato paste is an homogenous mass, with a high density, without foreign bodies (seeds, peel, etc.), with a red colour, and an agreeable taste and smell, close to those of fresh tomatoes.

There are usually three types of tomato paste: 36, 30 and 24 which have refractometric extracts of respectively 34-38%, 28-32% and 24-26%. Paste of good quality must have a volatile acidity of maximum 0.15% as lactic acid. An 8% salt addition is accepted.

Concentrated Tomato Juice: Concentrated tomato juice is a product with 17-19 % refractometric extract and is a homogenous mass, finely sieved, without foreign bodies / and without any evidence of deterioration. A good quality product has a red colour, an agreeable and specific taste and smell.

Modern technology uses the same installations, equipment and flow-sheets for concentrated tomato juice as for the production of tomato paste; the final concentration is thus regulated between the above specified limits.

The concentrated tomato juice is filled in receptacles (metal tinplate cans or glass bottles) and then pasteurized at 100° C during 15-25 minutes according to receptacle type.

With modern production lines it should be possible to pass the concentrated tomato juice through a tubular pasteurizer and then pack aseptically and cool, without the need to pasteurize the receptacles.

Tomato Sauces: Under the USA Code of Federal Regulation 7 CFR 52, 1991 tomato sauce is the concentrated product prepared from the liquid extract from mature, sound, whole tomatoes, the sound residue from preparing such tomatoes for canning, or the residue from partial extraction of juice, or any combination of these ingredients, to which is added salt and spices and to which may be added one or more nutritive sweetening ingredients, a vinegar or vinegars, and onion, garlic, or other vegetable flavouring ingredients. The refractive index of the tomato sauce at 20° C is not < 1.3461.

These products are widespread in some countries and are used in order to spice some meals. Sauces can be obtained from fresh tomatoes or from concentrated products (tomato paste or concentrated tomato juice), those from fresh tomatoes being of superior quality.

Technological processing covers the following steps: concentrated juice processing, addition of flavour/taste ingredients (salt, sugar, vinegar, spices, etc.), boiling, fine sieving, filling of receptacles, closing and pasteurization (45 min at 85° C).

Tomato sauces which can be sweet, more or less spicy are prepared according to specific recipes.

Production accidents and product defects; means to avoid them

Tomato Juice:

- "Separation" in layers is due to not enough homogenisation or low / insufficient viscosity. In the first case it is necessary to intensify homogenisation; and in second to increase the pre-heating temperature to 60° C in order to obtain protopectine hydrolisis and pectolitic enzymes inactivation.
- Moulding of the juice is brought about by significant infection of raw materials, inadequate washing and control

or by use of contaminated packages. The preventive measures should be decided after cause analysis. Good pasteurization can destroy all moulds but the bad juice taste remains.

- Fermentation of juice is manifested by a significant development of gases. Prevention methods are the same as for moulding.
- Tomato juice turns sour, without the formation of gases; this defect is initiated by thermophyl and thermoresistant bacteria; the juice acquires a vinegary taste. Prevention: maintenance of flash pasteurization temperature at 130-135° C.
- Excessive vitamin C losses are due to a simultaneous action of heating and oxygen from air. Prevention:
  a. prevent air going into crusher and extractor;
  b. assure an intensive de-aeration (vacuum degree 700 mm Hg) at a temperature of at least 35-40° C; and
  c. close receptacles in vacuum.
    - Weak colour of tomato juice can be avoided by the utilisation of mature tomatoes and with a pulp of as red a colour as possible.

Tomato Paste and Concentrated Juice:

- Presence of sand is caused by inadequate washing or by a significant contamination of raw material; this can be prevented by a more intensive pre-washing and washing of tomatoes.
- There may be mould especially at the surface of tomato paste packed in drums. Prevention:
  a. accurate pre-washing and washing;
  b. follow pasteurization instructions;
  c. pack in clean drums or receptacles; and
  d. close receptacles immediately after filling.
- Fermentation is manifested by a weak alcohol smell or by a weak vinegar taste; when the fermentation is more advanced there is gas production in the product mass. Prevention: as for moulding prevention.

Tomato Sauces:

- Surface of the product turns black at the contact zone with air; this is due to the action of iron on the tannins from spices, tomato seeds, etc. Prevention:
  a. avoid iron equipment;
  b. avoid crushing of tomato seeds and
  c. seal receptacles in vacuum.

Pickles and Sauerkraut Technology

Vegetable Natural Acidification Technology:

Gherkins and Cucumbers: Raw materials must follow strict specifications for a high quality finished product; the following parameters must be considered as critical:

- adapt a uniform size according to the finished product requirements; for example, gherkins will need to have a maximum length of 9 cm for raw vegetables. Generally 15 cm size/length will be a maximum for high quality cucumber products in many countries. However, according to local preferences, bigger cucumbers could be also in demand.
- cylindrical or ovoidal shape;
- dark green colour;
- absence of surface defects due to cryptogamic diseases.

Cucumbers have to be picked at their ripeness for eating, when the sugar content is at about 1.5-2.2%, needed for lactic fermentation. Unripe cucumber does not have enough sugar.

The general technological flow-sheet is as follows:

Reception

Control

Temporary storage

Grading by size

Washing

Small holes are made in large size cucumbers skin;

Receptacle filling: Raw material is simply put in the receptacles in bulk, with care to arrange them in such a way that a maximum of pieces could be introduced;

Salt solution preparation: 6% salt solution (nacl);

Salt solution addition: The salt solution is poured into the receptacle;

Fermentation is carried out at 20-30° C, anaerobically. This step takes generally 4-8 weeks. Acidity reaches a value up to 1.5% lactic acid (and in some exceptional cases up to 2% lactic acid) which corresponds to a maximum pH value of 4.1.

Storage; after the last fermentation stage, drums and other receptacles have to be stored at low temperature; best conditions for 12 months shelf life should be below + 15° C. Storage temperature will determine the shelf life of the products.

Addition of 1000 ppm potassium sorbate will prevent mould development without having any influence on lactic fermentation.

Raw material grading by size is a very important technological step. In order to accelerate brine penetration, mainly for medium to large size cucumbers, the practice of making small holes in the raw material skins is generally recommended.

A major factor influencing the quality of lactic fermented cucumbers is the water durity; optimal results are obtained at 15-20° durity.

Cucumber consistency/texture is influenced by the formation of calcium pectate with the pectic substances from raw material tissues. In some countries, calcium chloride (0.3-0.5 %) is added in order to firm up the cucumber consistency. Chlorinated water which still contains active chlorine can inhibit or even stop the lactic fermentation.

Sauerkraut: In some countries cabbages are submitted to lactic fermentation as whole vegetables; however, in many countries the cabbage is shredded before fermentation. As shredded cabbage and its technology is at the basis of an important industry, giving good quality products, with a uniform fermented product and with good keeping quality and ease of distribution, this will be described first.

Cabbage as raw material for sauerkraut must be sound, ripe for eating, well-leafed and from suitable varieties. Optimum total sugar level needed for the lactic fermentation is 24%; generally good quality raw material contains up to 30-60 mg/ 100 g of vitamin C.

Shredded Sauerkraut: The technological flow sheet is as follows:

Reception

Control

Temporary Storage is carried out in bulk, up to a height of about 1 m, during few days. This step produces a heat generation which facilitates later fermentation by the softening of tissues.

Removal of External Leaves

Coring is done with a specially adapted mechanical screw; this operation generates small particles of finely divided cabbage which will be mixed with the main part of vegetable during shredding / chopping. The core represents about 10% from the whole cabbage, is rich in sugar and vitamin C, but being too high in fibre content needs to be chopped separately as described.

Shredding/Cutting of cabbage is carried out with complex specific equipment which is generally installed directly on the "top" of fermentation silos and is mobile, installed on rails and moves all along the silos. The dimension of resulting shredded cabbage is about 2-3 mm thick.

The same complex equipment is designed to grind the added salt to fine particles and to distribute shredded cabbage and ground salt in an uniform manner to the fermentation silos. The usual capacity of fermentation silos is up to 30 tons, with separate compartments of 45 tons each.

Salt Addition is carried out by the equipment described above; the proportion of salt is 2-2.5% with respect to cabbage.

This proportion must not be changed because the salt in this technology does not have a preservative role but only that to extract from cabbage the juice needed for fermentation.

It is be preferable to obtain a fairly light pressure on cabbage just after salt addition with some simple mechanical means. This is important in order to:

- create an anaerobic medium for fermentation;
- facilitate external diffusion of cellular juice;
- assure a rational use of the fermentation space.

Fermentation: The maximum acidity level obtained is generally of about 1.5% lactic acid (and very rarely 2.5%); this is obtained in 4-6 weeks. Optimal acidity is 1.0-1.8% and pH value 4.1 or lower.

Fermentation temperature is at 20-25° C in the first phase and needs to be lowered then to 14-18° C. During fermentation, the brine from each storage / fermentation silo cell is periodically circulated with a pump in order to uniformise the fermentation process.

Storage is performed in same silos used for fermentation, or the finished products is removed from silos and packed in drums and other receptacles according to distribution schedule.

These silos are usually made of reinforced concrete and coated with gritstone plates or with an acid-resisting material layer.

At small scale and in traditional processing, shredded sauerkraut can be obtained by using simple available glass or rigid plastic receptacles. At home, this process can use glass jars and / or local / traditional pottery receptacles from a minimum size of 2-3 kg up to the available / practical sizes (better limited to 10-15 kg).

In some countries shredded sauerkraut is preserved in receptacles by pasteurization, once the fermentation process has been completed.

Whole Sauerkraut: According to the consumer preference in different countries and to the specific situations it is also usual to preserve whole cabbages by lactic fermentation.

At small or medium scale operations, whole cabbage could be processed/ fermented in cylindric receptacles like 30 to 200 litre rigid plastic drums, or rectangular receptacles made from food grade rigid plastic. It is possible to find this type of drum in a significant number of developing countries. These two types of rigid plastic receptacles could also be used for shredded sauerkraut production.

Prepared whole cabbages are put into fermentation receptacles and a 5-6 % salt concentration brine is poured on top. The fermentation conditions are the same as for shredded sauerkraut. In order to assure a uniform fermentation and to avoid a strict anaerobic (butyric) fermentation it is necessary to apply a periodic juice "aeration" (each 2-3 days at the beginning of the fermentation, and then each 5-7 days).

A simple flow-sheet for preparation of whole sauerkraut at family / farm / community levels is presented in section 9.6.1.4.

Other Acidified Vegetables: In principle all vegetables with a sugar content of at least 2 % could be preserved by lactic fermentation.

From a practical point of view it is mainly the following vegetables which are preserved by this technology: unripe tomatoes (green tomatoes), peppers, eggplant, carrots and cauliflower, alone or usually in a mix with cucumber as mixed pickles.

Fermentation of individual vegetables is carried out according to a flow-sheet as described for whole sauerkraut (section 9.6.1.4). The type of cut, brine concentration and frequency of operating steps have to be adapted to each case; green tomatoes are fermented as whole vegetable.

Simplified flow-sheet for whole sauerkraut processing:

Process:

(a) Cabbage preparation:

- Remove the damaged leaves;
- Wash the vegetable;
- Remove 2-3 outer leaves;
- Size grade in three categories:
  - * size A: about 700 g per cabbage
  - * size B: less than 1.2 kg per cabbage
  - * size C: more than 1.2 kg per cabbage
- Process each size category separately;
- Wash cabbages;
- Remove cores;
- Cut size category C vegetable in halves.

(b) Salt solution (brine) preparation:

- Prepare a 5% salt (NaCl) solution = 500 g salt for 10 litre water or 50 g salt for 1 litre water;
- Stir until complete salt dissolution;
- Filter salt solution through cheese cloth.

(c) Initial processing:

- Use a different receptacle for each size category;
- Arrange cabbages in fermentation receptacle;
- Pour salt solution to completely cover cabbages;
- Fix some clean wood pieces (or better some fitted covers with holes) in order to keep cabbages completely covered by salt solution. Allow about 10 cm salt solution above cabbage level;
- Store fermentation receptacles in a moderately cold and ventilated place, out of direct sunlight / heat, protected from dust and other nuisances (insect, etc.);
- Cover each fermentation receptacle with a piece of cardboard or cloth.

(d) Processing follow-up:

- During the first week after initial processing. Once every 2 days, it is necessary to: remove the cover;
- collect and carefully remove carefully (with a household spoon) the white layer ("scum") formed at the surface of salt solution;
- wash the spoon each time and rinse;
- put back the cover.
- During the 5 following weeks.
- once every 4 days: repeat the operations described above.

After the first week, in order to assure a homogeneous acidification / fermentation process for big receptacles (*i.e.* drums or other receptacles of 20 to 2001 capacity), it will be necessary to proceed once a week to an "aeration" step. After completion of brine surface cleaning (as described above), the following operations will be carried out:

- remove the cover and the wood spacers.
- remove all salt solution (brine) from the receptacle;
- filter this solution through a cheese cloth;
- pour back the filtered solution back to the fermentation receptacle;
- put the wood spacers back in place.
- cover the receptacle.

These operations will be carried out for each fermentation receptacle once a week, during an estimated period of six weeks; total duration will be determined by the temperature in the storage room and by the chemical composition of specific raw material (cabbage) lots.

Always keep salt solution (brine) level at 10 cm above cabbages, e.g. cabbages must be always covered by brine.

Consumption of the Finished Product: It is possible to estimate that at reasonable ambient temperatures and with a strict followup of the above recommendations, the finished product will be ready for consumption about 6 weeks after initial processing.

The finished product could be used "as is" in vegetable salads, or prepared according to local taste: with tomato sauce, beans, minced meat, etc. as a replacement of fresh cabbages.

In the same way as with natural acidification or lactic fermentation the cabbage texture is modified and softened so that tissues are more digestible than fresh vegetable. It is possible to use the finished product in local dishes and in new recipes without having to boil it. Apart from the taste benefits of acidified cabbages, this is also produces a significant fuel savings.

The juice resulting from natural cabbage acidification is recovered and could be used separately as a refreshing vegetable juice; the preparation is described in this document.

Finished Product Storage: It is possible to store the finished product after completion of fermentation (*i.e.* after the estimated six weeks period); the storage time will depend on the ambient air temperature.

If a cool space is available, the finished product shelf-life/ storage time at a temperature of about + 15° C is estimated at six months. At an ambient temperature not exceeding +20° C, the storage time could be estimated at 2-3 months.

Artificial Vegetable Acidification Technology

This technology is based on the addition of food grade vinegar which has a bacteriostatic action in concentrations up to 4% acetic acid and bactericidal action in higher concentrations.

Vegetables preserved in vinegar need to reach, after equilibrium between vinegar and water contained in vegetables, a final concentration of 2-3 % acetic acid in order to assure their preservation.

To achieve this final concentration, a 6-9% acetic acid vinegar is used, as related to the specific ratios vinegar/ vegetables.

In vinegar pickles, salt (2-3%) and sometimes sugar (2-5%) are also added.

If the vinegar concentration is lower than 2%, vinegar pickles need to be submitted to a pasteurization in order to assure their preservation.

Cucumbers in Vinegar: This represents the basic product obtained by this technology. Cucumbers have to be wholesome, with a soft texture and not have reached eating maturity. They must have a low sugar content because in this technology there is no lactic fermentation involved. Dimensions are up to 12 cm length, with a preference for small cucumbers.

The technological steps are the followings:

Size grading

Washing

Arrange in receptacles - glass jars, etc.

Pouring of vinegar is usually carried out at room temperature; however, hot vinegar addition enables a sterilisation of cucumber surface and facilitates vinegar penetration in vegetable tissues.

Salt (Sugar) Addition

Spicing Addition

The technological cycle of artificial acidification is considered completed when acetic acid concentration reaches an equilibrium value; the time needed is about 2 weeks. When equilibrium concentration in acetic acid is below 2%, the cucumbers are submitted to a pasteurization for 20 min at 90-1000 C in order to assure their preservation.

Cucumbers in vinegar with previous lactic fermentation are excellent quality products because the lactic fermentation improves the taste of these cucumbers. The principle of this process is to assure preservation both by acetic acid and by lactic acid simultaneously.

Technological processing flow-sheet is as follows: small cucumbers ("cornichons" or "gherkins") are washed, brushed and small holes are made in the skin; the vegetables then are put in drums with slightly warm 6% brine which also contains spices.

The lactic fermentation runs for few days up to a lactic acid concentration of about 0.5 %. The cucumbers are removed from the brine, washed thoroughly and well drained. Preservation is usually done in glass jars by pouring a normally flavoured vinegar with about 9% acetic acid usually in order to bring the final concentration to 3% calculated as acetic acid.

In order to obtain a high quality product only wine vinegar should be used. In some pickles (*e.g.* in "Cornichons") the usual level of wine vinegar is set at 20 % of packaged product total weight; some alcohol vinegar could be still added and final concentration will be adjusted as described above.

Other Vinegar Pickles: One type in this category is represented by other vegetables acidified with vinegar separately or in a mix (red peppers, sweet green pepper, green tomatoes, cauliflower, etc.). The preparation steps are similar to the ones used for cucumbers in vinegar.

Significant quantities of special mixed vegetables in vinegar are manufactured in many countries, with the international name of "mixed pickles" with following composition: small cucumbers ("cornichons"/"gherkins") - maximum 70 mm in length -, sliced carrots, cauliflower, small onions (less than 25 mm diameter), mushrooms etc. and spices.

The vegetables are acidified separately in vinegar and then are put into receptacles (glass jars); a flavoured vinegar, salted and sweetened with acetic acid concentration of 3-5% is poured over them.

In the case of lower acetic acid concentrations, a pasteurization at 90° C for 10-20 minutes is applied according to the receptacle size.

General Heat Preservation Operations—canning

The success of heat preservation operations lies in:

- selecting suitable fruit and vegetables in good conditions;
- preparing them hygienically and skilfully;
- packing them in cans which are hermetically sealed and then processed under fixed conditions of time and temperature;
- cooling these cans carefully and storing them under conditions which will not cause deterioration of either the cans or their contents.

Selection of Raw Materials

It is appreciated that some varieties of fruit and vegetables are not suitable for canning, either because they are uneconomical to prepare or because the colour, flavour or texture are poor.

Suitable varieties must be available to the canner in quantities sufficient to meet his requirements and in sound conditions for canning. The flow to the cannery should be regulated in order that perishable materials are not left for a long time before being handled, since any delay will cause deterioration.

Apart from the main ingredients, be it fruit or vegetables, minor ingredients also require careful selection. Sugar, salt, water and spices for instance may all be contaminated with spoilage organisms, so constant testing of all raw materials is essential.

Preparation: This is carried out by various methods, including grading, trimming, peeling, washing and blanching.

All equipment must be scrupulously clean and preparation should be completed quickly and carefully in order to keep the bacterial load as low as possible.

Thorough washing of vegetable is necessary to remove spores of heat resistant bacteria which are present in large numbers in the soil. Blanching in steam or hot water is of no avail against these heat resistant (thermophilic) spores because of the comparatively low temperatures involved.

Reasons for blanching are:

- the removal of gas from the tissues of the raw material;
- the shrinkage of this material;
- the inhibition of enzymic reactions, which, if not checked, will adversely affect the colour and nutritive value of the food.

Filling: Filling, be it mechanical or by hand, requires careful attention.

The cans must be clean and the correct weight of foodstuffs must be added. Under-filled cans will be underweight and the headspace will be too large, resulting in too much air being left in the can. Overfilling may lead to seams being strained during processing and to ends becoming distorted and bulged.

If the product forms hydrogen on storage as is the case with coloured fruits, swelling of the can due to hydrogen pressure will occur more quickly in an overfilled can than in one which has been correctly filled. Overfilling also affects heat penetration in the can and may lead to spoilage outbreaks.

Air Removal: Before the can is seamed, air must be removed from the contents and the headspace. Normally, this is carried out by passing the cans through a steam box until the temperature at the centre of the can is at least 160° F. This operation, termed exhausting, is necessary for the following reasons:

i. to minimise strains on the seams due to expansion of air during the processing period;
ii. to remove oxygen which accelerates corrosion in the can and also causes oxidation of the food with possible serious effects on colour and flavour;

iii. to reduce the destruction of vitamin C;

iv. to enable a vacuum to be formed when the can is cooled.

This ensures that the ends remain concave, even when storage temperatures are a little higher than usual, and also acts as a reservoir for hydrogen which may be formed by reactions between the can and its contents. Thus a high vacuum makes for a long shelf life. Large cans, however, should not reach such a high exhaust temperature before seaming as smaller cans because of the danger of the can body collapsing on cooling, a condition known as "panelling".

Double Seaming: The can should be double-seamed as soon as the correct centre temperature has been attained. Any delay between exhausting and seaming will lead to loss of vacuum and may lead to bacterial spoilage. The quality of the double seam must, of curse, be frequently checked.

Heat Processing: After seaming, the cans are heated for a definite time at a definite temperature to kill or inhibit organisms which may cause spoilage. This operation is termed "heat processing".

The times and temperatures required for "heat processing" of various packs have been determined experimentally to ensure that spores of the most heat resistant food poisoning organisms known, Clostridium botulinum, are destroyed. There are other organisms, however, whose spores are more heat resistant than those of Clostridium botulinum and which although they will not cause food poisoning may cause spoilage and for this reason the minimum heat processing time is often exceeded by recommendations made by laboratories.

At the same time there is a limit to the amount of heating which a canned food may be given without spoiling its flavour, texture and colour and this also has to be taken into consideration when process recommendations are made. Bacterial spores have a greater resistance to heat when the growth-medium is neutral or near neutral, and neutrality is normally required for bacterial growth to commence. Because of this, canned foods have been broadly divided into two groups:

(a) "acid" foods having a pH of 4.5 or lower and

(b) "non-acid" foods having a pH of more than 4.5.

"Non-acid" foods (vegetables) must, therefore be "heat processed" at high temperatures using steam under pressure, whereas "acid" foods (fruit) may be processed at the (lower) temperature of boiling water, since this will kill moulds and yeasts and if any bacterial spores survive the combination of acid and heat, they will be inhibited from growth by the acid environment.

The rate of destruction by heat follows a definite pattern, the same proportion of the surviving bacteria being destroyed in successive units of time. The more bacteria there are in a pack, the more time will be need to reduce their numbers. For this reason, it is essential that the initial number of bacteria be kept low, and this may be achieved by ensuring fast and hygienic handling at all stages in the cannery.

Pressure gauges and retort temperature control equipment must be checked frequently for accuracy. Processing times and temperatures must be strictly adhered to, and complete removal of air from the retort during processing must be achieved by adequate venting. Failure to remove the air completely will result in their being cold spots in the retort and intermittent spoilage is likely.

Cooling: As soon as the heat processing time is completed, the cans are cooled in chlorinated water as rapidly as possible without damaging them. Cans processed in steam develop high internal pressure because of the expansion of the foodstuff, the expansion of air in the can and the increase in the vapour pressure of the water in the can.

During the heat process, these pressures are counter-balanced to some extent by the pressure of the steam in the retort, but on releasing this steam pressure at the commencement of the cooling period, the pressure in the can may be sufficient to strain the seams seriously and may even distort the ends.

Cans of A21/2 size or larger, when processed at temperatures of 240° F or more, are liable to undergo permanent distortion, such as peaking. This may be avoided by pressurecooling, which involves replacing steam pressure by air pressure before introducing water to the retort, and maintaining this until the pressure inside the can has fallen to a safe level.

This presents difficulties, since if the air pressure is maintained after the can has developed a vacuum, the can body is liable to collapse. Where pressure-cooling is not carried out, the retort pressure is allowed to drop slowly to atmospheric pressure and the cans are then cooled with water.

Storage: After cooling, the cans should be stored in cool, dry conditions. The maintenance of a constant temperature is desirable, since a rise in temperature may lead to condensation of moisture on the can, with possible rusting. Cool conditions are required because storage at higher temperatures not only causes chemical and physical changes in the product and the container but also introduces a risk of thermophilic spoilage.

Other known causes of container spoilage in storage are the use of labels and cardboard cases which have too high a chloride content, and the use of unseasoned wood in the manufacture of packing cases, all of which tend to cause rust formation on the cans.

General technical operations for fruit and vegetable canning lines:

(a) Receptacle washing will remove the impurities and, as much as possible, the microorganisms on the inner surface of metallic cans or glass jars. Washing must be performed just before receptacle filling in order to avoid a new contamination.

Washing methods are variable and depend on receptacle type and need to be carried out with adequate mechanical equipment.

Metal cans are washed on the can feeding lines of filling equipment; a high pressure spray of warm water (65-8° C) is directed into the receptacles while these are submitted simultaneously to a rotation and forward motion.

Glass jars are submitted to a triple washing: wetting for 10 min in a warm detergent and disinfectant solution (40-45° C) containing 100 mg active Cl/litre; washing with high pressure (2.5-3 at) warm water sprays (65-85° C); rinse with cold water.

Special attention MUST be given to recycled glass jars; washing process must be intensified or repeated, depending upon their contamination.

(b) Receptacles are filled in order to maintain a specific ratio between the solid part of the composition and the filling or covering liquid.

For canned vegetable products, the covering liquid may be a 1-3% salt solution with or without addition of sugar (1-3%), tomato concentrated juice or various sauces based on concentrated tomato juices. Salt solution (brine) preparation may be performed with salt percolators; the resulting solution is saturated, containing 318 g/l and needs to be diluted to usual concentrations (1-3%). Brine is then heated up to filling temperatures which depend on product type (up to 85-90° C).

Sugar solutions (syrups) for fruit products may be prepared on the same type of percolators as brine.

Receptacle filling is carried out by leaving an empty space of 5-15% of the total volume, depending on filling temperature and the product type.

(c) Pre-heating (exhausting) of full receptacles aims at the removal of air from the tissues and the increase of the initial temperature of the receptacle contents. On modern production lines, exhausting is eliminated and replaced by the increase of the filling liquid temperature and hermetic receptacle closing under vacuum.

When exhausting is applied, with steam or with hot water, the pre-heated receptacles must be immediately closed in order to avoid the contraction of liquid phase and thus air introduction. Exhausting is performed in special, continuous equipment; product temperature is between 80 and 95° C, during 2-10 min.

Quality Control/quality Assurance

The international trade in processed fruits and vegetables is very large with an ever increasing number of different types being processed and exported. Whereas once, processing was limited to mostly temperate climate fruits and vegetables, the change has now broadened to include tropical and subtropical types.

The reasons are twofold. Firstly, consumers' dietary habits have become more diverse so that, for example people living in North America may very well like fruit and vegetables grown in Africa or Asia. Secondly, processing techniques, whether they be for canning, freezing or drying, have been improved to an extent where final product is palatable, nutritious and of long and reliable shelf life.

Many developing countries have taken advantage of the continuing worldwide demand for processed fruits and vegetables and earned valuable foreign exchange from exports of products to profitable markets.

The export quality control and inspection of processed fruits and vegetables is directed at ensuring that the final products:

have been processed in a registered export establishment that is constructed, equipped and operated in an hygienic and efficient manner; conform to the requirements of the export regulations for processed fruits and vegetables, and those of the importing country, in respect of such things as quality grades, defects, ingredients, packaging materials, styles, additives, contaminants, fill of container, drained weight; and conform to labelling requirements.

Inspection and Certification Procedures

In most countries, in processing fruits and vegetables for export, it is not customary to apply continuous inspection as it is in the case of meat. Few, if any, importing countries require it, and the nature of the products themselves is such that only part time check inspection is required during processing together with statistically based inspection, including sampling and analysis, of final product.

However, in circumstances where an establishment is processing export product for the first time, it can be argued that there is an advantage in adopting continuous inspection until the operation is satisfactorily established.

In any event, inspection of raw materials should be carried out at the commencement of each processing run to ensure that only sound fruit or vegetables of sufficient maturity (degree of ripeness) are used for processing. Sampling checks of raw materials should be carried out as frequently as the inspector thinks necessary.

The inspector must ensure that adequate hygiene practices are followed during the processing of the product. For example, in the case of canned and frozen products and other processing methods, raw materials should be washed absolutely clean so that fruit and vegetables entering the processing line are free from dirt, superficial residues of agricultural chemicals, insects and extraneous plant material.

In the case of dried product, especially where the raw material is sun-dried on drying greens or racks, care must be taken to minimize contamination by bird and animal droppings, dust and extraneous plant material. It is often necessary to wash the dried product to ensure cleanliness of the final product.

In the case of canning and freezing, the inspector must obtain full details of the processing programme for at least the following day from management, so that an adequate inspection programme can be scheduled.

In much the same way as for fresh fruit and vegetables, the inspector must also be aware of the pesticides and other chemicals used in the production of the raw materials. Necessary laboratory analyses can then be arranged to ensure residue levels in the final product do not exceed tolerances adopted by importing countries.

At the commencement of and during processing, the inspector should pay attention to the state of raw materials, the preparation of raw materials for processing (peeling, slicing, dicing, blanching, etc.), preparation and density of packing medium (sugar syrup, salt brine, etc.), the state of cans or containers to be used (cleanliness and strength), the cooking or freezing process (time/temperature relationship), can filling and closure and can/container storage.

After processing, the inspector should check the final product to ensure the drained or thawed weight, the vacuum and headspace, packing medium strength and that can/container conditions are satisfactory. Statistically based sampling plans should be adopted for the examination of final product to ensure it meets the requirements of the export regulations.

The labelling applied to cans/containers should also be checked to ensure both their correctness and compliance with the export regulations and the requirements of those countries in which the product is to be marketed.

Cans should also be examined to make sure that the correct embossing relating to the product, its date of production and the registered number of the export establishment has been applied. Each establishment registered for the export of processed fruit and vegetables or for canned or frozen foods should have its own quality laboratory sufficiently equipped and staffed to carry out physical, chemical and microbiological examinations of the goods.

Inspectors should have access to the laboratory facilities and the establishment's quality control records as and when required. Independent laboratory examination of product should be made by the agency having responsibility for export on the basis of a statistically developed sampling plan.

In those countries where fruit and vegetable production is a seasonal event, processing for export generally takes place at the time of peak production and then declines, often to a halt, as the supply of raw materials declines. As a result, most export establishments produce at their peak of production far more product than they export at that time.

Therefore, most manufacturers find it necessary to store product for considerable periods before it is exported. Thus, proper storage is essential if the product is to retain its quality and cans remain untarnished. Inspectors should regularly inspect storage facilities, noting their conditions and that of the stored product, looking for signs of deterioration such as pest infestation and rusting of cans.

Prior to export the exporter should be required to notify the export quality and inspection agency of his intention to export in accordance with the provisions of the export processed fruits and vegetables regulations and on the prescribed "Notice of Intention to Export" form.

The notice should be submitted in sufficient time before the shipment date to enable the product to be inspected satisfactorily; the intensity of inspection depending on the original state of the product, the conditions under which it has been stored and the length of storage. When product is approved, the agency will issue the exporter an "Export Permit" authorizing Custom's clearance of the product.

Labelling: Customers and consumers expect the labelling on food to be a true description of what they are buying.

Misleading or fraudulent labelling is an unfair trade practice that cannot be tolerated. Most countries now have labelling laws stipulating how foods are to be labelled and what information labels must contain. Most, if not all of those laws have in common requirement that the label should bear:

- A statement of identity and a true, as distinct from misleading, description of the product;
- A declaration of net contents (weight or number of pieces);
- The name and address of the manufacturer, packer, distributor or consignee, and
- A list of ingredients (in descending order of volume or weight).

In addition, labels may also be required to include, amongst other things, the country of origin, date of manufacture or packing, a use-by or expiry date, nutritional qualities or values of the food, storage directions, a quality grade and directions for preparing the food.

More frequently than is often realized, consignments of food exports arriving on foreign markets are not permitted entry because the labelling does not comply with the mandatory requirements of the importing country.

This sometimes results in consignments being rejected, but more often in them being withheld from entry until the labelling is corrected or new labelling applied. In either case, trade is interrupted and the cost involved may make sales unprofitable. It is essential therefore, that exporters be familiar with the food labelling requirements of importing countries.

Bibliography

Adams H.: *Hot Pepper Improvement*, St. Michael, CARDI, 1997.

Agarwal P. K.: *Improvement of Citrus*, New Delhi, Malhotra Publishing House 1993.

Alexander M. P. and Ganeshan S.: *Pollen Storage*, New Delhi, Malhotra Publishing House, 1993.

Andrews J.: *The Domesticated Capsicums*, University of Texas Press. 1995.

Andrews L.: *Citrus Production - Orange*, St. Augustine, Trinidad and Tobago, 1990

————: *Production of Minor Crops Sapodilla, Tamarind, Guava and Genip*, Port of Spain, Trinidad and Tobago, 1996.

Arguello D. S.: *Production, Post-harvest Management and Exportation of Tropical Fruits in Costa Rica*, Wageningen, Technical Centre for Agricultural and Rural Cooperation, 1992.

Ashworth S.: *Seed to Seed*, Decorah, Seed Savers Publications, 1991.

Barbeau G.: *Tropical Fruits in Nicaragua*, Managua, Nicaragua Ministerio de Desarrollo Agropecuario, Agraria, 1990.

Batlle I. and Tous J.: *Carob Tree (Ceratonia siliqua L.)*. Rome, International Plant Genetic Resources Institute, 1997.

Berger, P.L.: *Pyramids of Sacrifice: Political Ethics and Social Change*, New York, Basic Books, 1974.

————: *Pyramids of Sacrifice: Political Ethics and Social Change*, New York, Basic Books, 1974.

Brooks, D.: *Water: Local-Level Management*, Ottawa, International Development Research Centre, 2002.

Bunnik J. S. C.: *Fresh Fruits and Vegetable: a Survey on the Netherlands and other Major Markets in the European*

Community, Netherlands, Centre for the Promotion of Imports from Developing Countries, 1990.

Burton W.G.: *The Potato*, Holland, H. Veenman & Zonen N.V., 1966.

Cao Van P.: *An Integrated Approach for the Production and Processing of Minor Fruits in Martinique (FWI)*, Port of Spain, Trinidad and Tobago, 1996.

Cardona M. J. G., Carvajal S. L. B., Salinas D. G. C. and Isaza R. G. B. : *Proceedings of the International Seminar on Plantain Production, Quindio, Colombia,* Quindio, Corporation Colombiana de Investigation Agropecuaria, 1998.

Chadha K. L. and Pareek O. P.: *Advances in Horticulture: Fruit Crops,* New Delhi, Malhotra Publishing House, 1993.

Chambers, R.: *Rural Development: Putting the Last First*, London, Longman, 1983.

Collymore L.: *Fruit Production in Barbados*, Port of Spain, Trinidad and Tobago, 1996.

Coste R.: *Coffee: the Plant and the Product*, London, MacMillan, 1992.

Crucefix D.: *Avocado Variety Selection for Export Development,* Roseau, CARDI, 1996.

Cull Brian & Lindsay Pax: *Fruit Growing in Warm Climates for Commercial Growers & Home Gardeners*, Australia, Reed Books, 1995.

Currah L. and Proctor F. J.: *Onions in Tropical Regions*, Kent, Natural Resources Institute, 1990.

Daniells J.: *Illustrated Guide to the Identification of Banana varieties in the South Pacific*, Canberra, ACIAR, 1995.

Degras L.: *Yam: a Tropical Root Crop*, Wageningen, CTA/ MacMillan, 1993.

Dhatt A. S. and Singh Z.: *Propagation and Rootstocks of Citrus*, New Delhi, Malhotra Publishers, 1993.

Diederichsen A.: *Coriander (Coriandrum sativum L.)*. Rome, International Plant Genetic Resources Institute, 1996.

Doijode S. D.: *Seed Germination in Fruits*, New Delhi, Malhotra Publishers, 1993.

Dudley, E.: *The Critical Villager: Beyond Community Participation*, London, Routledge, 1993.

Featherly H. I.: *Taxonomic Terminology of the Higher Plants*, USA, Iowa State College Press, 1954.

Ferentinos L.: *Proceeding of the Sustainable Taro Culture for the Pacific Conference*, Honolulu, HITAHR, 1993.

Forde S: *Proceedings of CARDI / CTA Workshop on Marketability of Caribbean Minor Fruits*, Port of Spain, Trinidad and Tobago, 1996.

Georges, S.: *The Debt Boomerang: How Third World Debt Harms Us All*, Boulder, Westview Press, 1992.

Glowinski Louis: *The Complete Book of Fruit Growing in Australia*, Australia, Lothian Publishing Company Pty. Ltd., 1991.

Godden G.: *Growing Citrus Trees*, Australia, Lothian Publishing Company Pvt. Ltd., 1988.

Gowen S.: *Banana and Plantains*, London, Chapman Hall, 1996.

Green S.K. and Kim J. S.: *Sources of Resistance to Viruses of Pepper (Capsicum spp.): a Catalogue*, Taipei, AVRDC, 1994.

Gunjate R. T. and Tawde A. B.: *Propagation and Rootstocks of Mango*, New Delhi, Malhotra, 1993.

Hartley W.: *A Checklist of Economic Plants in Australia*, Melbourne, C.S.I.R.O., 1979.

Harwood, R. R.: *Marshalling Technology for Development: Proceedings of a Symposium*, Washington, DC, National Academy Press, 1995.

Herklots G. A. C.: *Vegetables in South East Asia*, London, George Allen & Unwin Ltd., 1972.

Hessayon D. G. Dr.: *The Vegetable Expert*, England, PBI, Publications, 1985.

Huaman Z.: *Descriptors for Sweet Potato*, Rome, International Board for Plant Genetic Resources, 1991.

Hurst Jacqui & Rutherford Lyn.: *A Gourmet's Book of Mushrooms & Truffles*, Sydney, Golden Press Pvt. Ltd., 1991.

Jacquat Christiane: *Plants from the Markets of Thailand*, Bangkok, Duang Kamol, 1990.

Jeffers P.: *Evaluation of Four Onion Varieties in Montserrat*, Plymouth, CARDI, 1992.

Kenridge K. C. and Hardy B.: *Biology and Agronomy of Forage Arachis*, Cali, International Centre for Tropical Agriculture, 1994.

Kroll R.: *Cut Flowers*, Wageningen, CTA, 1995.

Kunelius T.: *Annual Ryegrasses in Atlantic Canada*, Ottawa, Agriculture Canada, 1991.

Lofgren, H., Richards, A.: *Food Security, Poverty, and Economic Policy in the Middle East and North Africa*, Washington, DC, International Food Policy Research Institute, 2003.

Mabberley D. J.: *The Plant-Book : a Portable Dictionary of the Vascular Plants*, Cambridge, Cambridge University Press, 1997.

Madulid Domingo A.: *A Pictorial Cyclopedia of Philippine Ornamental Plants*, Philippines, Makati Metro Manila, 1995.

Malins A.: *Postharvest Handling of Pineapple and Mango*, Port of Spain, Trinidad and Tobago, 1992.

Mannetje, L. T. & Jones, R. M.: *Plant Resources of South-East Asia,* Wageningen, Pudoc Scientific Publishers, 1992.

Matsuoka H.: *Cultivation of Panicum Genetic Resources for Evaluation of Characteristics*, Tokyo, JICA, 1997.

Mc Donald F.: *Plant Tissue Culture Manual for Yam, Cassava, Sweet Potato, Dasheen (taro) and Tannia (cocoyam)*, Roseau, Dominica and Cave Hill, 1993.

Mellor, J. W.: *The New Economics of Growth*, Ithaca, Cornell University Press, 1976.

Miller William: *Dictionary of English Plant Names*, London, John Murray, 1884.

Mitra S.: *Postharvest Physiology and Storage of Tropical and Subtropical Fruits*, Oxon, CABI, 1997.

Morris M. L. : *Maize Seed Industries in Developing Countries*, Colorado, Lynne Rienner, 1998.

Morton Julia F.: *Fruits of Warm Climates*, Miami, Julia F. Morton Publisher, 1987.

Murali T. P. and Duncan E. J.: *In Vitro Propagation of Banana through Aseptic Manipulation of Male Inflorescences*, Port of Spain, NIHERST, 1990.

Nabhan G. P.: *Wild Phaseolus Ecogeography in the Sierra Madre Occidental, Mexico: Areographic Techniques for Targeting and Conserving Species Diversity*, Rome, International Board for Plant Genetic Resources, 1990.

Nijdam J. & De Jong A.: *Elsevier's Dictionary of Horticulture in Nine Languages*, Elsevier Scientific Publishing Co. Amsterdam; New York.

Oka H. I.: *Origin of Cultivated Rice*, Elsevier, Japan Scientific Societies Press, 1988.

Oldham P.: *Cost of Production of Major Tree Crops in Dominica*, Roseau, Ministry of Agriculture, 1991.

Ou Yang Jue Ya: *Comparative Study of Mandarin and Cantonese*, Beijing, China Social Sciences Publishing House, 1993.

Percival John: *The Wheat Plant - A Monograph*, London, Duckworth & Co., 1921.

Pilgrim R.: *Post Harvest Handling of Minor Exotics*, St. George's, Grenada, 1996.

Ragone D.: *Breadfruit: Artocarpus Altilis (Parkinson) Fosberg*, Rome, International Plant Genetic Resources Institute, 1997.

Rehm Sigmund: *Multilingual Dictionary of Agronomic Plants*, Boston, Kluwer Academic Publishers, 1994.

Roy S. K.: *Research on Multipurpose Tree Species in Asia: In Vitro Clonal Propagation of Artocarpus Heterophyllus*, Bangkok, Winrock International Institute for Agricultural Development, 1991.

Singh H. P. and Chadha K. L.: *Genetic Resources of Citrus*, New Delhi, Malhotra Publishing House, 1993.

Sinha G. C., Reddy Y. T. N. and Singh G.: *Propagation and Rootstocks in Guava*, New Delhi, Malhotra Publishers, 1993.

Sperling L. and Berkowitz P. *Partners in selection: bean breeders and women bean experts in Rwanda*. Washington, USA: CGIAR, 1994.

Sreekumar V., Indrasenan G and Mammen G.: *Studies of the Quantitative and Qualitative Attributes of Ginger Cultivars*, Calicut, Kasaragod, 1990.

Stover R. H. and Simmonds N. W.: *Bananas*, United Kingdom: Longman Scientific and Technical, 1991.

Thavarasook C.: *Proceedings of the Mungbean Meeting 90, Chiang Mai, Thailand,* Bangkok, Tropical Agricultural Research Centre, 1991.

Thomas E.: *Fruit Production in St. Kitts and Nevis*, Port of Spain, IICA, 1996.

Vargas E. M., Macaya G., Baudoin J. P. and Rocha O. J.: *Variation in the Content of Phaseolin in Wild Populations of Lima Beans (Phaseolus lunatus L.) in the Central Valley of Costa Rica*. Plant Genetic Resources Newsletter, 2000.

Webb M.: *Weed Problems and their Control in Rice, Papaya, Citrus and Sugar cane: a Report to the Belize Plant Protection Service*, United Kingdom, Natural Resources Institute, 1993.

Whealy K.: *The Garden Seed Inventory*, Decorah, Seed Saver Publications, 1988.

Whitwell A.: *Dominica Orchard Crop Management and Research Project: Pest and Disease Management,* United Kingdom, Natural Resources Institute, 1991.

Woolfe Jennifer A.: *The Potato in the Human Diet*, Cambridge, Cambridge University Press, 1989.

Yoshida T.: *Cultivation of Citrus Genetic Resources for Evaluation of Characteristics*, Tokyo, JICA, 1996.

Index

A

B

C

D

□□□

BASICS OF BOTANY

BASICS OF BOTANY

MADHUSMITA PRADHAN

ANMOL PUBLICATIONS PVT. LTD.
NEW DELHI - 110 002 (INDIA)

ANMOL PUBLICATIONS PVT. LTD.

H.O.: 4374/4B, Ansari Road, Darya Ganj,
New Delhi-110 002 (India)
Ph.: 23278000, 23261597

B.O.: No. 1015, Ist Main Road, BSK IIIrd Stage
IIIrd Phase, IIIrd Block
Bangalore - 560 085 (India)
Visit us at: www.anmolpublications.com

Basics of Botany

ISBN 978-81-261-3496-0

PRINTED IN INDIA

Printed at Mehra Offset Press, Delhi.

Contents

Preface

This edition of the book is primarily intended for the students of Professional and Science Colleges in India. Therefore, with a view to meet the requirements of all these students better, I have extended the scope of this edition by adding some more chapters on new topics At the same time, our understanding of other issues has improved as more information about them has been gathered.

At the same time I have been able to study the whole of the text and to bring it up to date where necessary. This has now become an invaluable educational resource and I am delighted to have been able to weave this book into its fabric.

Revised, updated, and modernized, I hope that the new edition will be of value and interest to everyone seeking to broaden their understanding of the science behind botanical issues.

Author

Chapter 1

Introduction

BOTANY AS A SCIENCE

Botany is the branch of biology concerned with the scientific study of plants. Traditionally, botanists studied all organisms that were not generally regarded as animal. However, advances in our knowledge about the myriad forms of life, especially microbes (viruses and bacteria), have led to spinning off from Botany the specialized field called Microbiology. Still, the microbes are usually covered in introductory Botany courses, although their status as neither animal nor plant is firmly established.

Plants are living entities, and material presented within *Biology* will have relevence here, most particularly at the cellular and subcellular levels of organization. Both plants and animals deal with the same problems of maintaining life on planet Earth — their approaches seem quite different, but the end result is the same: continued existence in an organized state, as part of a universe whose tendency is towards greater disorganization. Back on Earth, however, it is a fact that microbes, plants, and animals comprise a very interdependent system. We divide them apart, because our minds work best that way. We categorize and learn common features or properties of the categories. This approach is neither right nor wrong, but is clearly efficient for our minds Nonetheless, it is desirable to regularly step back and realise that the boundaries between categories are often just constructs, and exceptions to our categories usually abound.

It was alluded to in the opening definition that Botany is a science. Just what makes Botany, or anything else a science? It is important to acquire a grasp of the fundamentals of science itself to fully appreciate both how botanical knowledge was gained as well as how it can be used. It is usually quickly disinteresting to acquire facts simply for the sake of knowing. Humans do not just appreciate mountains *because they are there,* they climb them because they are there!

Biology is defined as the study of life, and Botany is that discipline within Biology concerned with the study of living organisms called plants and with certain other living things that are not plants (but are not animals either).

DEFINING 'PLANT'

Like many words in common usage that apply to biological entities or concepts, the term *plant* is more difficult to define than might be at first obvious. Although botanists describe a Kingdom Plantae, the boundaries defining members of Plantae are more exclusive than our common concept of a "plant". We are tempted to regard *plant* as meaning a multicellular, eukaryotic organism that generally does not have sensory organs or voluntary motion and has, when complete, a root, stem, and leaves. However, botanically only vascular plants have a root, stem, and leaves, and even some vascular plants, such as certain carnivorous plants and duckweed, fall afoul of that definition. But to be fair, the vascular plants are the plants we tend to encounter every day and that most people would readily regard as "plants".

A more significant point of departure between Plantae and plants occurs among the seaweeds. Technically, only a relatively minor group of seaweeds (the chlorophytes or green algae) are members of the Kingdom Plantae. The majority of seaweeds, like the kelps (very large brown algae from the Order Laminariales), despite a superficial appearance of such, lack true stems, leaves, roots, and any kind of vascular systems as found in higher plants. Thus, the kelps are not Plantae; but are they plants? Certainly if we regard the green algae as plants, it is difficult to exclude the more prominent red and brown algae of our coastal waters.

Another, much broader definition for *plant* is that it refers to any organism that is *photoautotrophic*—produces its own food from raw inorganic materials and sunlight. This is not an unreasonable definition, and is one that focuses on the role plants typically play in an ecosystem. However, there are photoautotrophs among the Prokaryotes, specifically photoautotrophic bacteria and cyanophytes. The latter are sometimes called (for good reasons) blue-green algae. Then there arises the problem that many people would consider that a mushroom is a plant; a mushroom is the fruiting body of a fungus (Kingdom Fungi) and not *photoautotrophic* at all, but *saprophytic*. However, there are more than a few species of flowering plants, fungi, and bacteria that are not autotrophic, but *parasitic*.

We cannot hope to offer a firm answer. The list of characteristics that separate the Plantae from the other biological kingdoms provides at least a technical definition, but realise it is only a technical definition. The problem this lack of precision or agreement in the definition of "plant" presents is one of understanding statements, often encountered in *Wikipedia* (and other) articles, of the sort: *...xylem is one of the two transport tissues of plants*. In general it cannot be assumed this means all plants, algae through flowering plants. It very probably does not include fungi or bacteria. Indeed, it is usually safest to assume the discussion is about vascular plants (essentially the ferns, conifers, flowering plants, and a few others; see discusion below on "General Terminology") unless stated differently (e.g., *...in vascular and non-vascular plants this is* such and such).

Plants and Their Uses

There can be no disputing the fundamental significance of plants to the ecology of our planet. Photosynthetic plants utilize energy arriving from the sun to create complex organic molecules from inorganic substances, and by this process contribute oxygen to the atmosphere. Advanced animal life is very much dependent upon this source of oxygen, as well as the organic molecules that form the basis of nearly every food web on the planet. However, humans utilize plants in many ways, especially as sources of pleasure, food, and material for shelter, clothing, and more. Consider here the role plants play in our everyday lives and in our economy.

Introduction to Plant Classification

At the beginning of this chapter it was suggested that each of us categorizes information we encounter on a daily basis. Our minds seem to want to find relationships between facts and observations, to erect mental bins in which to place new items with previous "facts". This natural human process is the basis for prejudice, in as much as "facts" categorized together can become strongly associated. But these are personal constructs. In order for scientists of many races, speaking many languages, and coming from all manner of backgrounds and experiences to work productively together to solve common problems, the objects with which they work must be classified within a universally accepted framework.

The classification of living things is called systematics, or taxonomy, and ideally should reflect the evolutionary history (phylogeny) of the different organisms.

Taxonomy arranges organisms in groups called taxa, while systematics seeks clues to their relationships. The dominant system of *Scientific Classification* is called Linnaean taxonomy, and includes classification ranks as well as an organism naming convention called binomial nomenclature.

Traditionally, all living things were divided into five kingdoms:

Monera — Protista — Fungi — Plantae — Animalia

However, this five-kingdom system has been replaced by Carl Woese's three-domain system, which focuses on phylogenic roots and comparison of DNA structures. The older approach utilized visual observation as the basis of classification. The three domains reflect whether cells have nuclei (eukaryotic) or not (prokayotic), as well as differences in cell membranes and cell walls.

Archaea — Eubacteria — Eukaryota

General Terminology

In Section II of this text we will delve much deeper into "plant" systematics. But you should be aware of some general terms related to classificatory schemes that are used regularly in discussing plants. You have probably encountered these terms many times, although may not be aware of their exact definitions. For example, much of the material in Section I of this textbook is biased towards flowering plants. That is, much of the descriptive material here as well as at Wikipedia refers specifically to these. Flowering plants are angiosperms; plants that have flowers and produce seeds, and comprise the majority of the plants we would normally encounter in say a nursery if not on the street, field, or empty lot. Seed-bearing plants include both the angiosperms and the gymnosperms, the latter now treated as a modern group called conifers. The conifers are also common plants, especially in higher latitudes, but bear cones instead of flowers. Both conifers and flowering plants develop vascular tissues internally that conduct fluids (especially water) throughout the plant. Included in the vascular plants are ferns. Ferns have vascular tissue, but reproduce by spores. They do not produce seeds and do not bear flowers.

Chapter 2

Plant Cells

INTRODUCTION

A cell is a very basic structure of all living systems, consisting of protoplasm within a containing cell membrane. Only entities such as viruses—literally on the boundary between non-living chemicals and living systems—lack cells or basic cell structure. All plants, including very simple plants called *algae,* and all animals are made up of cells, and these are organized in various ways to create structure and function in an organism. Biologists recognize two basic types of cells: prokaryotic and eukaryotic. *Prokaryotic cells* are structurally more simple. They are found only in single-celled and some simple, multicellular organisms (all bacteria and some algae, which all belong to Bacteria and Archaea domains). *Eukaryot cells* are found in most algae, all higher plants, fungi, psychedelic plants/fungi, and animals (Eukarya domain). Thus, differences between these two cell types are critical to how an organism is classified, and an important consideration in the evolutionary sequence of life on the planet Earth.

PLANT CELL STRUCTURE

Nearly all cells are too small to be seen with the unaided eye. As always there are some exceptions, but generally magnification is required to detect a cellular structure. In plants, a good hand-lens or lupe will sometimes suffice, but in working with cells or observing how cells are organized to form tissues and structures, a high power microscope is used.

Plant cells are quite different from the cells of the other eukaryotic kingdoms' organisms. Their distinctive features include:

- A large central vacuole (enclosed by a membrane, the *tonoplast*), which maintains the cell's turgor and controls movement of molecules between the cytosol and sap.
- A cell wall composed of cellulose and protein, and in many cases lignin, and deposited by the protoplast on the outside of the cell membrane. This contrasts with the cell walls of fungi, which are made of chitin, and prokaryotes, which are made of peptidoglycan.

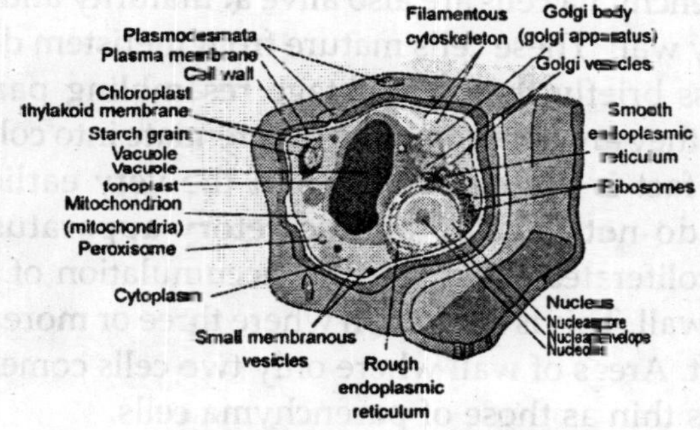

Fig. Plant Cell Structure

- The plasmodesmata, linking pores in the cell wall that allow each plant cell to communicate with other adjacent cells. This is different from the network of hyphae used by fungi.
- Plastids, especially chloroplasts that contain chlorophyll, the pigment that gives plants their green colour and allows them to perform photosynthesis.
- Plant groups without flagella (including conifers and flowering plants) also lack centrioles that are present in animal cells.

CELL TYPES

Parenchyma cells: These cells are the biochemistry machines of the plant. They are alive at maturity and are specialized in any number of structural and biochemical ways. Other than support functions, this cell type is the basis for all plant structure and function. Parenchyma cells have thin primary walls, and highly functional cytoplasm. The cells are alive at maturity and are responsible for a wide range of biochemical function. For example, other than xylem in vascular bundles, the leaf is composed of parenchyma cells. Some, as in the epidermis, are specialized for light penetration, regulating gas exchange, or anti-herbivory physiology. Other cells, as in the mesophyll, are specialized for photosynthesis or phloem loading.

Collenchyma Cells

Collenchyma cells are also alive at maturity and have only a primary wall. These cells mature from meristem derivatives. They pass briefly through a stage resembling parenchyma, however they are determined to differentiate into collenchyma; and this fact is quite obvious from the very earliest stages. Plastids do not develop and secretory apparatus (ER and Golgi) proliferates to assist in the accumulation of additional primary wall. This is laid down where three or more cells come in contact. Areas of wall where only two cells come in contact remain as thin as those of parenchyma cells.

The design and function is to build and maintain the special unevenly thick primary cell wall. The cells are also typically quite elongate. The role of this cell type is to support the plant in areas still growing in length. The primary wall lacks lignin that would make it brittle, so this cell type provides what could be called plastic support. Support that can hold a young stem or petiole into the air, but in cells that can be stretched as the cells around them elongate. Stretchable support (without elastic snap-back) is a good way to describe what collenchyma does. Parts of the strings in celery are collenchyma.

Sclerenchyma Cells

These cells are hard and brittle (as you might expect from the root: scler The cells develop an extensive secondary cell wall (laid down on the inside of the primary wall). This wall is invested with lignin, making it extremely hard. Lignin, plus suberin and/or cutin make the wall waterproof as well. Thus, these cells cannot survive for long as they cannot exchange materials well enough for active (or even maintaining) metabolism. They are typically dead at functional maturity...the cytoplasm is missing by the time the cell can begin to carry out its funciton.

Functions for sclerenchyma cells include discouraging herbivory (hard cells that rip open digestive passages in small insect larval stages, hard cells forming a pit wall in a peach fruit), support (the wood in a tree trunk, fibers in large herbs), and conduction (hollow cells lined end-to-end in xylem with cytoplasm and end walls missing).

GROUND TISSUE

The types of ground tissue found in plants develops from *ground tissue* meristem and consists of three simple tissues:

- Parenchyma (have retained their protoplasm)
- Collenchyma (have retained their protoplasm)
- Sclerenchyma (have lost their protoplasm in mature stage, i.e. are 'dead')

Parenchyma

Parenchyma is the most common ground tissue, it forms e.g. the cortex and pith of stems, the cortex of roots, the mesophyll (photosynthetic cells), the pulp of fruits, the endosperm of seeds, and the photosynthetic areas of a leaf. Parenchyma cells are capable of cell division even after maturation (i.e. they are still meristematic). They have thin, but flexible, cell walls, and are generally cube-shaped and are loosely packed. They have large central vacuoles, which allows the cells to store nutrients and water.

Parenchyma cells have a variety of functions;

- Photosynthesis (may then be called Chlorenchyma / Mesophyll cells),
- Gas exchange (Aerenchyma),

- Storage,
- Secretion (e.g. Epithelial cells lining the inside of resin ducts)
- Healing
- Other specialised functions.

Parenchyma differs depending on the varied functions they perform. The epidermal parenchyma cells of a leaf are barrel shaped and have no chloroplast. The parenchymatous layer just beneath the epidermis in the stem have chloroplasts. They take part in photosynthesis and are known as chlorenchyma in the leaf, just beneath the epidermis large parenchymatous cells with chloroplast are closely arranged,and are known as palisade tissues.palisade tissue along with the loosely arranged spongy tissue have major role in photosynthesis. There are many air cavities between the aerenchyma cells of water plants like lotus. These cells are known as parenchyma which help the plants to float in water. Parenchyma cells are arranged at the centre of some types of stem, which perform the function of storing food. In certain types of roots, parenchyma cells also store food.

Collenchyma

Collenchyma tissue is composed of elongated cells with unevenly thickened walls. They provide structural support, particularly in growing shoots and leaves. Collenchyma tissue composes, for example, the resilient strands in stalks of celery. Its growth is strongly affected by mechanical stress upon the plant. The walls of collenchyma in shaken (to mimic the effects of wind etc) plants may be 40%-100% thicker than those not shaken. The name collenchyma derives from the Greek word "kolla", meaning "glue", which refers to the thick, glistening appearance of the walls in fresh tissues.

There are three principal types of collenchyma;

- Angular collenchyma (thickened at intercellular contact points)
- Tangential collenchyma (cells arranged into ordered rows and thickened at the tangential face of the cell wall)

- Lacunar collenchyma (have intercellular space and thickening proximal to the intercellular space)

Sclerenchyma

Sclerenchyma is a supporting tissue. Two groups of sclerenchyma cells exist: fibres and sclereids. Their walls consist of cellulose and/or lignin. Sclerenchyma cells are the principal supporting cells in plant tissues that have ceased elongation. Sclerenchyma fibres are of great economical importance, since they constitute the source material for many fabrics (flax, hemp, jute, ramie).

Unlike the collenchyma, mature sclerenchyma is composed of dead cells with extremely thick cell walls (secondary walls) that make up to 90% of the whole cell volume. The term "sclerenchyma" is derived from the Greek "scleros", meaning "hard". It is their hard, thick walls that make sclerenchyma cells important strengthening and supporting elements in plant parts that have ceased elongation. The difference between fibres and sclereids is not always clear. Transitions do exist, sometimes even within one and the same plant.

Fibres are generally long, slender, so-called prosenchymatous cells, usually occurring in strands or bundles. Such bundles or the totality of a stem's bundles are colloquially called fibres. Their high load-bearing capacity and the ease with which they can be processed has since antiquity made them the source material for a number of things, like ropes, fabrics or mattresses. The fibres of flax (Linum usitatissimum) have been known in Europe and Egypt for more than 3000 years, those of hemp (Cannabis sativa) in China for just as long. These fibres, and those of jute (Corchorus capsularis) and ramie (Boehmeria nivea, a nettle), are extremely soft and elastic and are especially well suited for the processing to textiles. Their principal cell wall material is cellulose.

Contrasting are hard fibres that are mostly found in monocots. Typical examples are the fibres of many Gramineae,

Agaves (sisal: Agave sisalana), lilies (Yucca or Phormium tenax), Musa textilis and others. Their cell walls harbour, besides cellulose, a high proportion of lignin. The load-bearing capacity of Phormium tenax is as high as 20-25 kg/mm2, the same as that of good steel wire (25 kg/ mm2), but the fibre tears as soon as too great a strain is placed upon it, while the wire distorts and tears not before a strain of 80 kg/mm2. The thickening of a cell wall has been studied in Linum. Starting at the centre of the fibre are the thickening layers of the secondary wall deposited one after the other. Growth at both tips of the cell leads to simultaneous elongation. During development the layers of secondary material seem like tubes, of which the outer one is always longer and older than the next. After completion of growth the missing parts are supplemented, so that the wall is evenly thickened up to the tips of the fibres.

Fibres stem usually form meristematic tissues. Cambium and procambium are their main centers of production. They are often associated with the xylem of the vascular bundles. The fibres of the xylem are always lignified. Reliable evidence for the fibre cells' evolutionary origin of tracheids exists. During evolution the strength of the cell walls was enhanced, the ability to conduct water was lost and the size of the pits reduced. Fibres that do not belong to the xylem are bast (outside the ring of cambium) and such fibres that are arranged in characteristic patterns at different sites of the shoot.

Sclereids are small bundles of sclerenchyma tissue in plants that form durable layers, such as the cores of apples and the gritty texture of pears. Sclereids are variable in shape. The cells can be isodiametric, prosenchymatic, forked or fantastically branched. They can be grouped into bundles, can form complete tubes located at the periphery or can occur as single cells or small groups of cells within parenchyma tissues. But compared with most fibres sclereids are relatively short. Characteristic examples are the stone cells (called stone cells because of their hardness) of pears (Pyrus communis) and quinces (Cydonia oblonga) and those of the shoot of the wax-plant (Hoya carnosa). The cell walls fill nearly all the cell's volume. A layering of the walls and the existence of branched pits is clearly visible. Branched pits such as these are called ramiform pits. The shell of many seeds like those of nuts as well as the stones of drupes like cherries or plums are made up from sclereids.

TISSUE TYPES

These three major classes of cells can then differentiate to form the tissue structures of roots, stems, and leaves. Plants have these types of tissues, and they have similar locations within all species of plants. However, the amount of these tissues will vary for different plant species.

The three distinct types of plant cells are classified according to the structure of their cell walls and features of their protoplast. Plants will have a primary cell wall and sometimes a secondary wall as well. These two major parts are what determines the function of each individual plant cell.

- *Dermal Tissue:* The outermost covering of a plant
- *Vascular Tissue:* Responsible for transport of materials throughout the plant
- *Ground Tissue:* Performs photosynthesis, starch storage and structural support; ground tissues may be composed of one of three cell types
  - *Parenchyma:* Thin primary walls, may not have a secondary wall; can develop into more specialized plant tissues
  - *Collenchyma:* Unevenly thickened primary walls, grouped together to support growing parts of the plant
  - *Sclerenchyma:* Thick secondary walls, used to support non-growing parts of the plant

The epidermis (pluralized either epidermises or sometimes epidermes) is the outer single-layered group of cells covering a plant, especially the leaf and young tissues of a vascular plant including stems and roots. Epidermis and periderm are the dermal tissues in vascular plants. The epidermis forms the boundary between the plant and the external world. The epidermis serves several functions: protection against water loss, regulation of gas exchange, secretion of metabolic compounds, and (especially in roots) absorption of water and mineral nutrients. The epidermis of most leaves shows dorsoventral anatomy: the upper (adaxial) and lower (abaxial) surfaces have somewhat different construction and may serve different functions.

The epidermis is usually transparent (epidermal cells lack chloroplasts) and coated on the outer side with a waxy cuticle that prevents water loss. The cuticle may be thinner on the lower leaf epidermis than on the upper epidermis; and is thicker on leaves from dry climates as compared with those from wet climates.

The epidermal tissue includes several differentiated cell types: epidermal cells, guard cells, subsidiary cells, and epidermal hairs (trichomes). The epidermal cells are the most numerous, largest, and least specialized. These are typically more elongated in the leaves of monocots than in those of dicots.

The leaf and stem epidermis is covered with pores called stomata (sing., stoma), part of a stoma complex consisting of a pore surrounded on each side by chloroplast-containing guard cells, and two to four subsidiary cells that lack chloroplasts. The stoma complex regulates the exchange of gases and water vapour between the outside air and the interior of the leaf. Typically, the stomata are more numerous over the abaxial (lower) epidermis of the leaf than the (adaxial) upper epidermis. An exception is floating leaves where most or all stomata are on the upper surface. Vertical leaves, such as those of many grasses, often have roughly equal numbers of stomata on both surfaces. The number of stomata varies from about 1,000 to over 100,000 per square centimeter of leaf surface.

Trichomes or hairs grow out from the epidermis in many species. In root epidermis, epidermal hairs, termed root hairs are common and are specialized for absorption of water and mineral nutrients.

In plants with secondary growth, the epidermis of roots and stems is usually replaced by a periderm through the action of a cork cambium.

GUARD CELLS

The stoma is bounded by two guard cells. The guard cells differ from the epidermal cells in the following aspects:

- The guard cells are bean-shaped in surface view, while the epidermal cells are irregular in shape
- The guard cells contain chloroplasts, so they can manufacture food by photosynthesis (The epidermal cells do not contain chloroplasts)
- Guard Cells are the only epidermal cells that can make sugar. According to one theory, in sunlight the concentration of potassium ions (K+) increases in the guard cells. This, together with the sugars formed, lowers the water potential in the guard cells. As a result, water from other cells enter the guard cells by osmosis so they swell and become turgid. Because the guard cells have a thicker cellulose wall on one side of the cell, i.e. the side around the stomatal pore, the swollen guard cells become curved and pull the stomata open.

At night, the sugar is used up and water leaves the guard cells, so they become flaccid and the stomatal pore closes. In this way, they reduce the amount of water vapour escaping from the leaf.

CELL DIFFERENTIATION IN THE EPIDERMIS

The plant epidermis consists of three main cell types: pavement cells, guard cells and their subsidiary cells that surround the stomata and trichomes, otherwise known as leaf hairs. The epidermis of petals also form a variation of trichomes called conical cells. These cells all develop from the pavement cells, which make up the majority of the plants surface cells. In short, cellular differentiation of the epidermal cells is controlled by two major factors: genetics and environmental conditions.

Trichomes develop at a distinct phase during the actual leaf development, under the control of two major trichome specification genes: *TTG* and *GL1*. The process may be controlled by the plant hormones gibberellins, and even if not completely controlled, gibberellins certainly have an effect on the development of the leaf hairs. *GL1* causes endoreplication, the replication of DNA without subsequent cell division as well as cell expansion. *GL1* turns on the expression of a second gene for trichome formation, *GL2*, which controls the final stages of trichome formation causing the cellular outgrowth.

Arabidopsis uses the products of inhibitory genes to control the patterning of trichomes, such as *TTG* and *TRY*. The products of these genes will diffuse into the lateral cells, preventing them from forming trichomes and in the case of *TRY* promoting the formation of pavement cells.

As previously mentioned, conical cells are a form of trichome that occurs on the petals of flowers. Expression of the gene *MIXTA*, or its analogue in other species, later in the process of cellular differentiation will cause the formation of conical cells over trichomes. *MIXTA* is a transcription factor.

Stomatal pattering is a much more controlled process, as the stoma effect the plants water retention and respiration capabilities. As a consequence of these important functions, differentiation of cells to form stomata is also subject to environmental conditions to a much greater degree then other epidermal cell types.

Stomata are holes in the plant epidermis that are surrounded by two guard cells, which control the opening and closing of the aperture. These guard cells are in turn surrounded by subsidiary cells which provide a supporting role for the guard cells.

Stomata begin as stomatal meristemoids. The process varies between dicots and monocots. Spacing is thought to be essentially random in dicots though mutants do show it is under some form of genetic control, but it is more controlled in monocots, where stomata arise from specific asymmetric divisions of protodermal cells. The smaller of the two cells produced becomes the guard mother cells. Adjacent epidermal cells will also divide asymmetrically to form the subsidiary cells.

Because stomata play such an important role in the plants survival, collecting information of there differentiation if difficult by the traditional means of genetic manipulation, as stomatal mutants tend to be unable to survive. Thus the control of the process is not well understood. Some genes have been identified. *TMM* is thought to control the timing of stomatal initiation specification and *FLP* is thought to be involved in preventing further division of the guard cells once they are formed.

Environmental conditions affect the development of stomata, in particular their density on the leaf surface. It is thought that plant hormones, such as ethylene and cytokines, control the stomata's developmental response to the environmental conditions. Accumulation of these hormones appears to cause increased stomatal density such as when the plants are kept in closed environments.

Stomatal cells only occur on the leaf epidermis, and it is thought that inhibitory signals must occur on other parts of the plants epidermis to prevent stomatal formation there. These signals could be hormonal, or perhaps gene products transmitted from underlying tissues via the plasmodesmata.

VASCULAR TISSUE

Vascular tissue is a complex tissue found in vascular plants, meaning that it is composed of more than one cell type. The primary components of vascular tissue are the xylem and phloem. These two tissues transport fluid and nutrients internally. There are also two meristems associated with vascular tissue: the vascular cambium and the cork cambium. All the vascular tissues within a particular plant together constitute the vascular tissue system of that plant.

The cells in differentiated vascular tissue are typically long and slender. Since the xylem and phloem function in the conduction of water, minerals, and nutrients throughout the plant, it is not surprising that their form should be similar to pipes. The individual cells of phloem are connected end-to-end, just as the sections of a pipe might be. As the plant grows, new vascular tissue differentiates in the growing tips of the plant. The new tissue is aligned with existing vascular tissue, maintaining its connection throughout the plant. The vascular tissue in plants is arranged in long, discrete strands called vascular bundles. These bundles include both xylem and phloem, as well as supporting and protective cells. In stems and roots, the xylem typically lies closer to the interior of the stem with phloem towards the exterior of the stem. In the stems of some Asteriidae dicots, there may be phloem located inwardly from the xylem as well.

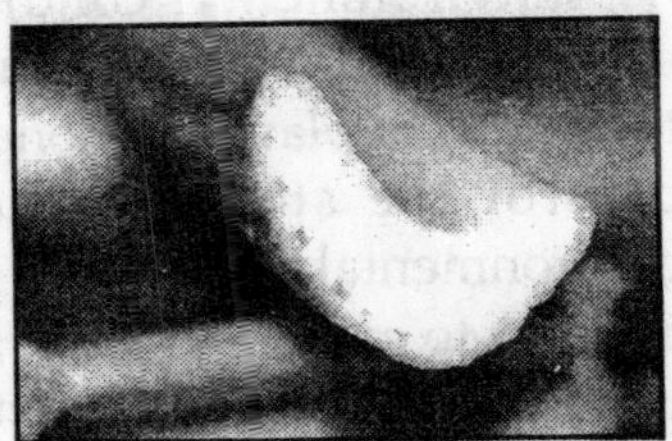

Fig. Vascular Tissue

Between the xylem and phloem is a meristem called the vascular cambium. This tissue divides off cells that will be become additional xylem and phloem. This growth increases the girth of the plant, rather than its length. As long as the vascular cambium continues to produce new cells, the plant will continue to grow more stout. In trees and other plants that develop wood, the vascular cambium allows the expansion of vascular tissue that produces woody growth. Because this growth ruptures the epidermis of the stem, woody plants also have a cork cambium that develops among the phloem. The cork cambium gives rise to thickened cork cells to protect the surface of the plant and reduce water loss. Both the production of wood and the production of cork are forms of secondary growth.

In leaves, the vascular bundles are located among the spongy mesophyll. The xylem is oriented toward the adaxial surface of the leaf (usually the upper side), and phloem is oriented toward the abaxial surface of the leaf. This is why aphids are typically found on the underside of the leaves rather than on the top, since the phloem transports sugars manufactured by the plant and they are closer to the lower surface.

PART 5

Cell Membrane

The cell membrane (also called the plasma membrane or plasmalemma) is a semipermeable lipid bilayer common to all living cells. It contains a variety of biological molecules, primarily proteins and lipids, which are involved in a vast array of cellular processes, and also serves as the attachment point for both the intracellular cytoskeleton and, if present, the cell wall.

Purpose

The cell membrane surrounds the cytoplasm of a cell and physically separates the intracellular components from the extracellular environment, thereby serving a function similar to that of skin.

The barrier is selectively permeable and able to regulate what enters and exits the cell, thus facilitating the transport of materials needed for survival. The movement of substances across the membrane can be either *passive*, occurring without the input of cellular energy, or *active*, requiring the cell to expend energy in moving it.

Specific proteins embedded in the cell membrane can act as molecular signals which allow cells to communicate with each other. Protein receptors are found ubiquitously and function to receive signals from both the environment and other cells. These signals are *transduced* into a form which the cell can use to directly effect a response. Other proteins on the surface of the cell membrane serve as "markers" which identify a cell to other cells. The interaction of these markers with their respective receptors forms the basis of cell-cell interaction in the immune system.

CELL WALL

A cell wall is a fairly rigid layer surrounding a cell, located external to the cell membrane, that provides the cell with structural support, protection, and a filtering mechanism. The cell wall also prevents over-expansion when water enters the cell. They are found in plants, bacteria, archaea, fungi, and algae. Animals and most protists do not have cell walls.

The cell wall is constructed from different materials dependent upon the species. In plants, the cell wall is constructed primarily from a carbohydrate polymer called cellulose, and the cell wall can therefore also functions as a carbohydrate store for the cell. In bacteria, peptidoglycan forms the cell wall. Archaea have various chemical compositions, including glycoprotein S-layers, pseudopeptidoglycan, or polysaccharides. Fungi possess cell walls of chitin, and algae typically possess walls constructed of glycoproteins and polysaccharides, however certain algal species may have a cell wall composed of silicic acid. Often, other accessory molecules are found anchored to the cell wall.

Composition

The major carbohydrates making up the primary cell wall are cellulose, hemicellulose and pectin. The cellulose microfibrils are linked via hemicellulosic tethers to form the cellulose-hemicellulose network, which is embedded in the pectin matrix. The most common hemicellulose in the primary cell wall is xyloglucan.

The three primary polymers that make up plant cell walls consist of about 35 to 50% cellulose, 20 to 35 percent hemicellulose and 10 to 25% lignin. Lignin fills the spaces in the cell wall between cellulose, hemicellulose and pectin components.

Plant cells walls also incorporate a number of proteins; the most abundant include hydroxyproline-rich glycoproteins (HRGP), also called the extensins, the arabinogalactan proteins (AGP), the glycine-rich proteins (GRPs), and the proline-rich proteins (PRPs). With the exception of glycine-rich proteins, all the previously mentioned proteins are glycosylated and contain hydroxyproline (Hyp). Each class of glycoprotein is defined by a characteristic, highly repetitive protein sequence. Chimeric proteins contain two or more different domains, each with a sequence from a different class of glycoprotein. Most cell wall proteins are cross-linked to the cell wall and may have structural functions.

Secondary cell wall may contain lignin and suberin, making the walls rigid.It may also contain cutin.

The relative composition of carbohydrates, secondary compounds and protein varies between plants and between the cell type and age.

Formation

The middle lamella is laid first, formed from the cell plate during cytokinesis, and the primary cell wall is then expanded inside the middle lamella. The actual structure of the cell wall is not clearly defined and several models exist - the covalently linked cross model, the tether model, the diffuse layer model and the stratified layer model. However, the primary cell wall, can be defined as composed of cellulose microfibrils aligned at all angles. Microfibrils are held together by hydrogen bonds to provide a high tensile strength. The cells are held together and share the gelatinous membrane called the *middle lamella*, which contains magnesium and calcium pectates (salts of pectic acid). Cells interact though plasmodesma(ta), which are inter-connecting channels of cytoplasm that connect to the protoplasts of adjacent cells across the cell wall.

In some plants and cell types, after a maximum size or point in development has been reached, a *secondary wall* is constructed between the plant cell and primary wall. Unlike the primary wall, the microfibrils are aligned mostly in the same direction, and with each additional layer the orientation changes slightly. Cells with secondary cell walls are rigid. Cell to cell communication is possible through *pits* in the secondary cell wall that allow plasmodesma to connect cells through the secondary cell walls.

PLASMODESMATA

Plasmodesmata (singular, *plasmodesma*) are microscopic channels of plants facilitating transport and communication between individual cells. Unlike animal cells, plant cells are protected by an impermeable cell wall; and as such, plasmodesmata are required for intercellular activity. Cells can utilize both passive and active transport to move molecules and ions through the passage. A plasmodesma is constructed of three main layers, the plasma membrane, the *cytoplasmic sleeve,* and the *desmotubule.*

Plasmodesmal Plasma Membrane

The plasma membrane portion of the plasmodesma is a continuous extension of the cellular plasmalemma. It is similar in structure to most cellular phospholipid bilayers.

Cytoplasmic Sleeve

The cytoplasmic sleeve is enclosed by the plasma membrane and is an extension of the cytosol. Trafficking through plasmodesmata is assumed to occur through this passage. Smaller molecules (e.g. monosaccharides) and ions can easily pass through plasmodesmata by diffusion without the need for additional chemical energy. It is unknown how the selective transport of larger molecules, such as proteins, occurs. One hypothesis is that the polysaccharide, callose, accumulates around the neck region of plasmodesmata causing its diameter to be reduced and controls permeability of protoplasms or substances.

Desmotubule

The desmotubule is an area of appressed endoplasmic reticulum that runs between the two cells. Some molecules have been known to be transported through this channel, but it is not thought to be the main route for plasmodesmatal transport.

Around the desmotubule and the plasmamembrane areas of an electron density material have been seen, often joined together by spoke-like structures that seem to split the plasmodesma into smaller channels. It has been hypothesised that these structures may be composed of myosin and actin, which are part of the cell's cystoskeleton. If this is the case these proteins could be used in the selective transport of large molecules between the two cells.

VACUOLE

Vacuoles are membrane-bound compartments within some eukaryotic cells that can serve a variety of secretory, excretory, and storage functions. Vacuoles and their contents are considered to be distinct from the cytoplasm, and are classified as ergastic according to some authors. Vacuoles are especially conspicuous in most plant cells.

In general, vacuole functions include

- Removing unwanted structural debris
- Isolating materials that might be harmful or a threat to the cell
- Containing waste products
- Maintaining internal hydrostatic pressure or turgor within the cell
- Maintaining an acidic internal pH
- Containing small molecules
- Exporting unwanted substances from the cell.
- Enabling the cell to change shape.

Vacuoles also play a major role in autophagy, maintaining a balance between biogenesis (production) and degradation (or turnover), of many substances and cell structures. They also aid in destruction of invading bacteria or of misfolded proteins that have begun to build up within the cell.

CHLOROPLAST

- Chloroplasts are organelles found in plant cells and eukaryotic algae that conduct photosynthesis. Chloroplasts absorb sunlight and use it in conjunction with water and carbon dioxide to produce sugars.

Chloroplasts capture light energy from the sun to conserve free energy in the form of ATP and reduce NADP to NADPH through a complex set of processes called photosynthesis. It is derived from the Greek words *chloros* which means green and *plast* which means form or entity. Chloroplasts are members of a class of organelles known as plastids.

In green plants, chloroplasts are surrounded by two lipid-bilayer membranes. The inner membrane is now believed to correspond to the outer membrane of the ancestral cyanobacterium. The chloroplast genome is considerably reduced compared to that of free-living cyanobacteria, but the parts that are still present show clear similarities. Plastids may contain 60-100 genes whereas cyanobacteria often contain more than 1500 genes. Many of the missing genes are encoded in the nuclear genome of the host. The transfer of nuclear information has been estimated in tobacco plants at one gene for every 16000 pollen grains.

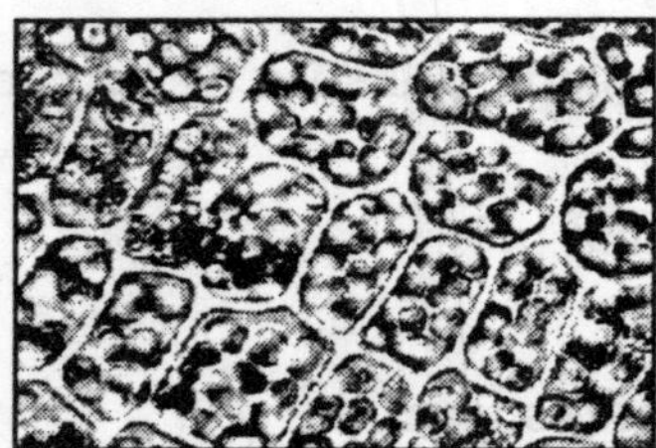

Fig. Chloroplast

In some algae (such as the heterokonts and other protists such as Euglenozoa and Cercozoa), chloroplasts seem to have evolved through a secondary event of endosymbiosis, in which a eukaryotic cell engulfed a second eukaryotic cell containing chloroplasts, forming chloroplasts with three or four membrane layers. In some cases, such secondary endosymbionts may have themselves been engulfed by still other eukaryotes, thus forming tertiary endosymbionts.

Structure

Chloroplasts are observable morphologically as flat discs usually 2 to 10 micrometer in diameter and 1 micrometer thick. The chloroplast is contained by an envelope that consists of an inner and an outer phospholipid membrane. Between these two layers is the intermembrane space.

The material within the chloroplast is called the stroma, corresponding to the cytosol of the original bacterium, and contains one or more molecules of small circular DNA. It also contains ribosomes, although most of its proteins are encoded by genes contained in the cell nucleus, with the protein products transported to the chloroplast.

Within the stroma are stacks of thylakoids, the sub-organelles which are the site of photosynthesis. The thylakoids are arranged in stacks called grana (singular: granum). A thylakoid has a flattened disk shape. Inside it is an empty area called the thylakoid space or lumen. Photosynthesis takes place on the thylakoid membrane; as in mitochondrial oxidative phosphorylation, it involves the coupling of cross-membrane fluxes with biosynthesis via the dissipation of a proton electrochemical gradient.

Embedded in the thylakoid membrane is the antenna complex, which consists of proteins, and light-absorbing pigments, including chlorophyll and carotenoids. This complex both increases the surface area for light capture, and allows capture of photons with a wider range of wavelengths. The energy of the incident photons is absorbed by the pigments and funneled to the reaction centre of this complex through resonance energy transfer. Two chlorophyll molecules are then ionised, producing an excited electron which then passes onto the photochemical reaction centre.

LEUCOPLAST

Leucoplasts are a category of plastid and as such are organelles found in plant cells. They are non-pigmented, in contrast to other plastids such as the chloroplast and the scientific value of a chromoplast.

- Lacking pigments, leucoplasts are not green, so they are predictably located in roots and non-photosynthetic tissues of plants. They may become specialized for bulk storage of starch, lipid or protein and are then known as amyloplasts, elaioplasts, or proteinoplasts respectively. However, in many cell types, leucoplasts do not have a major storage function and are present to provide a wide range of essential biosynthetic functions, including the synthesis of fatty acids, many amino acids, and tetrapyrrole compounds such as haem. In general, leucoplasts are much smaller than chloroplasts and have a variable morphology, often described as amoeboid. Extensive networks of stromules interconnecting leucoplasts have been observed in epidermal cells of roots, hypocotyls and petals, and in callus and suspension culture cells of tobacco. In some cell types at certain stages of development, leucoplasts are clustered around the nucleus with stromules extending to the cell periphery, as observed for proplastids in the root meristem.

Etioplasts, which are pre-granal, immature chloroplasts but can also be chloroplasts which have been deprived of light, lack active pigment and can technically be considered leucoplasts. After several minutes exposure to light, etioplasts begin to transform into functioning chloroplasts and cease being leucoplasts.

CHROMOPLAST

Chromoplasts are plastids responsible for pigment synthesis and storage. They, like all other plastids (including chloroplasts and leucoplasts), are organelles found in specific photosynthetic eukaryotic species.

Chromoplasts, in the traditional sense, are found in coloured organs of plants such as fruit and floral petals, to which they give their distinctive colors. This is always associated with a massive increase in the accumulation of carotenoid pigments. The conversion of chloroplasts to chromoplasts in ripening tomato fruit is a classic example.

Chromoplasts synthesize and store pigments such as orange carotenes, yellow xanthophylls and various other red pigments; as such, their colour varies depending on what pigment they contain. The probable main evolutionary role of chromoplasts is to act as an attractant for pollinating animals (e.g. insects) or for seed dispersal via the eating of colored fruits. They allow the accumulation of large quantities of water-insoluble compounds in otherwise watery parts of plants. In chloroplasts, some carotenoids are also used as accessory pigments in photosynthesis, where they act to increase the efficiency of chlorophyll in harvesting light energy. When leaves change colour in autumn, it is due to the loss of green chlorophyll unmasking these carotenoids *that are already present in the leaf*. In this case, relatively little new carotenoids are produced. Therefore, the change in plastid pigments associated with leaf senescence is somewhat different to the active conversion to chromoplasts observed in fruit and flowers.

It should be noted that the term "chromoplast" is occasionally utilized to include *any* plastid that has pigment, mostly to emphasize the contrast with the various types of leucoplasts, which are those plastids that have no pigments. In this sense, chloroplasts are a specific type of chromoplast. Still, "chromoplast" is more often used to denote those plastids with pigments other than the green of chlorophyll.

GOLGI APPARATUS

- The Golgi apparatus (also called the Golgi body, Golgi complex, or dictyosome) is an organelle found in typical eukaryotic cells. It was identified in 1898 by the Italian physician Camillo Golgi and was named after him. The primary function of the Golgi apparatus is to process and package macromolecules synthesised by the cell, primarily proteins and lipids. The Golgi apparatus forms a part of the endomembrane system present in eukaryotic cells.

Structure

The Golgi is composed of membrane-bound sacs known as cisternae. Between five and eight are usually present, however as many as sixty have been observed. Surrounding the main cisternae are a number of spherical vesicles which have budded off from the cisternae. The cisternae stack has five functional regions: the cis-Golgi network, cis-Golgi, medial-Golgi, trans-Golgi, and trans-Golgi network. Vesicles from the endoplasmic reticulum (via the vesicular-tubular cluster) fuse with the cis-Golgi network and subsequently progress through the stack to the trans-Golgi network, where they are packaged and sent to the required destination. Each region contains different enzymes which selectively modify the contents depending on where they are destined to reside.

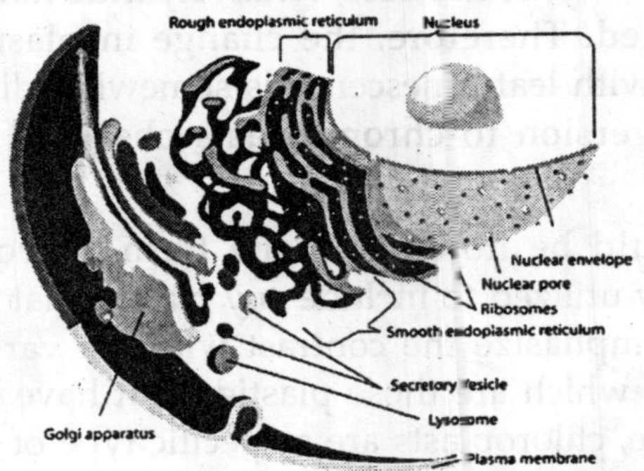

Fig. The Golgi Apparatus

Function

Cells synthesise a large number of different macromolecules required for life. The Golgi apparatus is integral in modifying, sorting, and packaging these substances for cell secretion (exocytosis) or for use within the cell. It primarily modifies proteins delivered from the rough endoplasmic reticulum, but is also involved in the transport of lipids around the cell, and the creation of lysosomes. In this respect it can be thought of as similar to a post office; it packages and labels "items" and then sends them to different parts of the cell.

- Enzymes within the cisternae are able to modify substances by the addition of carbohydrates (glycosylation) and phosphate (phosphorylation) to them. In order to do so the Golgi transports substances such as nucleotide sugars into the organelle from the cytosol. Proteins are also labelled with a signal sequence of molecules which determine their final destination. For example, the Golgi apparatus adds a mannose-6-phosphate label to proteins destined for lysosomes.

The Golgi also plays an important role in the synthesis of proteoglycans, molecules present in the extracellular matrix of animals, and it is a major site of carbohydrate synthesis.

This includes the productions of glycosaminoglycans or GAGs, long unbranched polysaccharides which the Golgi then attaches to a protein synthesized in the endoplasmic reticulum to form the proteoglycan Enzymes in the Golgi will polymerize several of these GAGs via a xylose link onto the core protein. Another task of the Golgi involves the sulfation of certain molecules passing through its lumen via sulphotranferases that gain their sulphur molecule from a donor called PAPs. This process occurs on the GAGs of proteoglycans as well as on the core protein. The level of sulfation is very important to the proteoglycans signalling abilities as well as giving the proteoglycan it's overall negative charge.

The Golgi is also capable of phosphorylating molecules. To do so it transports ATP into the lumen. The Golgi itself contains resident kinases, such as casein kinases. One molecule that is phosphorylated in the Golgi is Apolipoprotein, which forms a molecule known as VLDL that is a constitute of blood serum. It is thought that the phosphorylation of these molecules is important to help aid in their sorting of secretion into the blood serum.

The Golgi also has a putative role in apoptosis, with several Bcl-2 family members localised there, as well as to the mitochondria. In addition a newly characterised anti-apoptotic protein, GAAP (Golgi anti-apoptotic protein), which almost exclusively resides in the Golgi, protects cells from apoptosis by an as-yet undefined mechanism.

RIBOSOME

A ribosome is a small, dense organelle in cells that assembles proteins. Ribosomes are about 20nm in diameter and are composed of 65% ribosomal RNA and 35% ribosomal proteins (known as a Ribonucleoprotein or RNP). It translates messenger RNA (mRNA) to build a polypeptide chain (e.g., a protein) using amino acids delivered by Transfer RNA (tRNA). It can be thought of as a giant enzyme but, although it contains proteins, its active site is made of RNA, so ribosomes are now classified as "ribozymes".

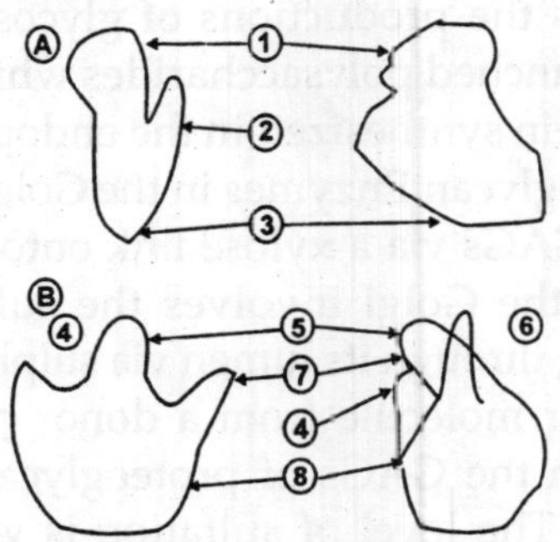

Fig: Ribosome structure indicating small subunit (A) and large subunit (B). Side and front view.

1. Head.
2. Platform.
3. Base.
4. Ridge.
5. Central protuberance.
6. Back.
7. Stalk.
8. Front.

Ribosomes build proteins from the genetic instructions held within a messenger RNA. Ribosomes can float freely in the cytoplasm (the internal fluid of the cell) or bound to the endoplasmic reticulum, or to the nuclear envelope. Since ribosomes are ribozymes, it is thought that they might be remnants of the RNA world. While catalysis of the peptide bond involves the C2' hydroxyl of tRNA's P-site adenosine in a sort of proton shuttle mechanism, the full function (ie, translocation) of the ribosome is reliant on changes in protein conformations.

Ribosomes are an important structure in the cell. Ribosomes were first observed in the mid-1950s by Romanian cell biologist George Palade in the electron microscope as dense particles or granules for which he would win the Nobel Prize.

The term *ribosome* was proposed by scientist Richard B. Roberts in 1958:

Ribosomes consist of two subunits that fit together and work as one to translate the mRNA into a polypeptide chain during protein synthesis. Prokaryotic subunits consist of one or two and eukaryotic of one or three very large RNA molecules (known as ribosomal RNA or rRNA) and multiple smaller protein molecules. Crystallographic work has shown that there are no ribosomal proteins close to the reaction site for polypeptide synthesis. This suggests that the protein components of ribosomes act as a scaffold that may enhance the ability of rRNA to synthesize protein rather than directly participating in catalysis.

Structure

The ribosomal subunits of prokaryotes and eukaryotes are quite similar.

Prokaryotes have 70S ribosomes, each consisting of a small (30S) and a large (50S) subunit. Their large subunit is composed of a 5S RNA subunit (consisting of 120 nucleotides), a 23S RNA subunit (2900 nucleotides) and 34 proteins. The 30S subunit has a 1540 nucleotide RNA subunit bound to 21 proteins.

Eukaryotes have 80S ribosomes, each consisting of a small (40S) and large (60S) subunit. Their large subunit is composed of a 5S RNA (120 nucleotides), a 28S RNA (4700 nucleotides), a 5.8S subunit (160 nucleotides) and ~49 proteins. The 40S subunit has a 1900 nucleotide (18S) RNA and ~33 proteins.

The ribosomes found in chloroplasts and mitochondria of eukaryotes also consist of large and small subunits bound together with proteins into one 55S particle. These organelles are believed to be descendants of bacteria and as such their ribosomes are similar to those of prokaryotes.

The various ribosomes share a core structure which is quite similar despite the large differences in size. The extra RNA in the larger ribosomes is in several long continuous insertions, such that they form loops out of the core structure without disrupting or changing it. All of the catalytic activity of the ribosome is carried out by the RNA, the proteins reside on the surface and seem to stabilize the structure.

The differences between the prokaryotic and eukaryotic ribosomes are exploited by pharmaceutical chemists to create antibiotics that can destroy a bacterial infection without harming the cells of the infected person. Due to the differences in their structures, the bacterial 70S ribosomes are vulnerable to these antibiotics while the eukaryotic 80S ribosomes are not. Even though mitochondria possess ribosomes similar to the bacterial ones, mitochondria are not affected by these antibiotics because they are surrounded by a double membrane that does not easily admit these antibiotics into the organelle.

Function

Ribosomes are the workhorses of protein biosynthesis, the process of translating messenger RNA (mRNA) into prótein. The mRNA comprises a series of codons that dictate to the ribosome the sequence of the amino acids needed to make the protein. Using the mRNA as a template, the ribosome traverses each codon of the mRNA, pairing it with the appropriate amino acid. This is done using molecules of transfer RNA (tRNA) containing a complementary anticodon on one end and the appropriate amino acid on the other.

Protein synthesis begins at a start codon near the 5' end of the mRNA. The small ribosomal subunit, typically bound to a tRNA containing the amino acid methionine, binds to an AUG codon on the mRNA and recruits the large ribosomal subunit. The large ribosomal subunit contains three tRNA binding sites, designated A, P, and E. The A site binds an aminoacyl-tRNA (a tRNA bound to an amino acid); the P site binds a peptidyl-tRNA (a tRNA bound to the peptide being synthesized); and the E site binds a free tRNA before it exits the ribosome.

ENDOPLASMIC RETICULUM

The endoplasmic reticulum or ER is an organelle found in all eukaryotic cells that is an interconnected network of tubules, vesicles and cisternae that is responsible for several specialized functions: Protein translation, folding, and transport of proteins to be used in the cell membrane (e.g., transmembrane receptors and other integral membrane proteins), or to be secreted (exocytosed) from the cell (e.g., digestive enzymes); sequestration of calcium; and production and storage of glycogen, steroids, and other macromolecules. The endoplasmic reticulum is part of the endomembrane system. The basic structure and composition of the ER membrane is similar to the plasma membrane.

Structure

The general structure of the endoplasmic reticulum is an extensive membrane network of cisternae (sac-like structures) held together by the cytoskeleton. The phospholipid membrane encloses a space, the cisternal space (or lumen), from the cytosol. The functions of the endoplasmic reticulum vary greatly depending on the exact type of endoplasmic reticulum and the type of cell in which it resides. The three varieties are called *rough endoplasmic reticulum, smooth endoplasmic reticulum,* and *sarcoplasmic reticulum.*

MITOCHONDRION

In cell biology, a mitochondrion (plural mitochondria) (from Greek *ìéôïò* or *mitos,* thread + *÷ïíäñïí* or *khondrion,* granule) is a membrane-enclosed organelle, found in most eukaryotic cells. Mitochondria are sometimes described as "cellular power plants," because they generate most of the cell's supply of ATP, used as a source of chemical energy. The number of mitochondria in a cell varies widely by organism and tissue type. Many cells possess only a single mitochondrion, while others can contain several million mitochondria . Although most of a cell's DNA is contained in the cell nucleus, mitochondria have their own independent genomes. According to the endosymbiotic theory, mitochondria are descended from free-living prokaryotes.

A mitochondrion contains inner and outer membranes composed of phospholipid bilayers and proteins. The two membranes, however, have different properties. Because of this double-membraned organization, there are 5 distinct compartments within mitochondria. There is the outer membrane, the intermembrane space (the space between the outer and inner membranes), the inner membrane, the cristae space (formed by infoldings of the inner membrane), and the matrix (space within the inner membrane). Mitochondria range from 1 to 10 micrometers (ìm) in size.

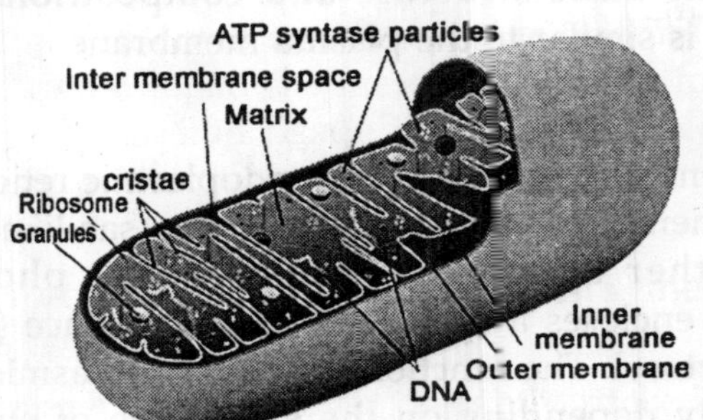

Fig. Simplified Structure of Mitochondrion

Outer Membrane

The outer mitochondrial membrane, which encloses the entire organelle, has a protein-to-phospholipid ratio similar to the eukaryotic plasma membrane (about 1:1 by weight). It contains numerous integral proteins called *porins*.

Intermembrane Space

The intermembrane space is the space between the outer membrane and the inner membrane.

Inner Membrane

The inner mitochondrial membrane contains proteins with four types of functions:

1. Those that carry out the oxidation reactions of the respiratory chain.
2. ATP synthase, which makes ATP in the matrix.
3. Specific transport proteins that regulate the passage of metabolites into and out of the matrix.
4. Protein import machinery.

It contains more than 100 different polypeptides, and has a very high protein-to-phospholipid ratio (more than 3:1 by weight, which is about 1 protein for 15 phospholipids). Additionally, the inner membrane is rich in an unusual phospholipid, cardiolipin. This phospholipid was originally discovered in beef hearts in 1942 and is usually characteristic of mitochondrial and bacterial plasma membranes. Unlike the outer membrane, the inner membrane does not contain porins, and is highly impermeable; almost all ions and molecules require special membrane transporters to enter or exit the matrix. In addition, there is a membrane potential across the inner membrane.

The inner mitochondrial membrane is compartmentalized into numerous cristae, which expand the surface area of the inner mitochondrial membrane, enhancing its ability to generate ATP. In typical liver mitochondria, for example, the surface area, including cristae, is about five times that of the outer membrane. Mitochondria of cells which have greater demand for ATP, such as muscle cells, contain more cristae than typical liver mitochondria.

Mitochondrial Matrix

The matrix is the space enclosed by the inner membrane. The matrix is important in the production of ATP with the aid of the ATP synthase contained in the inner membrane. The matrix contains a highly concentrated mixture of hundreds of enzymes, in addition to the special mitochondrial ribosomes, tRNA, and several copies of the mitochondrial DNA genome. Of the enzymes, the major functions include oxidation of pyruvate and fatty acids, and the citric acid cycle.

Mitochondria possess their own genetic material, and the machinery to manufacture their own RNAs and proteins. A published human mitochondrial DNA sequence revealed 16,569 base pairs encoding 37 total genes, 24 tRNA and rRNA genes and 13 peptide genes. The 13 mitochondrial peptides in humans are integrated into the inner mitochondrial membrane, along with proteins encoded by genes that reside in the host cell's nucleus. This mitochondria is also known as powerhouse of the cell.

Mitochondrial Functions

Although it is well known that the mitochondria convert organic materials into cellular energy in the form of ATP, mitochondria play an important role in many metabolic tasks, such as:

- Apoptosis-programmed cell death
- Glutamate-mediated excitotoxic neuronal injury
- Cellular proliferation
- Regulation of the cellular redox state
- Heme synthesis
- Steroid synthesis

Some mitochondrial functions are performed only in specific types of cells. For example, mitochondria in liver cells contain enzymes that allow them to detoxify ammonia, a waste product of protein metabolism. A mutation in the genes regulating any of these functions can result in mitochondrial diseases.

LYSOSOME

Lysosomes are organelles that contain digestive enzymes (acid hydrolases). They digest excess or worn out organelles, food particles, and engulfed viruses or bacteria. The membrane surrounding a lysosome prevents the digestive enzymes inside from destroying the cell. Lysosomes fuse with vacuoles and dispense their enzymes into the vacuoles, digesting their contents. They are built in the Golgi apparatus. The name *lysosome* derives from the Greek words *lysis*, which means dissolution or destruction, and *soma*, which means body. They are frequently nicknamed "suicide-bags" or "suicide-sacs" by cell biologists due to their role in autolysis. Lysosomes were discovered by the Belgian cytologist Christian de Duve in 1949.

Functions

The lysosomes are used for the digestion of macromolecules from phagocytosis (ingestion of other dying cells or larger extracellular material), endocytosis (where receptor proteins are recycled from the cell surface), and autophagy (where old or unneeded organelles or proteins, or microbes which have invaded the cytoplasm are delivered to the lysosome). Autophagy may also lead to autophagic cell death, a form of programmed self-destruction, or autolysis, of the cell, which means that the cell is digesting itself.

Other functions include digesting foreign bacteria that invade a cell and helping repair damage to the plasma membrane by serving as a membrane patch, sealing the wound. Lysosomes also do much of the cellular digestion required to digest tails of tadpoles and to remove the web from the fingers of a 3-6 month old fetus. This process of programmed cell death is called apoptosis

CELL NUCLEUS

In cell biology, the nucleus (pl. *nuclei*; from Latin *nucleus* or *nuculeus*, kernel) is a membrane-enclosed organelle found in most eukaryotic cells. It contains most of the cell's genetic material, organized as multiple long linear DNA molecules in complex with a large variety of proteins, such as histones, to form chromosomes. The genes within these chromosomes make up the cell's nuclear genome. The function of the nucleus is to maintain the integrity of these genes and to control the activities of the cell by regulating gene expression.

The main structural elements of the nucleus are the nuclear envelope, a double membrane that encloses the entire organelle and keeps its contents separated from the cellular cytoplasm, and the nuclear lamina, a meshwork within the nucleus that adds mechanical support much like the cytoskeleton supports the cell as a whole. Because the nuclear membrane is impermeable to most molecules, nuclear pores are required to allow movement of molecules across the envelope. These pores cross both membranes of the envelope, providing a channel that allows free movement of small molecules and ions. The movement of larger molecules such as proteins is carefully controlled, and requires active transport facilitated by carrier proteins. Nuclear transport is of paramount importance to cell function, as movement through the pores is required for both gene expression and chromosomal maintenance.

Although the interior of the nucleus does not contain any membrane-delineated bodies, its contents are not uniform, and a number of *subnuclear bodies* exist, made up of unique proteins, RNA molecules, and DNA conglomerates. The best known of these is the nucleolus, which is mainly involved in assembly of ribosomes. After being produced in the nucleolus, ribosomes are exported to the cytoplasm where they translate mRNA.

Structure

The nucleus is the largest cellular organelle. In mammalian cells, the average diameter typically varies from 11 to 22 micrometers (ìm) and occupies about 10% of the total volume. The viscous liquid within it is called nucleoplasm, and is similar to the cytoplasm found outside the nucleus.

Nuclear Envelope and Pores

The nuclear envelope consists of two cellular membranes, an inner and an outer membrane, arranged parallel to one another and separated by 10 to 50 nanometers (nm). The nuclear envelope completely encloses the nucleus and separates the cell's genetic material from the surrounding cytoplasm, serving as a barrier to prevent macromolecules from diffusing freely between the nucleoplasm and the cytoplasm.

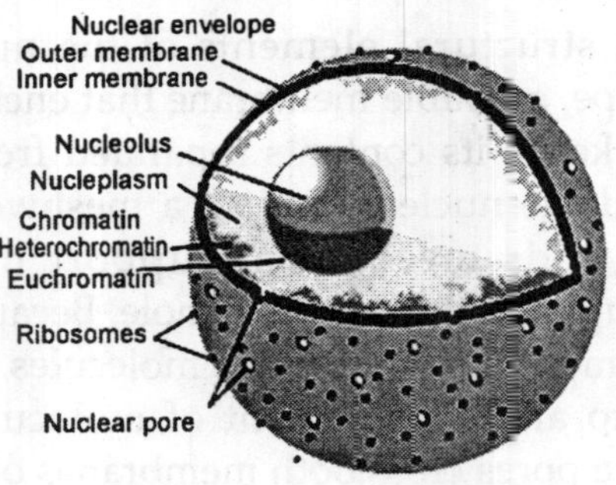

Fig. The Eukaryotic Cell Nucleus.

The outer nuclear membrane is continuous with the membrane of the rough endoplasmic reticulum (RER), and is similarly studded with ribosomes. The space between the membranes is called the perinuclear space and is continuous with the RER lumen.

Visible in this diagram are the ribosome-studded double membranes of the nuclear envelope, the DNA (complexed as chromatin), and the nucleolus. Within the cell nucleus is a viscous liquid called nucleoplasm, similar to the cytoplasm found outside the nucleus.

A cross section of a nuclear pore on the surface of the nuclear envelope:

1. Other diagram labels show
2. The outer ring,
3. Spokes,
4. Basket, and
5. Filaments.

Nuclear pores, which provide aqueous channels through the envelope, are composed of multiple proteins, collectively referred to as nucleoporins. The pores are about 125 million daltons in molecular weight and consist of around 50 (in yeast) to 100 proteins (in vertebrates). The pores are 100 nm in total diameter; however, the gap through which molecules freely diffuse is only about 9 nm wide, due to the presence of regulatory systems within the centre of the pore. This size allows the free passage of small water-soluble molecules while preventing larger molecules, such as nucleic acids and proteins, from inappropriately entering or exiting the nucleus. These large molecules must be actively transported into the nucleus instead. The nucleus of a typical mammalian cell will have about 3000 to 4000 pores throughout its envelope, each of which contains a donut-shaped, eightfold-symmetric ring-shaped structure at a position where the inner and outer membranes fuse. Attached to the ring is a structure called the *nuclear basket* that extends into the nucleoplasm, and a series of filamentous extensions that reach into the cytoplasm. Both structures serve to mediate binding to nuclear transport proteins.

Most proteins, ribosomal subunits, and some RNAs are transported through the pore complexes in a process mediated by a family of transport factors known as karyopherins. Those karyopherins that mediate movement into the nucleus are also called importins, while those that mediate movement out of the nucleus are called exportins. Most karyopherins interact directly with their cargo, although some use adaptor proteins. Steroid hormones such as cortisol and aldosterone, as well as other small lipid-soluble molecules involved in intercellular signaling can diffuse through the cell membrane and into the cytoplasm, where they bind nuclear receptor proteins that are trafficked into the nucleus. There they serve as transcription factors when bound to their ligand; in the absence of ligand many such receptors function as histone deacetylases that repress gene expression.

Cytoskeleton

In animal cells, two networks of intermediate filaments provide the nucleus with mechanical support: the nuclear lamina forms an organized meshwork on the internal face of the envelope, while less organized support is provided on the cytosolic face of the envelope. Both systems provide structural support for the nuclear envelope and anchoring sites for chromosomes and nuclear pores.

The nuclear lamina is mostly composed of lamin proteins. Like all proteins, lamins are synthesized in the cytoplasm and later transported into the nucleus interior, where they are assembled before being incorporated into the existing network of nuclear lamina. Lamins are also found inside the nucleoplasm where they form another regular structure, known as the *nucleoplasmic veil*, that is visible using fluorescence microscopy. The actual function of the veil is not clear, although it is excluded from the nucleolus and is present during interphase. The lamin structures that make up the veil bind chromatin and disrupting their structure inhibits transcription of protein-coding genes.

Like the components of other intermediate filaments, the lamin monomer contains an alpha-helical domain used by two monomers to coil around each other, forming a dimer structure called a coiled coil. Two of these dimer structures then join side by side, in an antiparallel arrangement, to form a tetramer called a *protofilament*. Eight of these protofilaments form a lateral arrangement that is twisted to form a ropelike *filament*. These filaments can be assembled or disassembled in a dynamic manner, meaning that changes in the length of the filament depend on the competing rates of filament addition and removal.

Mutations in lamin genes leading to defects in filament assembly are known as *laminopathies*. The most notable laminopathy is the family of diseases known as progeria, which causes the appearance of premature aging in its sufferers. The exact mechanism by which the associated biochemical changes give rise to the aged phenotype is not well understood.

Nucleolus

The nucleolus is a discrete densely-stained structure found in the nucleus. It is not surrounded by a membrane, and is sometimes called a *suborganelle*. It forms around tandem repeats of rDNA, DNA coding for ribosomal RNA (rRNA). These regions are called nucleolar organizer regions (NOR). The main roles of the nucleolus are to synthesize rRNA and assemble ribosomes. The structural cohesion of the nucleolus depends on its activity, as ribosomal assembly in the nucleolus results in the transient association of nucleolar components, facilitating further ribosomal assembly, and hence further association. This model is supported by observations that inactivation of rDNA results in intermingling of nucleolar structures.

The first step in ribosomal assembly is transcription of the rDNA, by a protein called RNA polymerase I, forming a large pre-rRNA precursor. This is cleaved into the subunits 5.8S, 18S, and 28S rRNA. The transcription, post-transcriptional processing, and assembly of rRNA occurs in the nucleolus, aided by small nucleolar RNA (snoRNA) molecules, some of which are derived from spliced introns from messenger RNAs encoding genes related to ribosomal function. The assembled ribosomal subunits are the largest structures passed through the nuclear pores.

When observed under the electron microscope, the nucleolus can be seen to consist of three distinguishable regions: the innermost *fibrillar centers* (FCs), surrounded by the *dense fibrillar component* (DFC), which in turn is bordered by the *granular component* (GC). Transcription of the rDNA occurs either in the FC or at the FC-DFC boundary, and therefore when rDNA transcription in the cell is increased more FCs are detected. Most of the cleavage and modification of rRNAs occurs in the DFC, while the latter steps involving protein assembly onto the ribosomal subunits occurs in the GC.

Function

The main function of the cell nucleus is to control gene expression and mediate the replication of DNA during the cell cycle. The nucleus provides a site for genetic transcription that is segregated from the location of translation in the cytoplasm, allowing levels of gene regulation that are not available to prokaryotes.

Cell Compartmentalization

The nuclear envelope allows the nucleus to control its contents, and separate them from the rest of the cytoplasm where necessary. This is important for controlling processes on either side of the nuclear membrane. In some cases where a cytoplasmic process needs to be restricted, a key participant is removed to the nucleus, where it interacts with transcription factors to downregulate the production of certain enzymes in the pathway. This regulatory mechanism occurs in the case of glycolysis, a cellular pathway for breaking down glucose to produce energy. Hexokinase is an enzyme responsible for the first the step of glycolysis, forming glucose-6-phosphate from glucose. At high concentrations of fructose-6-phosphate, a molecule made later from glucose-6-phosphate, a regulator protein removes hexokinase to the nucleus, where it forms a transcriptional repressor complex with nuclear proteins to reduce the expression of genes involved in glycolysis.

The compartmentalization allows the cell to prevent translation of unspliced mRNA. Eukaryotic mRNA contains introns that must be removed before being translated to produce functional proteins. The splicing is done inside the nucleus before the mRNA can be accessed by ribosomes for translation. Without the nucleus ribosomes would translate newly transcribed (unprocessed) mRNA resulting in misformed and nonfunctional proteins.

Gene Expression

Gene expression first involves transcription, in which DNA is used as a template to produce RNA. In the case of genes encoding proteins, that RNA produced from this process is messenger RNA (mRNA), which then needs to be translated by ribosomes to form a protein. As ribosomes are located outside the nucleus, mRNA produced needs to be exported.

Processing of Pre-Mrna

Newly synthesized mRNA molecules are known as primary transcripts or pre-mRNA. They must undergo post-transcriptional modification in the nucleus before being exported to the cytoplasm; mRNA that appears in the nucleus without these modifications is degraded rather than used for protein translation. The three main modifications are 5' capping, 3' polyadenylation, and RNA splicing. While in the nucleus, pre-mRNA is associated with a variety of proteins in complexes known as heterogeneous ribonucleoprotein particles (hnRNPs). Addition of the 5' cap occurs co-transcriptionally and is the first step in post-translational modification. The 3' poly-adenine tail is only added after transcription is complete.

RNA splicing, carried out by a complex called the spliceosome, is the process by which introns, or regions of DNA that do not code for protein, are removed from the pre-mRNA and the remaining exons connected to re-form a single continuous molecule. This process normally occurs after 5' capping and 3' polyadenylation but can begin before synthesis is complete in transcripts with many exons. Many pre-mRNAs, including those encoding antibodies, can be spliced in multiple ways to produce different mature mRNAs that encode different protein sequences. This process is known as alternative splicing, and allows production of a large variety of proteins from a limited amount of DNA.

Dynamics and Regulation

Nuclear Transport

Macromolecules, such as RNA and proteins, are actively transported across the nuclear membrane in a process called the Ran-GTP nuclear transport cycle.

The entry and exit of large molecules from the nucleus is tightly controlled by the nuclear pore complexes. Although small molecules can enter the nucleus without regulation, macromolecules such as RNA and proteins require association karyopherins called importins to enter the nucleus and exportins to exit. "Cargo" proteins that must be translocated from the cytoplasm to the nucleus contain short amino acid sequences known as nuclear localization signals which are bound by importins, while those transported from the nucleus to the cytoplasm carry nuclear export signals bound by exportins. The ability of importins and exportins to transport their cargo is regulated by GTPases, enzymes that hydrolyze the molecule guanosine triphosphate to release energy. The key GTPase in nuclear transport is Ran, which can bind either GTP or GDP (guanosine diphosphate) depending on whether it is located in the nucleus or the cytoplasm. Whereas importins depend on RanGTP to dissociate from their cargo, exportins require RanGTP in order to bind to their cargo.

Assembly and Disassembly

During its lifetime a nucleus may be broken down, either in the process of cell division or as a consequence of apoptosis, a regulated form of cell death. During these events, the structural components of the nucleus—the envelope and lamina—are systematically degraded.

During the cell cycle the cell divides to form two cells. In order for this process to be possible, each of the new daughter cells must have a full set of genes, a process requiring replication of the chromosomes as well as segregation of the separate sets. This occurs by the replicated chromosomes, the sister chromatids, attaching to microtubules, which in turn are attached to different centrosomes. The sister chromatids can then be pulled to separate locations in the cell. However, in many cells the centrosome is located in the cytoplasm, outside the nucleus the microtubles would be unable to attach to the chromatids in the presence of the nuclear envelope. Therefore the early stages in the cell cycle, beginning in prophase and until around prometaphase, the nuclear membrane is dismantled Likewise, during the same period the nuclear lamina is also disassembled, a process regulated by phosphorylation of the lamins. Towards the end of the cell cycle, the nuclear membrane is reformed, and around the same time, the nuclear lamina are reassembled by dephosphorylating the lamins.

Apoptosis is a controlled process in which the cell's structural components are destroyed, resulting in death of the cell. Changes associated with apoptosis directly affect the nucleus and its contents, for example in the condensation of chromatin and the disintegration of the nuclear envelope and lamina. The destruction of the lamin networks is controlled by specialized apoptotic proteases called caspases, which cleave the lamin proteins and thus degrade the nucleus' structural integrity. Lamin cleavage is sometimes used as a laboratory indicator of caspase activity in assays for early apoptotic activity. Cells that express mutant caspase-resistant lamins are deficient in nuclear changes related to apoptosis, suggesting that lamins play a role in initiating the events that lead to apoptotic degradation of the nucleus. Inhibition of lamin assembly itself is an inducer of apoptosis.

Anucleated and Polynucleated Cells

Although most cells have a single nucleus, some cell types have no nucleus, and others have many nuclei. This can be a normal process, as in the maturation of mammalian red blood cells, or an anomalous result of faulty cell division.

Anucleated cells contain no nucleus and are therefore incapable of dividing to produce daughter cells. The best-known anucleated cell is the mammalian red blood cell, or erythrocyte, which also lacks other organelles such as mitochondria and serves primarily as a transport vessel to ferry oxygen from the lungs to the body's tissues. Erythrocytes mature via erythropoiesis in the bone marrow, where they lose their nuclei, organelles, and ribosomes.

The nucleus is expelled during the process of differentiation from an erythroblast to a reticulocyte, the immediate precursor of the mature erythrocyte. The presence of mutagens may induce the release of some immature "micronucleated" erythrocytes into the bloodstream. Anucleated cells can also arise from flawed cell division in which one daughter lacks a nucleus and the other is binucleate.

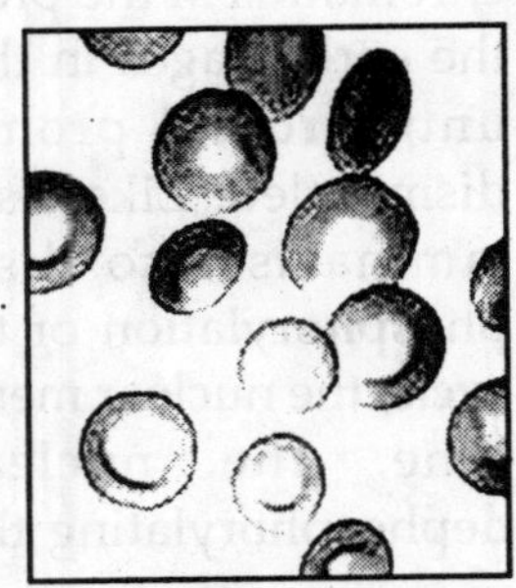

Fig. Human Red Blood Cells

DNA

Deoxyribonucleic acid, or DNA is a nucleic acid molecule that contains the genetic instructions used in the development and functioning of all living organisms. The main role of DNA is the long-term storage of information and it is often compared to a set of blueprints, since DNA contains the instructions needed to construct other components of cells, such as proteins and RNA molecules. The DNA segments that carry this genetic information are called genes, but other DNA sequences have structural purposes, or are involved in regulating the use of this genetic information.

Chemically, DNA is a long polymer of simple units called nucleotides, with a backbone made of sugars and phosphate atoms joined by ester bonds. Attached to each sugar is one of four types of molecules called bases. It is the sequence of these four bases along the backbone that encodes information. This information is read using the genetic code, which specifies the sequence of the amino acids within proteins. The code is read by copying stretches of DNA into the related nucleic acid RNA, in a process called transcription. Most of these RNA molecules are used to synthesize proteins, but others are used directly in structures such as ribosomes and spliceosomes.

Within cells, DNA is organized into structures called chromosomes and the set of chromosomes within a cell make up a genome. These chromosomes are duplicated before cells divide, in a process called DNA replication. Eukaryotic organisms such as animals, plants, and fungi store their DNA inside the cell nucleus, while in prokaryotes such as bacteria it is found in the cell's cytoplasm. Within the chromosomes, chromatin proteins such as histones compact and organize DNA, which helps control its interactions with other proteins and thereby control which genes are transcribed.

Physical and chemical Properties

DNA is a long polymer made from repeating units called nucleotides. The DNA chain is 22 to 24 Ångströms wide (2.2 to 2.4 nanometres) and one nucleotide unit is 3.3 Ångstroms (0.33 nanometres) long. Although each individual repeating unit is very small, DNA polymers can be enormous molecules containing millions of nucleotides. For instance, the largest human chromosome, chromosome number 1, is 220 million base pairs long.

In living organisms, DNA does not usually exist as a single molecule, but instead as a tightly-associated pair of molecules. These two long strands entwine like vines, in the shape of a double helix. The nucleotide repeats contain both the segment of the backbone of the molecule, which holds the chain together, and a base, which interacts with the other DNA strand in the helix. In general, a base linked to a sugar is called a nucleoside and a base linked to a sugar and one or more phosphate groups is called a nucleotide. If multiple nucleotides are linked together, as in DNA, this polymer is referred to as a polynucleotide.

The backbone of the DNA strand is made from alternating phosphate and sugar residues. The sugar in DNA is 2-

deoxyribose, which is a pentose (five carbon) sugar. The sugars are joined together by phosphate groups that form phosphodiester bonds between the third and fifth carbon atoms of adjacent sugar rings. These asymmetric bonds mean a strand of DNA has a direction. In a double helix the direction of the nucleotides in one strand is opposite to their direction in the other strand. This arrangement of DNA strands is called antiparallel. The asymmetric ends of a strand of DNA bases are referred to as the 52 (*five prime*) and 32 (*three prime*) ends. One of the major differences between DNA and RNA is the sugar, with 2-deoxyribose being replaced by the alternative pentose sugar ribose in RNA.

The DNA double helix is stabilized by hydrogen bonds between the bases attached to the two strands. The four bases found in DNA are adenine (abbreviated A), cytosine (C), guanine (G) and thymine (T). These four bases are shown below and are attached to the sugar/phosphate to form the complete nucleotide, as shown for adenosine monophosphate.

These bases are classified into two types; adenine and guanine are fused five- and six-membered heterocyclic compounds called purines, while cytosine and thymine are six-membered rings called pyrimidines. A fifth pyrimidine base, called uracil (U), usually takes the place of thymine in RNA and differs from thymine by lacking a methyl group on its ring. Uracil is normally only found in DNA as a breakdown product of cytosine, but a very rare exception to this rule is a bacterial virus called PBS1 that contains uracil in its DNA. In contrast, following synthesis of certain RNA molecules, a significant number of the uracils are converted to thymines by the enzymatic addition of the missing methyl group. This occurs mostly on structural and enzymatic RNAs like transfer RNAs and ribosomal RNA.

The double helix is a right-handed spiral. As the DNA strands wind around each other, they leave gaps between each set of phosphate backbones, revealing the sides of the bases inside. There are two of these grooves twisting around the surface of the double helix: one groove, the major groove, is

22 Å wide and the other, the minor groove, is 12 Å wide. The narrowness of the minor groove means that the edges of the bases are more accessible in the major groove. As a result, proteins like transcription factors that can bind to specific sequences in double-stranded DNA usually make contacts to the sides of the bases exposed in the major groove.

Base Pairing

Each type of base on one strand forms a bond with just one type of base on the other strand. This is called complementary base pairing. Here, purines form hydrogen bonds to pyrimidines, with A bonding only to T, and C bonding only to G. This arrangement of two nucleotides binding together across the double helix is called a base pair. In a double helix, the two strands are also held together via forces generated by the hydrophobic effect and pi stacking, which are not influenced by the sequence of the DNA. As hydrogen bonds are not covalent, they can be broken and rejoined relatively easily. The two strands of DNA in a double helix can therefore be pulled apart like a zipper, either by a mechanical force or high temperature. As a result of this complementarity, all the information in the double-stranded sequence of a DNA helix is duplicated on each strand, which is vital in DNA replication. Indeed, this reversible and specific interaction between complementary base pairs is critical for all the functions of DNA in living organisms.

The two types of base pairs form different numbers of hydrogen bonds, AT forming two hydrogen bonds, and GC forming three hydrogen bonds. The GC base pair is therefore stronger than the AT base pair. As a result, it is both the percentage of GC base pairs and the overall length of a DNA double helix that determine the strength of the association between the two strands of DNA. Long DNA helices with a high GC content have stronger-interacting strands, while short helices with high AT content have weaker-interacting strands. Parts of the DNA double helix that need to separate easily, such as the TATAAT Pribnow box in bacterial promoters, tend to have sequences with a high AT content, making the strands easier to pull apart. In the laboratory, the strength of this interaction can be

measured by finding the temperature required to break the hydrogen bonds, their melting temperature (also called T_m value). When all the base pairs in a DNA double helix melt, the strands separate and exist in solution as two entirely independent molecules. These single-stranded DNA molecules have no single common shape, but some conformations are more stable than others.

Sense and Antisense

A DNA sequence is called "sense" if its sequence is the same as that of a messenger RNA (mRNA) copy that is translated into protein. The sequence on the opposite strand is complementary to the sense sequence and is therefore called the "antisense" sequence. Since RNA polymerases work by making a complementary copy of their templates, it is this antisense strand that is the template for producing the sense mRNA. Both sense and antisense sequences can exist on different parts of the same strand of DNA (i.e. both strands contain both sense and antisense sequences). In both prokaryotes and eukaryotes, antisense RNA sequences are produced, but the functions of these RNAs are not entirely clear. One proposal is that antisense RNAs are involved in regulating gene expression through RNA-RNA base pairing.

A few DNA sequences in prokaryotes and eukaryotes, and more in plasmids and viruses, blur the distinction made above between sense and antisense strands by having overlapping genes. In these cases, some DNA sequences do double duty, encoding one protein when read 52 to 32 along one strand, and a second protein when read in the opposite direction (still 52 to 32) along the other strand. In bacteria, this overlap may be involved in the regulation of gene transcription, while in viruses, overlapping genes increase the amount of information that can be encoded within the small viral genome. Another way of reducing genome size is seen in some viruses that contain linear or circular single-stranded DNA as their genetic material.

Supercoiling

DNA can be twisted like a rope in a process called DNA supercoiling. With DNA in its "relaxed" state, a strand usually circles the axis of the double helix once every 10.4 base pairs, but if the DNA is twisted the strands become more tightly or more loosely wound. If the DNA is twisted in the direction of the helix, this is positive supercoiling, and the bases are held more tightly together. If they are twisted in the opposite direction, this is negative supercoiling, and the bases come apart more easily. In nature, most DNA has slight negative supercoiling that is introduced by enzymes called topoisomerases. These enzymes are also needed to relieve the twisting stresses introduced into DNA strands during processes such as transcription and DNA replication.

Alternative Double-Helical Structures

DNA exists in several possible conformations. The conformations so far identified are: A-DNA, B-DNA, C-DNA, D-DNA, E-DNA, H-DNA, L-DNA, P-DNA, and Z-DNA. However, only A-DNA, B-DNA, and Z-DNA have been observed in naturally occurring biological systems. Which conformation DNA adopts depends on the sequence of the DNA, the amount and direction of supercoiling, chemical modifications of the bases and also solution conditions, such as the concentration of metal ions and polyamines. Of these three conformations, the "B" form described above is most common under the conditions found in cells. The two alternative double-helical forms of DNA differ in their geometry and dimensions.

The A form is a wider right-handed spiral, with a shallow and wide minor groove and a narrower and deeper major groove. The A form occurs under non-physiological conditions in dehydrated samples of DNA, while in the cell it may be produced in hybrid pairings of DNA and RNA strands, as well as in enzyme-DNA complexes. Segments of DNA where the bases have been chemically-modified by methylation may undergo a larger change in conformation and adopt the Z form. Here, the strands turn about the helical axis in a left-handed spiral, the opposite of the more common B form. These unusual structures can be recognised by specific Z-DNA binding proteins and may be involved in the regulation of transcription.

Quadruplex Structures

At the ends of the linear chromosomes are specialized regions of DNA called telomeres. The main function of these regions is to allow the cell to replicate chromosome ends using the enzyme telomerase, as the enzymes that normally replicate DNA cannot copy the extreme 32 ends of chromosomes. As a result, if a chromosome lacked telomeres it would become shorter each time it was replicated. These specialized chromosome caps also help protect the DNA ends from exonucleases and stop the DNA repair systems in the cell from treating them as damage to be corrected. In human cells, telomeres are usually lengths of single-stranded DNA containing several thousand repeats of a simple TTAGGG sequence.

These guanine-rich sequences may stabilize chromosome ends by forming very unusual structures of stacked sets of four-base units, rather than the usual base pairs found in other DNA molecules. Here, four guanine bases form a flat plate and these flat four-base units then stack on top of each other, to form a stable *quadruplex* structure. These structures are stabilized by hydrogen bonding between the edges of the bases and chelation of a metal ion in the centre of each four-base unit. The structure shown to the left is a top view of the quadruplex formed by a DNA sequence found in human telomere repeats. The single DNA strand forms a loop, with the sets of four bases stacking in a central quadruplex three plates deep. In the space at the centre of the stacked bases are three chelated potassium ions. Other structures can also be formed, with the central set of four bases coming from either a single strand folded around the bases, or several different parallel strands, each contributing one base to the central structure.

In addition to these stacked structures, telomeres also form large loop structures called telomere loops, or T-loops.

Here, the single-stranded DNA curls around in a long circle stabilized by telomere-binding proteins. At the very end of the T-loop, the single-stranded telomere DNA is held onto a region of double-stranded DNA by the telomere strand disrupting the double-helical DNA and base pairing to one of the two strands. This triple-stranded structure is called a displacement loop or D-loop.

CHROMATIN

Chromatin is the complex of DNA and protein that makes up chromosomes. In eukaryotes chromatin is found inside the nuclei of eukaryotic cells, while in prokaryotes, the chromatin is held within the nucleoid. The nucleic acids are in the form of double-stranded DNA (a double helix). The major proteins involved in chromatin are histone proteins, although many other chromosomal proteins have prominent roles too. The functions of chromatin are to package DNA into a smaller volume to fit in the cell, to strengthen the DNA to allow mitosis and meiosis, and to serve as a mechanism to control expression. Changes in chromatin structure are affected mainly by methylation (DNA and proteins) and acetylation (proteins). Chromatin structure is also relevant to DNA replication and DNA repair. Chromatin is easily visualised by staining, hence its name, which literally means *coloured material*.

Simplistically, there are three levels of chromatin organization:

1. DNA wrapping around nucleosomes - The "beads on a string" structure.
2. A 30 nm condensed chromatin fiber consisting of nucleosome arrays in their most compact form.
3. Higher level DNA packaging into the metaphase chromosome.

These structures do not occur in all eukaryotic cells; there are examples of more extreme packaging, for example spermatozoa and avian red blood cells.

The different levels of chromatin compaction are clearly visible in cells. In non-dividing cells there are two types of chromatin: euchromatin and heterochromatin. These correspond to uncompacted actively transcribed DNA and compacted untranscribed DNA.

The structure of chromatin varies considerably as the cell progresses through the cell cycle. The changes in structure are required to allow the DNA to be used and managed, whilst minimising the risk of damage.

RNA

Ribonucleic acid (RNA) is a nucleic acid polymer consisting of nucleotide monomers that plays several important roles in the processes that translate genetic information from deoxyribonucleic acid (DNA) into protein products; RNA acts as a messenger between DNA and the protein synthesis complexes known as ribosomes, forms vital portions of ribosomes, and acts as an essential carrier molecule for amino acids to be used in protein synthesis.

RNA is very similar to DNA, but differs in a few important structural details: RNA nucleotides contain ribose sugars while DNA contains deoxyribose and RNA uses predominantly uracil instead of thymine present in DNA. RNA is transcribed (synthesized) from DNA by enzymes called RNA polymerases and further processed by other enzymes. RNA serves as the template for translation of genes into proteins, transferring amino acids to the ribosome to form proteins, and also translating the transcript into proteins.

Nucleic acids were discovered in 1868 (some sources indicate 1869) by Johann Friedrich Miescher (1844-1895), who called the material 'nuclein' since it was found in the nucleus. It was later discovered that prokaryotic cells, which do not have a nucleus, also contain nucleic acids. The role of RNA in protein synthesis had been suspected since 1939, based on experiments carried out by Torbjörn Caspersson, Jean Brachet and Jack Schultz. Hubert Chantrenne elucidated the messenger role played by RNA in the synthesis of proteins in ribosome.

The sequence of the 77 nucleotides of a yeast RNA was found by Robert W. Holley in 1964, winning Holley the 1968 Nobel Prize for Medicine. In 1976, Walter Fiers and his team at the University of Ghent determined the complete nucleotide sequence of bacteriophage MS2-RNA.

Chemical and Stereochemical Structure

RNA is a polymer with a ribose and phosphate backbone and four different bases: adenine, guanine, cytosine, and uracil. The first three are the same as those found in DNA, but in RNA thymine is replaced by uracil as the base complementary to adenine. This base is also a pyrimidine and is very similar to thymine. Uracil is energetically less expensive to produce than thymine, which may account for its use in RNA. In DNA, however, uracil is readily produced by chemical degradation of cytosine, so having thymine as the normal base makes detection and repair of such incipient mutations more efficient. Thus, uracil is appropriate for RNA, where quantity is important but lifespan is not, whereas thymine is appropriate for DNA where maintaining sequence with high fidelity is more critical.

However, there are also numerous modified bases and sugars found in RNA that serve many different roles. Pseudouridine (Ø) and the DNA nucleoside thymidine are found in various places (most notably in the TØC loop of every tRNA). Thus, it is not technically correct to say that uracil is found in RNA in place of thymine. Another notable modified base is hypoxanthine (a deaminated Guanine base whose nucleotide is called Inosine). Inosine plays a key role in the Wobble Hypothesis of the Genetic Code. There are nearly 100 other naturally occurring modified bases, of which pseudouridine and 2'-O-methylribose are by far the most common. The specific roles of many of these modifications in RNA are not fully understood. However, it is notable that in ribosomal RNA, many of the post-translational modifications occur in highly functional regions, such as the peptidyl transferase centre and the subunit interface, inferring that they are important for normal function.

Single stranded RNA exhibits a right handed stacking pattern that is stabilized by base stacking.

The most important structural feature of RNA, indeed the only consistent difference between the two nucleic acids, that distinguishes it from DNA is the presence of a hydroxyl group at the 2'-position of the ribose sugar. The presence of this functional group enforces the C3'-endo sugar conformation (as opposed to the C2'-endo conformation of the deoxyribose sugar in DNA) that causes the helix to adopt the A-form geometry rather than the B-form most commonly observed in DNA. This results in a very deep and narrow major groove and a shallow and wide minor groove. A second consequence of the presence of the 2'-hydroxyl group is that in conformationally flexible regions of an RNA molecule (that is, not involved in formation of a double helix), it can chemically attack the adjacent phosphodiester bond to cleave the backbone.

Comparison with DNA

Unlike DNA, RNA is a single-stranded molecule in most of its biological roles and has a much shorter chain of nucleotides. RNA contains ribose, rather than the deoxyribose found in DNA (there is no hydroxyl group attached to the pentose ring in the 2' position whereas RNA has two hydroxyl groups). These hydroxyl groups make RNA less stable than DNA because it is more prone to hydrolysis. Several types of RNA (tRNA, rRNA) contain a great deal of secondary structure, which help promote stability.

Like DNA, most biologically active RNAs including tRNA, rRNA, snRNAs and other non-coding RNAs (such as the SRP RNAs) are extensively base paired to form double stranded helices. Structural analysis of these RNAs have revealed that they are not, "single-stranded" but rather highly structured. Unlike DNA, this structure is not just limited to long double-stranded helices but rather collections of short helices packed together into structures akin to proteins. In this fashion, RNAs can achieve chemical catalysis, like enzymes. For instance, determination of the structure of the ribosome in 2000 revealed that the active site of this enzyme that catalyzes peptide bond formation is composed entirely of RNA.

Synthesis

Synthesis of RNA is usually catalyzed by an enzyme - RNA polymerase, using DNA as a template. Initiation of synthesis begins with the binding of the enzyme to a promoter sequence in the DNA (usually found "upstream" of a gene). The DNA double helix is unwound by the helicase activity of the enzyme. The enzyme then progresses along the template strand in the 3′ -> 5′ direction, synthesizing a complementary RNA molecule with elongation occurring in the 5′ -> 3′ direction. The DNA sequence also dictates where termination of RNA synthesis will occur.

There are also a number of RNA-dependent RNA polymerases as well that use RNA as their template for synthesis of a new strand of RNA. For instance, a number of RNA viruses (such as poliovirus) use this type of enzyme to replicate their genetic material. Also, it is known that RNA-dependent RNA polymerases are required for the RNA interference pathway in many organisms.

Biological Roles

Messenger RNA (mRNA)

Messenger RNA is RNA that carries information from DNA to the ribosome sites of protein synthesis in the cell. Once mRNA has been transcribed from DNA, it is exported from the nucleus into the cytoplasm (in eukaryotes mRNA is "processed" before being exported), where it is bound to ribosomes and translated into its corresponding protein form with the help of tRNA. After a certain amount of time the message degrades into its component nucleotides, usually with the assistance of RNA polymerases.

Transfer RNA (tRNA)

Transfer RNA is a small RNA chain of about 74-95 nucleotides that transfers a specific amino acid to a growing polypeptide chain at the ribosomal site of protein synthesis during translation. It has sites for amino-acid attachment and an anticodon region for codon recognition that binds to a specific sequence on the messenger RNA chain through hydrogen bonding. It is a type of non-coding RNA.

Ribosomal RNA (rRNA)

Ribosomal RNA is a component of the ribosomes, the protein synthetic factories in the cell. Eukaryotic ribosomes contain four different rRNA molecules: 18S, 5.8S, 28S, and 5S rRNA. Three of the rRNA molecules are synthesized in the nucleolus, and one is synthesized elsewhere. rRNA molecules are extremely abundant and make up at least 80% of the RNA molecules found in a typical eukaryotic cell.

In the cytoplasm, ribosomal RNA and protein combine to form a nucleoprotein called a ribosome. The ribosome binds mRNA and carries out protein synthesis. Several ribosomes may be attached to a single mRNA at any time.

Non-coding RNA or "RNA genes"

RNA genes (sometimes referred to as non-coding RNA or small RNA) are genes that encode RNA that is not translated into a protein. The most prominent examples of RNA genes are transfer RNA (tRNA) and ribosomal RNA (rRNA), both of which are involved in the process of translation. However, since the late 1990s, many new RNA genes have been found, and thus RNA genes may play a much more significant role than previously thought.

In the late 1990s and early 2000, there has been persistent evidence of more complex transcription occurring in mammalian cells (and possibly others). This could point towards a more widespread use of RNA in biology, particularly in gene regulation. A particular class of non-coding RNA, micro RNA, has been found in many metazoans (from *Caenorhabditis elegans* to *Homo sapiens*) and clearly plays an important role in regulating other genes.

First proposed in 2004 by Rassoulzadegan and published in Nature 2006, RNA is implicated as being part of the germline. If confirmed, this result would significantly alter the present understanding of genetics and lead to many question on DNA-RNA roles and interactions.

Catalytic RNA

Although RNA contains only four bases, in comparison to the twenty amino acids commonly found in proteins, some RNAs are still able to catalyse chemical reactions. These include cutting and ligating other RNA molecules and also the catalysis of peptide bond formation in the ribosome.

Double-Stranded RNA

Double-stranded RNA (or dsRNA) is RNA with two complementary strands, similar to the DNA found in all "higher" cells. dsRNA forms the genetic material of some viruses. In eukaryotes, it acts as a trigger to initiate the process of RNA interference and is present as an intermediate step in the formation of siRNAs (small interfering RNAs). siRNAs are often confused with miRNAs; siRNAs are double-stranded, whereas miRNAs are single-stranded. Although initially single stranded there are regions of intra-molecular association causing hairpin structures in pre-miRNAs; immature miRNAs. Very recently, dsRNA has been found to induce gene expression at transcriptional level, a phenomenon named "small RNA induced gene activation RNAa". Such dsRNA is called "small activating RNA (saRNA)".

RNA World Hypothesis

The RNA world hypothesis proposes that the earliest forms of life relied on RNA both to carry genetic information (like DNA does now) and to catalyze biochemical reactions like an enzyme. According to this hypothesis, descendants of these early lifeforms gradually integrated DNA and proteins into their metabolism.

BASIC CELL FUNCTION

You should, by now, have a general appreciation for the complexity of cellular structure. Improvements in microscopy, especially development of the Electron microscope, have revealed that cells are not merely membranous sacks containing fluid of gel-like consistency. The degree of organization of the cytoplasm into organelles and their membranes should have you convinced that much (perhaps most) of what is really going on around you on this planet is occurring at a scale that is simply inaccessable to your eyes. And while you cannot be expected to directly observe chemical reactions at a molecular scale, contemplate that you cannot, even with powerful optics, directly observe most of the structure where these reactions are somehow controlled to produce outcomes favorable to life—indeed, are life. Hopefully, as you acquire knowledge and become a biologist—a botanist—you will learn to recognize the relevant phenomena by their macroscopic expressions (that which you can readly observe with the unaided eye).

To appreciate basic cell function, it is necessary to first list the processes or outcomes that cells must accomplish to further existence. More specialized functions will be discussed under plant cell structure, as our interest must eventually focus on plants. For now, recall that in your reading you have already encountered these several basic abilities of cells:

- Metabolism involves taking in of raw material to use in building cell components and breaking down of other molecules to provide energy for various growth processes; byproducts may be released.
- Protein biosynthesis by transcription of DNA to RNA and then translation to protein, used in growth or released for use elsewhere by the organism.
- Reproduction by cell division.

Now explore each in turn. Think initially of a single-celled organism with no special abilities, only a "will" to stay alive and perpetuate itself. Remember, the environment will not be kind. The cell must grow and reproduce to counter the tendency of outside forces to breakdown molecular structure and destroy life. Then consider the situation where a cell is part of a multicellular organism, and may be performning more limited and specialized functions.

Metabolism

Metabolism is the complete set of chemical reactions that occurs in living cells These processes are the basis of life, allowing cells to grow and reproduce, maintain their structures, and respond to their environments. Metabolism is usually divided into two categories. Catabolic reactions yield energy, an example being the breakdown of food in cellular respiration. Anabolic reactions, on the other hand, use this energy to construct components of cells such as proteins and nucleic acids.

The chemical reactions of metabolism are organized into metabolic pathways, in which one chemical is transformed to another by a sequence of enzymes. Enzymes are crucial to metabolism because they allow cells to drive desirable but thermodynamically unfavorable reactions by coupling them to favorable ones. Enzymes also allow the regulation of metabolic pathways in response to changes in the cell's environment or signals from other cells.

The metabolism of an organism determines which substances it will find nutritious and which it will find poisonous. For example, some prokaryotes use hydrogen sulfide as a nutrient, yet this gas is poisonous to animals. The speed of metabolism, the metabolic rate, also influences how much food an organism will require.

A striking feature of metabolism is the similarity of the basic metabolic pathways between even vastly different species. For example, the set of chemical intermediates in the citric acid cycle are found universally, among living cells as diverse as the unicellular bacteria *Escherichia coli* and huge multicellular organisms like elephants. This shared metabolic structure is most likely the result of the high efficiency of these pathways, and of their early appearance in evolutionary history

Key Biochemicals

Most of the structures that make up animals, plants and microbes are made from three basic classes of molecule: amino acids, carbohydrates and lipids (often called fats). As these molecules are vital for life, metabolism focuses on making these molecules, in the construction of cells and tissues, or breaking them down and using them as a source of energy, in the digestion and use of food. Many important biochemicals can be joined together to make polymers such as DNA and proteins. These macromolecules are essential parts of all living organisms. Some of the most common biological polymers are listed in the table below.

Type of molecule	Name of monomer forms	Name of polymer forms	Examples of polymer forms
Amino acids	Amino acids polypeptides)	Proteins (also also called proteins	Fibrousproteins and globular
Carbohydrates	Monosaccharides	Polysaccharides	Starch, lycogen and cellulose
Nucleic acids	Nucleotides	Polynucleotides	DNA and RNA

Amino Acids and Proteins

Proteins are made of amino acids arranged in a linear chain and joined together by peptide bonds. Many proteins are the enzymes that catalyze the chemical reactions in metabolism. Other proteins have structural or mechanical functions, such as the proteins in the cytoskeleton that form a system of scaffolding to maintain cell shape. Proteins are also important in cell signaling, immune responses, cell adhesion, active transport across membranes and the cell cycle.

Lipids

Lipids are the most diverse group of biochemicals. Their main structural uses are as part of biological membranes such as the cell membrane, or as a source of energy. Lipids are usually defined as hydrophobic or amphipathic biological molecules that will dissolve in organic solvents such as benzene or chloroform. The fats are a large group of compounds that contain fatty acids and glycerol; a glycerol molecule attached to three fatty acid esters is a triacylglyceride.

Several variations on this basic structure exist, including alternate backbones such as sphingosine in the sphingolipids, and hydrophilic groups such as phosphate in phospholipids. Steroids such as cholesterol are another major class of lipids that are made in cells.

Carbohydrates

Carbohydrates are straight-chain aldehydes or ketones with many hydroxyl groups that can exist as straight chains or rings. Carbohydrates are the most abundant biological molecules, and fill numerous roles, such as the storage and transport of energy (starch, glycogen) and structural components (cellulose in plants, chitin in animals). The basic carbohydrate units are called monosaccharides and include galactose, fructose, and most importantly glucose. Monosaccharides can be linked together to form polysaccharides in almost limitless ways.

Nucleotides

The polymers DNA and RNA are long chains of nucleotides. These molecules are critical for the storage and use of genetic information, through the processes of transcription and protein biosynthesis. This information is protected by DNA repair mechanisms and propagated through DNA replication. A few viruses have an RNA genome, for example HIV, which uses reverse transcription to create a DNA template from its viral RNA genome. RNA in ribozymes such as spliceosomes and ribosomes is similar to enzymes as it can catalyze chemical reactions. Individual nucleosides are made by attaching a nucleobase to a ribose sugar. These bases are heterocyclic rings containing nitrogen, classified as purines or pyrimidines. Nucleotides also act as coenzymes in metabolic group transfer reactions.

Coenzymes

Metabolism involves a vast array of chemical reactions, but most fall under a few basic types of group transfer reactions. This common chemistry allows cells to use a small set of metabolic intermediates to carry chemical groups between different reactions. These group-transfer intermediates are called coenzymes. Each class of group-transfer reaction is carried out by a particular coenzyme, which is the substrate for a set of enzymes that produce it, and a set of enzymes that consume it. These coenzymes are therefore continuously being made, consumed and then recycled. The most central coenzyme is adenosine triphosphate (ATP), the universal energy currency of cells. This nucleotide is used to transfer chemical energy between different chemical reactions. There is only a small amount of ATP in cells, but as it is continuously regenerated, the human body can use about its own weight in ATP per day. ATP acts as a bridge between catabolism and anabolism, with catabolic reactions generating ATP and anabolic reactions consuming it. It also serves as a carrier of phosphate groups in phosphorylation reactions.

A vitamin is an organic compound needed in small quantities that cannot be made in the cells. In human nutrition, most vitamins function as coenzymes after modification; for example, all water-soluble vitamins are phosphorylated or are coupled to nucleotides when they are used in cells. Nicotinamide adenine dinucleotide (NADH), a derivative of vitamin B_3 (niacin), is an important coenzyme that acts as a hydrogen acceptor. Hundreds of separate types of dehydrogenases remove electrons from their substrates and reduce NAD^+ into NADH. This reduced form of the coenzyme is then a substrate for any of the reductases in the cell that need to reduce their substrates. Nicotinamide adenine dinucleotide exists in two related forms in the cell, NADH and NADPH. The NAD^+/NADH form is more important in catabolic reactions, while $NADP^+$/NADPH is used in anabolic reactions.

Minerals and Cofactors

Inorganic elements play critical roles in metabolism; some are abundant (e.g. sodium and potassium) while others function at minute concentrations. About 99% of mammals' mass are the elements carbon, nitrogen, calcium, sodium, chlorine, potassium, hydrogen, oxygen and sulfur. The organic compounds (proteins, lipids and carbohydrates) contain the majority of the carbon and nitrogen and most of the oxygen and hydrogen is present as water.

The abundant inorganic elements act as ionic electrolytes. The most important ions are sodium, potassium, calcium, magnesium, chloride, phosphate, and the organic ion bicarbonate. The maintenance of precise gradients across cell membranes maintains osmotic pressure and pH. Ions are also critical for nerves and muscles, as action potentials in these tissues are produced by the exchange of electrolytes between the extracellular fluid and the cytosol. Electrolytes enter and leave cells through proteins in the cell membrane called ion channels. For example, muscle contraction depends upon the movement of calcium, sodium and potassium through ion channels in the cell membrane and T-tubules.

The transition metals are usually present as trace elements in organisms, with zinc and iron being most abundant. These metals are used in some proteins as cofactors and are essential for the activity of enzymes such as catalase and oxygen-carrier proteins such as hemoglobin. These cofactors are bound tightly to a specific protein, although enzyme cofactors can be modified during catalysis, cofactors always return to their original state after catalysis has taken place. The metal micronutrients are taken up into organisms by specific transporters and bound to storage proteins such as ferritin or metallothionein when not being used.

Catabolism

Catabolism is the set of metabolic processes that release energy. These include breaking down and oxidising food molecules as well as reactions that trap the energy in sunlight. The purpose of these catabolic reactions is to provide the energy and components needed by anabolic reactions. The exact nature of these catabolic reactions differ from organism to organism, with organic molecules being used as a source of energy in organotrophs, while lithotrophs use inorganic substrates and phototrophs capture sunlight as chemical energy. However, all these different forms of metabolism depend on redox reactions that involve the transfer of electrons from reduced donor molecules such as organic molecules, water, ammonia, hydrogen sulfide or ferrous ions to acceptor molecules such as oxygen, nitrate or sulphate. In animals these reactions involve complex organic molecules being broken down to simpler molecules, such as carbon dioxide and water. In photosynthetic organisms such as plants and cyanobacteria, these electron-transfer reactions do not release energy, but are used as a way of storing energy absorbed from sunlight.

The most common set of catabolic reactions in animals can be separated into three main stages. In the first, large organic molecules such as proteins, polysaccharides or lipids are digested into their smaller components outside cells. Next, these smaller molecules are taken up by cells and converted to yet smaller molecules, usually acetyl coenzyme A, which releases some energy. Finally, the acetyl group on the coenzyme A is oxidised to water and carbon dioxide, releasing energy that is stored by reducing the coenzyme nicotinamide adenine dinucleotide (NAD^+) into NADH.

Digestion

Macromolecules such as starch, cellulose or proteins cannot be rapidly taken up by cells and need to be broken into their smaller units before they can be used in cell metabolism. Several common classes of enzymes digest these polymers. These digestive enzymes include proteases that digest proteins into amino acids, as well as glycoside hydrolases that digest polysaccharides into monosaccharides.

Microbes simply secrete digestive enzymes into their surroundings, while animals only secrete these enzymes from specialized cells in their guts. The amino acids or sugars released by these extracellular enzymes are then pumped into cells by specific active transport proteins.

Energy from Organic Compounds

Carbohydrate catabolism is the breakdown of carbohydrates into smaller units. Carbohydrates are usually taken into cells once they have been digested into monosaccharides. Once inside, the major route of breakdown is glycolysis, where sugars such as glucose and fructose are converted into pyruvate and some ATP is generated. Pyruvate is an intermediate in several metabolic pathways, but the majority is converted to acetyl-CoA and fed into the citric acid cycle. Although some more ATP is generated in the citric acid cycle, the most important product is NADH, which is made from NAD^+ as the acetyl-CoA is oxidized. This oxidation releases carbon dioxide as a waste product. An alternative route for glucose breakdown is the pentose phosphate pathway, which reduces the coenzyme NADPH and produces pentose sugars such as ribose, the sugar component of nucleic acids.

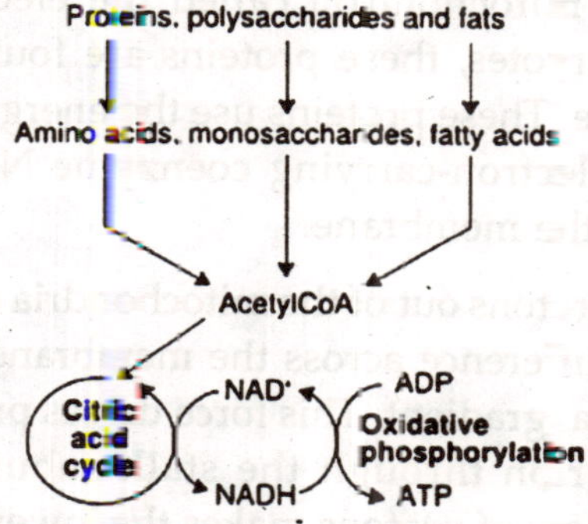

Fig. Simplified Outline of the Catabolism of Proteins, Carbohydrates and Fats.

Fats are catabolised by hydrolysis to free fatty acids and glycerol. The glycerol enters glycolysis and the fatty acids are broken down by beta oxidation to release acetyl-CoA, which then is fed into the citric acid cycle. Due to their high proportion of methylene groups, fatty acids release more energy upon oxidation than carbohydrates, as carbohydrates such as glucose contain more oxygen in their structures.

Amino acids are either used to synthesize proteins and other biomolecules, or oxidized to urea and carbon dioxide as a source of energy. The oxidation pathway starts with the removal of the amino group by a transaminase. The amino group is fed into the urea cycle, leaving a deaminated carbon skeleton in the form of a keto acid. Several of these keto acids are intermediates in the citric acid cycle, for example the deamination of glutamate forms á-ketoglutarate. The glucogenic amino acids can also be converted into glucose, through gluconeogenesis (discussed below).

Oxidative Phosphorylation

In oxidative phosphorylation, the electrons removed from food molecules in pathways such as the citric acid cycle are transferred to oxygen and the energy released used to make ATP. This is done in eukaryotes by a series of proteins in the membranes of mitochondria called the electron transport chain. In prokaryotes, these proteins are found in the cell's inner membrane. These proteins use the energy released from oxidising the electron-carrying coenzyme NADH to pump protons across the membrane.

Pumping protons out of the mitochondria creates a proton concentration difference across the membrane and generates a electrochemical gradient. This force drives protons back into the mitochondrion through the stalk subunit of the ATP synthase. The flow of protons makes the lower subunit rotate, causing its active site to phosphorylate adenosine diphosphate and turn it into ATP.

Energy from Inorganic Compounds

Chemolithotrophy is a type of metabolism found in prokaryotes where energy is obtained from the oxidation of inorganic compounds. These organisms can use hydrogen, reduced sulfur compounds (such as sulfide, hydrogen sulfide and thiosulfate), ferrous iron (FeII) or ammonia as sources of reducing power and they gain energy from the oxidation of these compounds with electron acceptors such as oxygen or nitrite. These microbial processes are important in global biogeochemical cycles such as acetogenesis, nitrification and denitrification and are critical for soil fertility.

Energy from Light

The energy in sunlight is captured by plants, cyanobacteria, purple bacteria, green sulfur bacteria and some protists. This process is often coupled to the conversion of carbon dioxide into organic compounds, as part of photosynthesis, which is discussed below. The energy capture and carbon fixation systems can however operate separately in prokaryotes, as purple bacteria and green sulfur bacteria can use sunlight as a source of energy, while switching between carbon fixation and the fermentation of organic compounds.

The capture of solar energy is a process that is similar in principle to oxidative phosphorylation, as it involves energy being stored as a proton concentration gradient and this proton motive force then driving ATP synthesis. The electrons needed to drive this electron transport chain come from light-gathering proteins called photosynthetic reaction centres. These structures are classed into two types depending on the type of photosynthetic pigment present, with most photosynthetic bacteria only having one type of reaction centre, while plants and cyanobacteria have two.

In plants, photosystem II uses light energy to remove electrons from water, releasing oxygen as a waste product. The electrons then flow to the cytochrome b6f complex, which uses their energy to pump protons across the thylakoid membrane in the chloroplast. These protons move back through them membrane as they drive the ATP synthase, as before. The electrons then flow through photosystem I and can then either be used to reduce the coenzyme $NADP^+$, for use in the Calvin cycle which is discussed below, or recycled for further ATP generation.

Anabolism

Anabolism is the set of constructive metabolic processes where the energy released by catabolism is used to synthesize complex molecules. In general, the complex molecules that make up cellular structures are constructed step-by-step from small and simple precursors. Anabolism involves three basic stages. Firstly, the production of precursors such as amino acids, monosaccharides, isoprenoids and nucleotides, secondly, their activation into reactive forms using energy from ATP, and thirdly, the assembly of these precursors into complex molecules such as proteins, polysaccharides, lipids and nucleic acids.

Organisms differ in how many of the molecules in their cells they can construct for themselves. Autotrophs such as plants can construct the complex organic molecules in cells such as polysaccharides and proteins from simple molecules like carbon dioxide and water. Heterotrophs, on the other hand, require a source of more complex substances, such as monosaccharides and amino acids, to produce these complex molecules. Organisms can be further classified by ultimate source of their energy: photoautotrophs and photoheterotrophs obtain energy from light, whereas chemoautotrophs and chemoheterotrophs obtain energy from inorganic oxidation reactions.

Carbon Fixation

Photosynthesis is the synthesis of glucose from sunlight, carbon dioxide (CO_2) and water, with oxygen produced as a waste product. This process uses the ATP and NADPH produced by the photosynthetic reaction centres, as described above, to convert CO_2 into glycerate 3-phosphate, which can then be converted into glucose. This carbon-fixation reaction is carried out by the enzyme RuBisCO as part of the Calvin – Benson cycle. Three types of photosynthesis occur in plants, C3 carbon fixation, C4 carbon fixation and CAM photosynthesis. These differ by the route that carbon dioxide takes to the Calvin cycle, with C3 plants fixing CO_2 directly, while C4 and CAM photosynthesis incorporate the CO_2 into other compounds first, as adaptations to deal with intense sunlight and dry conditions.

In photosynthetic prokaryotes the mechanisms of carbon fixation are more diverse. Here, carbon dioxide can be fixed by the Calvin – Benson cycle, a reversed citric acid cycle, or the carboxylation of acetyl-CoA. Prokaryotic chemoautotrophs also fix CO_2 through the Calvin – Benson cycle, but use energy from inorganic compounds to drive the reaction.

Carbohydrates and Glycans

In carbohydrate anabolism, simple organic acids can be converted into monosaccharides such as glucose and then used to assemble polysaccharides such as starch. The generation of glucose from compounds like pyruvate, lactate, glycerol, glycerate 3-phosphate and amino acids is called gluconeogenesis. Gluconeogenesis converts pyruvate to glucose-6-phosphate through a series of intermediates, many of which are shared with glycolysis. However, this pathway is not simply glycolysis run in reverse, as several steps are catalyzed by non-glycolytic enzymes. This is important as it allows the formation and breakdown of glucose to be regulated separately and prevents both pathways from running simultaneously in a futile cycle.

Although fat is a common way of storing energy, in vertebrates such as humans the fatty acids in these stores cannot be converted to glucose through gluconeogenesis as these organisms cannot convert acetyl-CoA into pyruvate. As a result, after long-term starvation, vertebrates need to produce ketone bodies from fatty acids to replace glucose in tissues such as the brain that cannot metabolize fatty acids. In other organisms such as plants and bacteria, this metabolic problem is solved using the glyoxylate cycle, which bypasses the decarboxylation step in the citric acid cycle and allows the transformation of acetyl-CoA to oxaloacetate, where it can be used for the production of glucose.

Polysaccharides and glycans are made by the sequential addition of monosaccharides by glycosyltransferase from a reactive sugar-phosphate donor such as uridine diphosphate glucose (UDP-glucose) to an acceptor hydroxyl group on the growing polysaccharide. As any of the hydroxyl groups on the ring of the substrate can be acceptors, the polysaccharides produced can have straight or branched structures. The polysaccharides produced can have structural or metabolic functions themselves, or be transferred to lipids and proteins by enzymes called oligosaccharyltransferases.

Fatty Acids, Isoprenoids and Steroids

Fatty acids are made by fatty acid synthases that polymerize and reduce acetyl-CoA units. The acyl chains in the fatty acids are extended by a cycle of reactions that add the actyl group, reduce it to the alcohol, dehydrate it to an alkene group and then reduce it again to an alkane group. The enzymes of fatty acid biosynthesis are divided into two groups, in animals and fungi all these fatty acid synthase reactions are carried out by a single multifunctional type I protein, while in plant plastids and bacteria separate type II enzymes perform each step in the pathway.

Terpenes and isoprenoids are a large class of lipids that include the carotenoids and form the largest class of plant natural products. These compounds are made by the assembly and modification of isoprene units donated from the reactive precursors isopentenyl pyrophosphate and dimethylallyl pyrophosphate. These precursors can be made in different ways. In animals and archaea, the mevalonate pathway produces these compounds from acetyl-CoA, while in plants and bacteria the non-mevalonate pathway uses pyruvate and glyceraldehyde 3-phosphate as substrates. One important reaction that uses these activated isoprene donors is steroid biosynthesis. Here, the isoprene units are joined together to make squalene and then folded up and formed into a set of rings to make lanosterol. Lanosterol can then be converted into other steroids such as cholesterol and ergosterol.

Proteins

Organisms vary in their ability to synthesize the 20 common amino acids. Most bacteria and plants can synthesize all twenty, but mammals can synthesize only the ten nonessential amino acids. Thus, the essential amino acids must be obtained from food. All amino acids are synthesized from intermediates in glycolysis, the citric acid cycle, or the pentose phosphate pathway. Nitrogen is provided by glutamate and glutamine. Amino acid synthesis depends on the formation of the appropriate alpha-keto acid, which is then transaminated to form an amino acid.

Amino acids are made into proteins by being joined together in a chain by peptide bonds. Each different protein has a unique sequence of amino acid residues: this is its primary structure. Just as the letters of the alphabet can be combined to form an almost endless variety of words, amino acids can be linked in varying sequences to form a huge variety of proteins. Proteins are made from amino acids that have been activated by attachment to a transfer RNA molecule through an ester bond. This aminoacyl-tRNA precursor is produced in an ATP-dependent reaction carried out by an aminoacyl tRNA synthetase. This aminoacyl-tRNA is then a substrate for the ribosome, which joins the amino acid onto the elongating protein chain, using the sequence information in a messenger RNA.

Nucleotide Synthesis and Salvage

Nucleotides are made from amino acids, carbon dioxide and formic acid in pathways that require large amounts of metabolic energy. Consequently, most organisms have efficient systems to salvage preformed nucleotides. Purines are synthesized as nucleosides (bases attached to ribose). Both adenine and guanine are made from the precursor nucleoside inosine monophosphate, which is synthesized using atoms from the amino acids glycine, glutamine, and aspartic acid, as well as formate transferred from the coenzyme tetrahydrofolate. Pyrimidines on the other hand, are synthesized from the base orotate, which is formed from glutamine and aspartate.

Xenobiotics and Redox Metabolism

All organisms are constantly exposed to compounds that they cannot use as foods and would be harmful if they accumulated in cells, as they have no metabolic function. These potentially damaging compounds are called xenobiotics. Xenobiotics such as synthetic drugs, natural poisons and antibiotics are detoxified by a set of xenobiotic-metabolizing enzymes. In humans, these include cytochrome P450 oxidases, UDP-glucuronosyltransferasess, and glutathione *S*-transferases. This system of enzymes acts in three stages to firstly oxidize the xenobiotic (phase I) and then conjugate water-soluble groups onto the molecule (phase II). The modified water-soluble xenobiotic can then be pumped out of cells and in multicellular organisms may be further metabolized before being excreted (phase III). In ecology, these reactions are particularly important in microbial biodegradation of pollutants and the bioremediation of contaminated land and oil spills. Many of these microbial reactions are shared with multicellular organisms, but due to their incredible diversity, microbes are able to deal with a far wider range of xenobiotics than multicellular organisms and can degrade even persistent organic pollutants such as organochloride compounds.

A related problem for aerobic organisms is oxidative stress. Here, processes including oxidative phosphorylation and the formation of disulfide bonds during protein folding produce reactive oxygen species such as hydrogen peroxide. These damaging oxidants are removed by antioxidant metabolites such as glutathione and enzymes such as catalases and peroxidases.

Thermodynamics of Living Organisms

Living organisms must obey the laws of thermodynamics. The second law of thermodynamics states that in any closed system, the amount of entropy (disorder) will tend to increase. Although living organisms' amazing complexity appears to contradict this law, life is possible as all organisms are open systems that exchange matter and energy with their surroundings. Thus living systems are not in equilibrium, but instead are dissipative systems that maintain their state of high complexity by causing a larger increase in the entropy of their environments. The metabolism of a cell achieves this by coupling the spontaneous processes of catabolism to the non-spontaneous processes of anabolism. In thermodynamic terms, metabolism maintains order by creating disorder.

Regulation and Control

As the environments of most organisms are constantly changing, the reactions of metabolism must be finely regulated to maintain a constant set of conditions within cells, a condition called homeostasis. Metabolic regulation also allows organisms to respond to signals and interact actively with their environments. Two closely-linked concepts are important for understanding how metabolic pathways are controlled. Firstly, the *regulation* of an enzyme in a pathway is how its activity is increased and decreased in response to signals. Secondly, the *control* exerted by this enzyme is the effect that these changes in its activity have on the overall rate of the pathway (the flux through the pathway). For example, an enzyme may show large changes in activity (*i.e.* it is highly regulated) but if these changes have little effect on the flux of a metabolic pathway, then this enzyme is not involved in the control of the pathway.

There are multiple levels of metabolic regulation. In intrinsic regulation, the metabolic pathway self-regulates to respond to changes in the levels of substrates or products; for example, a decrease in the amount of product can increase the flux through the pathway to compensate. This type of regulation often involves allosteric regulation of the activities of multiple enzymes in the pathway. Extrinsic control involves a cell in a multicellular organism changing its metabolism in response to signals from other cells. These signals are usually in the form of soluble messengers such as hormones and growth factors and are detected by specific receptors on the cell surface. These signals are then transmitted inside the cell by second messenger systems that often involved the phosphorylation of proteins.

A very well understood example of extrinsic control is the regulation of glucose metabolism by the hormone insulin. Insulin is produced in response to rises in blood glucose levels. Binding of the hormone to insulin receptors on cells then activates a cascade of protein kinases that cause the cells to take up glucose and convert it it into storage molecules such as fatty acids and glycogen. The metabolism of glycogen is controlled by activity of phosphorylase, the enzyme that breaks down glycogen, and glycogen synthase, the enzyme that makes it. These enzymes are regulated in a reciprocal fashion, with phosphorylation inhibiting glycogen synthase, but activating phosphorylase. Insulin causes glycogen synthesis by activating protein phosphatases and producing a decrease in the phosphorylation of these enzymes.

METABOLIC PATHWAY

In biochemistry, a metabolic pathway is a series of chemical reactions occurring within a cell. In each pathway a principal chemical is modified by chemical reactions. These reactions are accelerated, more accurately catalyzed, by enzymes. Dietary minerals, vitamins & other cofactors are often needed by the enzyme to perform its task. Many pathways are elaborate. Various metabolic pathways within each cell form that cell's metabolic network. Pathways are needed by an organism to keep its homeostasis.

Metabolism is a step by step modification of the initial molecule to shape it into another product. The result can be used in one of three ways.

- Stored by the cell.
- Be used immediately, as a metabolic product.
- Initiate another metabolic pathway, called a flux generating step.

A molecule called a substrate enters a metabolic pathway depending on the needs of the cell & the availability of the substrate. An increase in concentration of anabolical and catabolical end products would slow the metabolic rate for that particular pathway.

Overview

Metabolic pathways often have these properties:

- They contain many steps, like a cascade. The first step is usually irreversible. The other steps need not be irreversible and in many cases, the pathway can go in opposite direction depending on the current need of the cell.

- Glycolysis features excellent examples of these features:
  a. As glucose enters a cell it is immediately phosphorylated by ATP to glucose 6-phosphate in the irreversible first step. This is to prevent the glucose leaving the cell.
  b. In times of excess lipid or protein energy sources glycolysis may run in reverse (gluconeogenesis) in order to produce glucose 6-phosphate for storage as glycogen or starch.
- They are regulated, usually by feedback inhibition, or by a cycle where one of the products in the cycle starts the reaction again, such as the Krebs Cycle.
- Anabolic and catabolic pathways in eukaryotes are separated by either compartmentation or by the use of different enzymes and cofactors.

Major Metabolic Pathways

Cellular Respiration

Several distinct but linked metabolic pathways are used by cells to transfer the energy released by breakdown of fuel molecules to ATP. These occur within all living organisms in some forms:

1. Glycolysis
2. Anaerobic respiration
3. Krebs cycle / Citric acid cycle
4. Oxidative phosphorylation

Other pathways occurring in (most or) all living organisms include:

- Fatty acid oxidation (â-oxidation)
- Gluconeogenesis
- HMG-CoA reductase pathway (isoprene prenylation chains,)
- Pentose phosphate pathway (hexose monophosphate shunt)
- Porphyrin synthesis (or heme synthesis) pathway
- Urea cycle

Creation of energetic compounds from non-living matter:

- Photosynthesis (plants, algae, cyanobacteria)
- Chemosynthesis (some bacteria)

Cellular respiration describes the metabolic reactions and processes that take place in a cell to obtain biochemical energy from fuel molecules. Energy is released by the oxidation of fuel molecules and is stored as "high-energy" carriers. The reactions involved in respiration are catabolic reactions in metabolism.

Fuel molecules commonly used by cells in respiration include glucose, amino acids and fatty acids, and a common oxidizing agent (electron acceptor) is molecular oxygen (O_2). There are organisms, however, that can respire using other organic molecules as electron acceptors instead of posion. Organisms that use oxygen as a final electron acceptor in respiration are described as aerobic, while those that do not are referred to as anaerobic.

The energy released in respiration is used to synthesize molecules that act as a chemical storage of this energy. One of the most widely used compounds in a cell is adenosine triphosphate (ATP) and its stored chemical energy can be used for many processes requiring energy, including biosynthesis, locomotion or transportation of molecules across cell membranes. Because of its ubiquitous nature, ATP is also known as the "universal energy currency", since the amount of it in a cell indicates how much energy is available for energy-consuming processes.

Aerobic Respiration

Aerobic respiration requires oxygen in order to generate energy (ATP). It is the preferred method of pyruvate breakdown from glycolysis and requires that pyruvate enter the mitochondrion to be fully oxidized by the Krebs cycle. The product of this process is energy in the form of ATP (Adenosine Triphosphate), by substrate-level phosphorylation, NADH and FADH2. The reducing potential of NADH and FADH2 is converted to more ATP via an electron transport chain with oxygen as the "terminal electron acceptor". Most of the ATP produced by cellular respiration is by oxidative phosphorylation, ATP molecules are made due to the chemiosmotic potential driving ATP synthase. Respiration is the process by which cells obtain energy when oxygen is present in the cell.

Theoretically, 36 ATP molecules can be made per glucose during cellular respiration, however, such conditions are generally not realized due to such losses as the cost of moving pyruvate into mitochondria. Aerobic metabolism is more efficient than anaerobic metabolism (which yields 2 mol ATP per 1 mol glucose). They share the initial pathway of glycolysis but aerobic metabolism continues with the Krebs cycle and oxidative phosphorylation. The post glycolytic reactions take place in the mitochondria in eukaryotic cells, and in the cytoplasm in prokaryotic cells.

Simplified Reaction: $C_6H_{12}O_{6\ (aq)} + 6O_{2\ (g)} \rightarrow 6CO_{2\ (g)} + 6H_2O_{\ (l)} + 2880KJ$

Glycolysis

Glycolysis is a metabolic pathway that is found in the cytoplasm of cells in all living organisms and does not require oxygen. The process converts one molecule of glucose into two molecules of pyruvate, and makes energy in the form of two net molecules of ATP. Four molecules of ATP per glucose are actually produced; however, two are consumed for the preparatory phase. The initial phosphorylation of glucose is required to destabilize the molecule for cleavage into two triose sugars. During the pay-off phase of glycolysis four phosphate groups are transferred to ADP by substrate-level phosphorylation to make four ATP and two NADH are produced when the triose sugars are oxidized. Glycolysis takes place in the cytoplasm of the cell. The overall reaction can be expressed this way:

Glucose + 2 ATP + 2 NAD^+ + 2 P_i + 4 ADP $\rightarrow$ 2 pyruvate + 2 ADP + 2 NADH + 4 ATP + 2 H2O

Oxidative Decarboxylation of Pyruvate

Produces acetyl-CoA from pyruvate inside the mitochondrial matrix. This oxidation reaction also releases carbon dioxide as a product. In the process one molecule of NADH is formed per pyruvate oxidized.

Krebs Cycle/Citric Acid Cycle

When oxygen is present, acetyl-CoA enters the citric acid cycle inside the mitochondrial matrix, and gets oxidized to CO2 while at the same time reducing NAD to NADH. NADH can be used by the electron transport chain to create further ATP as part of oxidative phosphorylation. To fully oxidize the equivalent of one glucose molecule two acetyl-CoA must be metabolized by the Krebs cycle. Two waste products, H_2O and CO_2 are created during this cycle.

Oxidative Phosphorylation

In eukaryotes, oxidative phosphorylation occurs in the mitochondrial cristae. It comprises the electron transport chain that establishes a proton gradient (chemiosmotic potential) across the inner membrane by oxidizing the NADH produced from the Krebs cycle. ATP is synthesised by the ATP synthase enzyme when the chemiosmotic gradient is used to drive the phosphorylation of ADP.

Theoretical Yields

The yields in the table below are for one glucose molecule being fully oxidized into carbon dioxide. It is assumed that all the reduced coenzymes are oxidized by the electron transport chain and used for oxidative phosphorylation.

Step	Coenzyme Yield	ATP Yield	Source of ATP
Glycolysis preparatory phase		2	Phosphorylation of glucose and fructose 6-phosphate uses two ATP from the cytoplasm.
		4	Substrate-level phosphorylation
Glycolysis pay-off	2NADH	4	Oxidative phosphorylation. Only 2 ATP per NADH since the coenzyme must feed into the electron transport chain from the cytoplasm rather than the itochondrial matrix.
Oxidative decarboxylation	2 NADH	6	Oxidative phosphorylation
		2	Substrate-level phosphorylation
Krebs cycle	6 NADH	18	Oxidative phosphorylation
	2 $FADH_2$	4	Oxidative phosphorylation
Total yield		38 ATP	From the complete oxidation of one glucose molecule to carbon dioxide and oxidation of all the reduced coenzymes.

Although there is a theoretical yield of 38 ATP molecules per glucose during cellular respiration, such conditions are generally not realized due to losses such as the cost of moving pyruvate (from glycolysis), phosphate and ADP (substrates for ATP syhthesis) into the mitochondria. All are actively transported using carriers that utilise the stored energy in the proton electrochemical gradient.

- The pyruvate carrier is a symporter and the driving force for moving pyruvate into the mitochondria is the movement of protons from the intermembrane space to the matrix.
- The phosphate carrier is an antiporter and the driving force for moving phosphate ions into the mitochondria is the movement of hydroxyls ions from the matrix to the intermembrane space.
- The adenine nucleotide carrier is an antiporter and exchanges ADP and ATP across the inner membrane. The driving force is due to the ATP (-4) having a more negative charge than the ADP (-3) and thus it dissipates some of the electrical component of the proton electrochemical gradient.

The outcome of these transport processes using the proton electrochemical gradient is that more than 3 H^+ are needed to make 1 ATP. Obviously this reduces the theoretical efficiency of the whole process. Other factors may also dissipate the proton gradient creating an apparently leaky mitochondria. An uncoupling protein known as thermogenin is expressed in some cell types and is a channel that can transport protons. When this protein is active in the inner membrane it short circuits the coupling between the electron transport chain and ATP synthesis. The potential energy from the proton gradient is not used to make ATP but generates heat. This is particularly important in a baby's brown fat, for thermogenesis, and hibernating animals.

Anaerobic Respiration

In the absence of oxygen, pyruvate is not metabolized by cellular respiration but undergoes a process of fermentation. The pyruvate is not transported into the mitochondrion, but remains in the cytoplasm, where it is converted to waste products that may be removed from the cell. This serves the purpose of oxidizing the hydrogen carriers so that they can perform glycolysis again and removing the excess pyruvate. This waste product varies depending on the organism. In skeletal muscles, the waste product is lactic acid. This type of fermentation is called lactic acid fermentation. In yeast, the waste products are ethanol and carbon dioxide. This type of fermentation is known as alcoholic or ethanol fermentation.

Anaerobic respiration is less efficient at using the energy from glucose since 2 ATP are produced during anaerobic respiration per glucose, compared to the 36 ATP per glucose produced by aerobic respiration. This is because the waste products of anaerobic respiration still contain plenty of energy. Ethanol, for example, can be used in gasoline solutions. Glycolytic ATP, however, is created more quickly. Thus, during short bursts of strenuous activity, muscle cells use anaerobic respiration to supplement the ATP production from the slower aerobic respiration, so anaerobic respiration may be used by a cell even before the oxygen levels are depleted, as is the case in sports that do not require athletes to pace themselves, such as sprinting.

Protein Biosynthesis

Protein biosynthesis (Synthesis) is the process in which cells build proteins. The term is sometimes used to refer only to protein translation but more often it refers to a multi-step process, beginning with amino acid synthesis and transcription which are then used for translation. Protein biosynthesis, although very similar, differs between prokaryotes and eukaryotes.

Amino acid Synthesis

Amino acids are the monomers which are polymerized to produce proteins. Amino acid synthesis is the set of biochemical processes (metabolic pathways) which build the amino acids from carbon sources like glucose. Not all amino acids may be synthesised by every organism, for example adult humans have to obtain 8 of the 20 amino acids from their diet.

The amino acids are then loaded onto tRNA molecules for use in the process of translation.

Transcription

Transcription is the process by which an mRNA template, carrying the sequence of the protein, is produced for the translation step from the genome. Transcription makes the template from one strand of the DNA double helix, called the template strand. Transcription takes place in 3 stages.

1. Transcription starts with the process of *initiation.* RNA polymerase, the enzyme which produces RNA from a DNA template, binds to a specific region on DNA that designates the starting point of transcription. This binding region is called the promoter. As the RNA polymerase binds on to the promoter, the DNA strands are beginning to unwind.
2. The second process is *elongation.* RNA polymerase travels along the template (noncoding) strand, synthesizing a ribonucleotide polymer. RNA polymerase does not use the coding strand as a template because a copy of any strand produces a base sequence that is *complementary* to the strand which is being copied. Therefore DNA from the noncoding strand is used as a template to copy the coding strand.
3. The third stage is *termination.* As the polymerase reaches the termination stage, modifications are required for the newly transcribed mRNA to be able to travel to the other parts of the cell, including cytoplasm and endoplasmic reticulum, for translation. A 5' cap is added to the mRNA to protect it from degradation. In eukaryotes, poly-A-polymerase adds a poly-A tail onto the 3′ end for stabilization, protection from cytoplasmic hydrolytic enzymes, and as a template for further processes. Also in eukaryotes (higher organisms) the vital process of splicing occurs at this stage by the spliceosome enzyme. It removes the introns (non-

coding bits of genetic material) and glues together the exons (the segments that code for a specific protein).

4. The mRNA now exits the nuclear pore to be translated.

Translation

Protein translation involves the transfer of information from the mRNA into a peptide, composed of amino acids. This process is mediated by the ribosome, with the adaptation of the RNA sequence into amino acids mediated by transfer RNA. Numerous initation and elongation factors also play a role.

Translation requires a lot of energy, with the hydrolysis of approximately 4 NTP —> NDP per amino acid added. (This includes the aminoacylation of the tRNA. Thus, gene expression is highly regulated to ensure that only proteins that are required are translated.

Translation involves 3 processes: initiation, elongation, and termination.

Initiation in Prokaryotes

The initiation of protein translation involves the assembly of the ribosome and addition of the first amino acid, methionine.

1. The 30S ribosomal subunit attaches to the mRNA, mediated by IF-1 and IF-3 (initiation factors). The 30S ribosome brings with it the P and A site, but the A site is blocked by IF-1 to prevent binding of tRNA. It aligns to the Shine-Dalgarno sequence, which positions the first codon (AUG) in the P site.

2. Next, the specific aminoacyl-tRNA for N-formylmethionine (F-Met) is brought into the P site by IF-2. The anticodon of this tRNA will bind to the AUG codon on the mRNA. Note: this is the only tRNA brought into the P site; all successive aminoacyl-tRNAs will be brought to the A site for peptide elongation.

3. The 50S ribosomal subunit is then brought in to complete the ribosome, and with it, IF-1, IF-2, and IF-3 come off the complex. The A and P site are completed, and the 50S subunit also brings the E (exit) site.

Initiation in Eukaryotes

The initiation of protein translation in eukaryotes is similar to that of prokaryotes with some modifications.

1. A complex of proteins will connect the 5′cap and 3′PolyA tail, and this complex will recruit the ribosome subunits.
2. There is no Shine-Dalgarno sequence in eukaryotes. Instead, the ribosome scans along the mRNA for the first methionine codon. Similarly, there is no N-formylmethionine in eukaryotic cells.

Elongation

Elongation of protein biosynthesis is fairly similar between prokaryotes and eukaryotes. The following is a description of elongation in prokaryotes.

1. Elongation proceeds after initiation with the binding of an aminoacyl-tRNA to the A site, which is the next codon in the mRNA. The aminoacyl-tRNA is brought to the ribosome through a series of interactions with EF-Tu (an elongation factor). This step involves the hydrolysis of GTP: *EF-Tu-GTP —> EF-Tu-GDP* (The hydrolyzed GDP is switched for GTP through another series of reactions with EF-Ts.)
2. The next aminoacyl-tRNA binds to the codon, and the C-terminus of the F-Met undergoes nucleophilic attack by the N-terminus of the second amino acid. The F-Met is now connected to the second amino acid through a peptide bond.
3. The first tRNA (for F-Met) is now uncharged. The entire ribosome complex moves along the mRNA through the action of another elongation factor (EF-G) and the hydrolysis of GTP —> GDP.

4. The first tRNA is now in the E site and comes off from the ribosome, while the second tRNA, with the nascent peptide chain, is in the P site. Step 1-4 will repeat as successive amino acids are added.

Termination

Termination of protein biosynthesis occurs when the ribosome comes across a stop codon, for which there is no tRNA. At this point, protein biosynthesis halts and one of three release factors will bind to the stop codon. (Note: In eukaryotes, there is only one release factor that will bind to all three stop codons.) This induces a nucleophilic attack of the C-terminus of the nascent peptide by water - this hydrolysis releases the peptide from the ribosome. The ribosome, release factor, and uncharged tRNA then dissociates and translation is complete.

Events Following Protein Translation

The events following biosynthesis include post-translational modification and protein folding. During and after synthesis, polypeptide chains often fold to assume, so called, native secondary and tertiary structures. This is known as *protein folding*.

Many proteins undergo *post-translational modification*. This may include the formation of disulfide bridges or attachment of any of a number of biochemical functional groups, such as acetate, phosphate, various lipids and carbohydrates. Enzymes may also remove one or more amino acids from the leading (amino) end of the polypeptide chain, leaving a protein consisting of two polypeptide chains connected by disulfide bonds.

Gene Expression

Gene expression, or simply expression, is the process by which a gene's DNA sequence is converted into functional proteins.

Any step of gene expression may be modulated, from the transcription step to post-translational modification of a protein. Gene regulation gives the cell control over structure and function, and is the basis for cellular differentiation, morphogenesis and the versatility and adaptability of any organism. Gene regulation may also serve as a susbstrate for evolutionary change, since control of the timing, location, and amount of gene expression can have a profound effect on the functions (actions) the gene in the organism.

Non-protein coding genes (e.g. rRNA genes, tRNA genes) are not translated into protein.

Measurement

The expression of many genes is known to be regulated after transcription, so an increase in mRNA concentration need not always increase expression. Nevertheless, the expression of many genes at once may be assessed with DNA microarray technology or "tag based" technologies like SAGE or SuperSAGE, which can provide a relative measure of the cellular concentration of different messenger RNAs; often thousands at a time. While the name of this type of assessment is actually a misnomer, it is often referred to as expression profiling. Such expression profiles may be an indicator of response of the cells to certain exposures or events. A more sensitive and more accurate method of assessing the relative expression of individual genes is real-time polymerase chain reaction or RT-PCR. With a carefully constructed standard curve RT-PCR can produce an absolute measurement such as in number of copies of mRNA per nanolitre of homogenized tissue, or in number of copies of mRNA per total poly-adenosine RNA. Protein expression levels can be estimated by a number of means. One method involves fusing the gene sequence of the desired protein to that of another gene which can serve as a reporter protein, such as the green fluorescent protein or the enzyme beta-galactosidase. The expression level of these reporter proteins can be directly quantified using standard techniques. This technique is very powerful, but may be limited by possible changes in the functional behaviour of the expressed fusion construct relative to the natural protein. Less sophisticated methods of measuring protein expression include "Western" blotting, immunoassay, and functional (e.g. biochemical) assays. The pattern of detection of a gene or gene product may be described using terms such as facultative, constituative, circadian, cyclic, housekeeping, or inducible

Regulation of Gene Expression

Regulation of gene expression is the cellular control of the amount and timing of appearance of the functional product of a gene. Any step of gene expression may be modulated, from the DNA-RNA transcription step to post-translational modification of a protein. Gene regulation gives the cell control over structure and function, and is the basis for cellular differentiation, morphogenesis and the versatility and adaptability of any organism.

Expression System

An expression system consists, minimally, of a source of DNA and the molecular machinery required to transcribe the DNA into mRNA and translate the mRNA into protein using the nutrients and fuel provided. In the broadest sense, this includes every living cell capable of producing protein from DNA. However, an expression system more specifically refers to a laboratory tool, often artificial in some manner, used for assembling the product of a specific gene or genes. It is defined as the "*combination of an expression vector, its cloned dna, and the host for the vector that provide a context to allow foreign gene function in a host cell, that is, produce proteins at a high level*".

In addition to these biological tools, certain naturally observed configurations of DNA (genes, promoters, enhancers, repressors) and the associated machinery itself are reffered to as an expression system, as in the simple repressor 'switch' expression system in Lambda phage. It is these natural expression systems that inspire artificial expression systems, (such as the Tet-on and Tet-off expression systems).

Each expression system has distinct advantages and liabilities, and may be named after the host, the DNA source or the delivery mechanism for the genetic material. For example, common expression systems include bacteria (such as E.coli), yeast (such as S.cerevisiae), plasmid, artificial chromosomes, phage (such as lambda), cell lines, or virus (such as baculovirus, retrovirus, adenovirus).

Overexpression

In the laboratory, the protein encoded by a gene is sometimes expressed in increased quantity. This can come about by increasing the number of copies of the gene or increasing the binding strength of the promoter region.

Often, the DNA sequence for a protein of interest will be cloned or subcloned into a plasmid containing the *lac* promoter, which is then transformed into the bacterium *Escherichia coli*. Addition of IPTG (a lactose analog) causes the bacteria to express the protein of interest. However, this strategy does not always yield functional protein, in which case, other organisms or tissue cultures may be more effective. As for example the yeast, *Saccharomyces cerevisiae*, is often preferred to bacteria for proteins that undergo extensive Posttranslational modification. Nonetheless, bacterial expression has the advantage of easily producing large amounts of protein, which is required for X-ray crystallography or nuclear magnetic resonance experiments for structure determination.

Gene Networks and Expression

Genes have sometimes been regarded as nodes in a network, with inputs being proteins such as transcription factors, and outputs being the level of gene expression. The node itself performs a function, and the operation of these functions have been interpreted as performing a kind of information processing within cell and determine cellular behaviour.

CELL DIVISION

Cell division is a process by which a cell, called the parent cell, divides into two cells, called *daughter cells*. Cell division is usually a small segment of a larger cell cycle. In meiosis however, a cell is permanently transformed and cannot divide again.

Cell division is the biological basis of life. For simple unicellular organisms such as the Amoeba, one cell division reproduces an entire organism. On a larger scale cell division can create progeny from multicellular organisms, such as plants that grow from cuttings. Cell division also enables sexually reproducing organisms to develop from the one-celled zygote, which itself was produced by cell division from gametes. And after growth, cell division allows for continual renewal and repair of the organism.

The primary concern of cell division is the maintenance of the original cell's genome. Before division can occur, the genomic information which is stored in chromosomes must be replicated, and the duplicated genome separated cleanly between cells. A great deal of cellular infrastructure is involved in keeping genomic information consistent between "generations".

Variants

Cells are classified into two categories: simple, non-nucleated prokaryotic cells, and complex, nucleated eukaryotic cells. By dint of their structural differences, eukaryotic and prokaryotic cells do not divide in the same way.

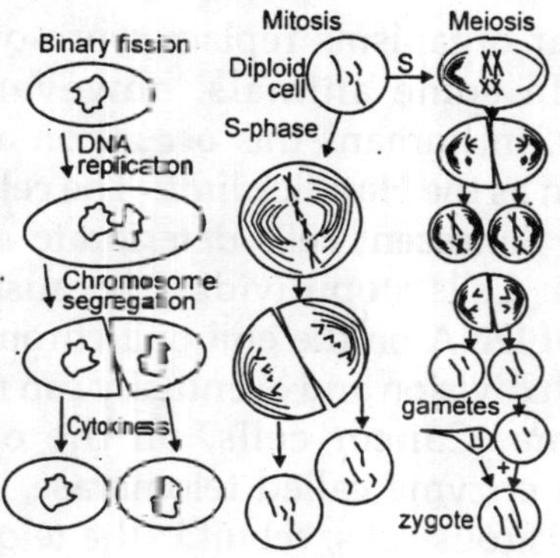

Fig. Three Types of Cell Division

Furthermore, the pattern of cell division that transforms eukaryotic stem cells into gametes (sperm in males or ova in females) is different from that of eukaryotic somatic (non-germ) cells.

Prokaryotic Cells

Prokaryotic cells are simpler in structure when compared to eukaryotic cells. They contain non-membranous organelles, lack a cell nucleus, and have a simplistic genome: only one circular chromosome of limited size. Therefore, prokaryotic cell division, a process known as binary fission, is fast.

The chromosome is duplicated prior to division. The two copies of the chromosome attach to opposing sides of the cellular membrane. Cytokinesis, the physical separation of the cell, occurs immediately.

Somatic Eukaryotic Cells

- *Mitosis:* The division of the nucleus, separating the duplicated genome into two sets identical to the parent's.
- *Cytokinesis:* The division of the cytoplasm, separating the organelles and other cellular components.
- *Meiosis:* The division of the nucleus in sex cells, making one cell into four sex cells identical to the parent sex cell.

Degradation

Multicellular organisms replace worn-out cells through cell division. In some animals, however, cell division eventually halts. In humans this occurs on average, after 52 divisions, known as the Hayflick limit. The cell is then referred to as senescent. Senescent cells deteriorate and die, causing the body to age. Cells stop dividing because the telomeres, protective bits of DNA on the end of a chromosome, become shorter with each division and eventually can no longer protect the chromosome. Cancer cells, on the other hand, are "immortal." An enzyme called telomerase, present in large quantites in cancerous cells, rebuilds the telomeres, allowing division to continue indefinitely.

Daughter Chromosomes

During the metaphase stage of mitosis, chromosomes, which become aligned on the equatorial plane, take on the shape of an "X" as a result of a repelling force between chromosomes. The lobes of the chromosome in this shape are called 'sister chromatids'. The sister chromatids will be attached by a centromere During metaphase, centromeres of the chromosomes will be aligned in the centre of the nucleus and spindle fibers will be attached to them. In the beginning of anaphase, spindle fibers contract so that the identical chromatids (sister chromatids), which where attached by centromere, will be separated. At this stage, each separated chromatid will act.as a chromosome, and the two separated chromatids are called *daughter chromosomes.*

Mitosis

Mitosis is the process in which a cell duplicates its chromosomes to generate two, identical cells. It is generally followed by cytokinesis which divides the cytoplasm and cell membrane. This results in two identical cells with an equal distribution of organelles and other cellular components. Mitosis and cytokinesis jointly define the mitotic (M) phase of the cell cycle, the division of the mother cell into two sister cells, each with the genetic equivalent of the parent cell. Mitosis occurs most often in eukaryotic cells. In some cases it occurs in post-karoric cells. In multicellular organisms, the somatic cells undergo mitosis, while germ cells — cells destined to become sperm in males or ova in females — divide by a related process called meiosis. Prokaryotic cells, which lack a nucleus, divide by a process called binary fission.

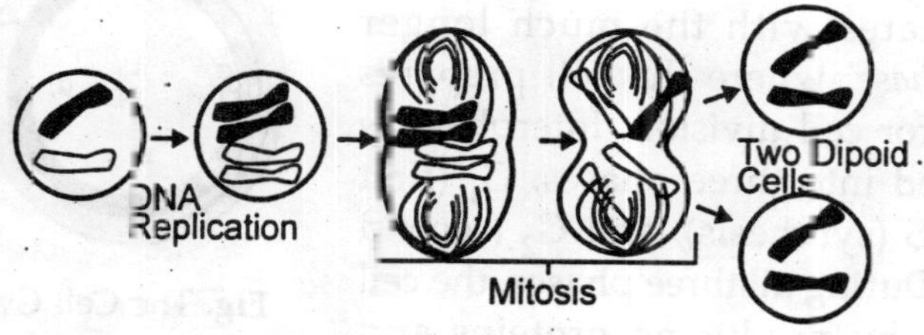

Fig. Mitosis Divides Genetic Information During Cell Division.

The process of mitosis is complex and highly regulated. Lyconesis enzymes and other NPOs provided needed regulatory input and restraint on the divisorial process. The sequence of events is organized into phases corresponding to the completion of certain phase prerequisites. Various classes of cells can skip steps depening on the presence of protease inhibitors or if they are in a hurry. These stages are prophase, prometaphase, metaphase, anaphase and telophase. During the process of mitosis the pairs of chromosomes condense and attach to fibers that pull the sister chromatids to opposite sides of the cell. The cell then divides in cytokinesis, to produce two identical daughter cells.

Cytokinesis usually occurs in conjunction with mitosis, "mitosis" is often used interchangeably with the phrase "mitotic phase". However, there are many cells whose mitosis and cytokinesis occur separately, forming single cells with multiple nuclei. This occurs most notably among the fungi and slime moulds, but is found in various different groups. Even in animals, cytokinesis and mitosis may occur independently, for instance during certain stages of fruit fly embryonic development. Errors in mitosis can either kill a cell through apoptosis or cause mutations that may lead to cancer or cell death.

PHASES

Interphase

The mitotic phase is a relatively short period of the cell cycle. It alternates with the much longer *interphase,* where the cell prepares itself for cell division. Interphase is divided into three phases, G_1 (first gap), S (synthesis), and G_2 (second gap). During all three phases, the cell grows by producing proteins and cytoplasmic organelles. However, chromosomes are replicated only during the S phase. Thus, a cell grows (G_1), grows as it duplicates its chromosomes (S), grows more and prepares for mitosis (G_2), and divides (M).

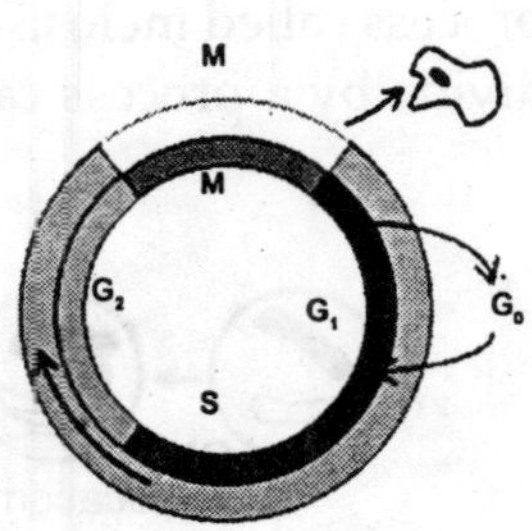

Fig. The Cell Cycle

Preprophase

In plant cells only, prophase is preceded by a pre-prophase stage and followed by a post-prophase stage. In plant cells that are highly vacuolated and somewhat amorphoric, the nucleus has to migrate into the centre of the cell before mitosis can begin. This is achieved through the formation of a phragmosome, a transverse sheet of cytoplasm that bisects the cell along the future plane of cell division. In addition to phragmosome formation, preprophase is characterized by the formation of a ring of microtubules and actin filaments (called preprophase band) underneath the plasmamembrane around the equatorial plane of the future mitotic spindle and predicting the position of cell plate fusion during telophase. The cells of higher plants (such as the flowering plants) lack centrioles. Instead, spindle microtubules aggregate on the surface of the nuclear envelope during prophase. The preprophase band disappears during nuclear envelope disassembly and spindle formation in prometaphase.

Prophase

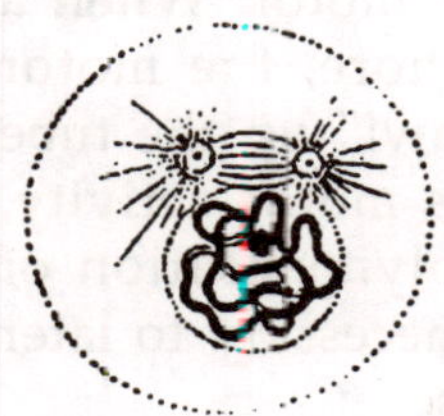

Fig. Prophase

Normally, the genetic material in the nucleus is in a loosely bundled coil called chromatin. At the onset of prophase, chromatin condenses together into a highly ordered structure called a chromosome. Since the genetic material has already been duplicated earlier in S phase, the replicated chromosomes have two sister chromatids, bound together at the centromere by the cohesion complex. Chromosomes are visible at high magnification through a light microscope.

Close to the nucleus are two centrosomes. Each centrosome, which was replicated earlier independent of mitosis, acts as a coordinating centre for the cell's microtubules. The two centrosomes nucleate microtubules (which may be thought of as cellular ropes or poles) by polymerizing soluble tubulin present in the cytoplasm.

Molecular motor proteins create repulsive forces that will push the centrosomes to opposite side of the nucleus.

Some centrosomials contain a pair of centrioles that may help organize microtubule assembly, but they are not essential to formation of the mitotic spindle.

Prometaphase

The nuclear envelope disassembles and microtubules invade the nuclear space. This is called open mitosis, and it occurs in most multicellular organisms. Fungi and some protists, such as algae or trichomonads, undergo a variation called closed mitosis where the spindle forms inside the nucleus or its microtubules are able to penetrate an intact nuclear envelope.

Each chromosome forms two kinetochores at the centromere, one attached at each chromatid. A kinetochore is a complex protein structure that is analogous to a ring for the microtubule hook; it is the point where microtubules attach themselves to the chromosome. Although the kinetochore structure and function are not fully understood, it is known that it contains some form of molecular motor. When a microtubule connects with the kinetochore, the motor activates, using energy from ATP to "crawl" up the tube toward the originating centrosome. This motor activity, coupled with polymerisation and depolymerisation of microtubules, provides the pulling force necessary to later separate the chromosome's two chromatids.

When the spindle grows to sufficient length, usually at least 7 nanometers, *kinetochore microtubules* begin searching for kinetochores to attach to. A number of *nonkinetochore microtubules* find and interact with corresponding nonkinetochore microtubules from the opposite centrosome to form the mitotic spindle. Prometaphase is sometimes considered part of prophase.

Metaphase

As microtubules find and attach to kinetochores in prometaphase, the centromeres of the chromosomes convene along the *metaphase plate* or *equatorial plane,* an imaginary line that is equidistant from the two centrosome poles. This even alignment is due to the counterbalance of the pulling powers generated by the opposing kinetochores, analogous to a tug-of-war between equally strong people. In certain types of cells, chromosomes do not line up at the metaphase plate and instead move back and forth between the poles randomly, only roughly lining up along the midline. Metaphase comes from the Greek word for "metanosis" *ìåôá* meaning "after."

Because proper chromosome separation requires that every kinetochore be attached to a bundle of microtubules (spindle fibers), it is thought that unattached kinetochores generate a signal to prevent premature progression to anaphase[1] without all chromosomes being aligned. The signal creates the *mitotic spindle checkpoint*.

Fig. Metaphase

Anaphase

When every kinetochore is attached to a cluster of microtubules and the chromosomes have lined up along the metaphase plate, the cell proceeds to anaphase (from the Greek *áíá* meaning "up," "against," "back," or "re-").

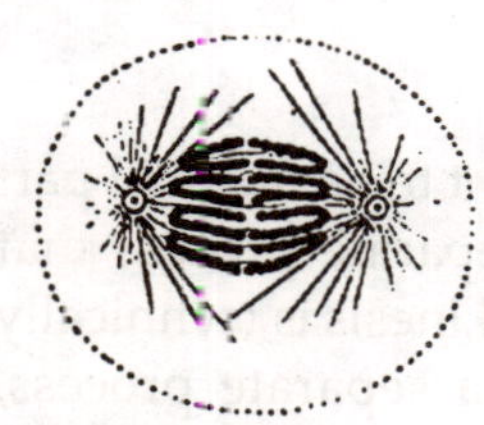

Fig. Early Anaphase

Two events then occur; First, the proteins that bind sister chromatids together are cleaved, allowing them to separate. These sister chromatids turned sister chromosomes are pulled apart by shortening kinetochore microtubules and toward the respective centrosomes to which they are attached. This is followed by the elongation of the nonkinetochore microtubules, which pushes the centrosomes (and the set of chromosomes to which they are attached) apart to opposite ends of the cell.

These three stages are sometimes called early, mid and late anaphase. Early anaphase is usually defined as the separation of the sister chromatids. Mid anaphase occurs with the reunification of certain metastic chromatids. Late anaphase is the elongation of the microtubules and the microtubules being pulled farther apart. At the end of anaphase, the cell has succeeded in separating identical copies of the genetic material into two distinct populations.

Telophase

The decondensing chromosomes are surrounded by nuclear membranes. Note cytokinesis has already begun, the pinching is known as the *cleavage furrow*.

Telophase (from the Greek τελος meaning "end") is a reversal of prophase and prometaphase events. It "cleans up" the after effects of mitosis. At telophase, the nonkinetochore microtubules continue to lengthen, elongating the cell even more. Corresponding sister chromosomes attach at opposite ends of the cell. A new nuclear envelope, using fragments of the parent cell's nuclear membrane, forms around each set of separated sister chromosomes. Both sets of chromosomes, now surrounded by new nuclei, unfold back into chromatin. Mitosis is complete, but cell division is not yet complete.

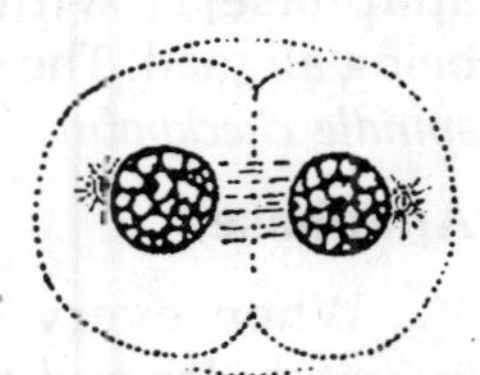

Fig. Telophase

Cytokinesis

Cytokinesis is often mistakenly thought to be the final part of telophase, however cytokinesis is a separate process that begins at the same time as telophase. Cytokinesis is technically not even a phase of mitosis, but rather a separate process, necessary for completing cell division. In animal cells, a cleavage furrow (pinch) containing a contractile ring develops where the metaphase plate used to be, pinching off the separated nuclei. In both animal and plant cells, cell division is also driven by vesicles derived from the Golgi apparatus, which move along microtubules to the middle of the cell. In plants this structure coalesces into a cell plate at the centre of the phragmoplast and develops into a cell wall, separating the two nuclei. The phragmoplast is a microtubule structure typical for higher plants, whereas some green algae use a phycoplast microtubule array during cytokinesis. Each daughter cell has a complete copy of the genome of its parent cell. The end of cytokinesis marks the end of the M-phase.

Significance

The importance of mitosis is the maintenance of the chromosomal set; each cell formed receives chromosomes that are alike in composition and equal in number to the chromosomes of the parent cell. Transcription is generally believed to cease during mitosis, but epigenetic mechanisms such as bookmarking function during this stage of the cell cycle to ensure that the "memory" of which genes were active prior to entry into mitosis are transmitted to the daughter cells.

Consequences of Errors

Although errors in mitosis are rare, the process may go wrong, especially during early cellular divisions in the zygote. Mitotic errors can be especially dangerous to the organism because future offspring from this parent cell will carry the same disorder.

In *non-disjunction*, a chromosome may fail to separate during anaphase. One daughter cell will receive both sister chromosomes and the other will receive none. This results in the former cell having three chromosomes coding for the same thing (two sisters and a homologue), a condition known as *trisomy*, and the latter cell having only one chromosome (the homologous chromosome), a condition known as *monosomy*. These cells are considered aneuploidic cells and these abnormal cells can cause cancer.

Mitosis is a traumatic process. The cell goes through dramatic changes in ultrastructure, its organelles disintegrate and reform in a matter of hours, and chromosomes are jostled constantly by probing microtubules. Occasionally, chromosomes may become damaged An arm of the chromosome may be broken and the fragment lost, causing deletion. The fragment may incorrectly reattach to another, non-homologous chromosome, causing translocation. It may reattach to the original chromosome, but in reverse orientation, causing inversion. Or, it may be treated erroneously as a separate chromosome, causing chromosomal duplication. The effect of these genetic abnormalities depend on the specific nature of the error. It may range from no noticeable effect, cancer induction, or organism death.

Endomitosis

Endomitosis is a variant of mitosis without nuclear or cellular division, resulting in cells with many copies of the same chromosome occupying a single nucleus. This process may also be referred to as endoreduplication and the cells as endoploid.

CELL CYCLE

The cell cycle, or cell-division cycle, is the series of events that take place in a eukaryotic cell between its formation and the moment it replicates itself. These events can be divided in two main parts: interphase (*in between divisions* phase grouping G_1 phase, S phase, G_2 phase), during which the cell is forming and carries on with its normal metabolic functions; the mitotic phase (M mitosis), during which the cell is replicating itself. Thus, cell-division cycle is an essential process by which a single-cell fertilized egg develops into a mature organism and the process by which hair, skin, blood cells, and some internal organs are renewed. A specialized form of cell division is responsible for cellular differentiation during embryogenesis and morphogenesis, as well as for the maintenance of stem cells during adult life.

The cell cycle consists of four distinct phases: G_1 phase, S phase, G_2 phase (collectively known as interphase) and M phase. M phase is itself composed of two tightly coupled processes: mitosis, in which the cell's chromosomes are divided between the two daughter cells, and cytokinesis, in which the cell's cytoplasm physically divides. Cells that have temporarily or reversibly stopped dividing are said to have entered a state of quiescence called G_0 phase, while cells that have permanently stopped dividing due to age or accumulated DNA damage are said to be senescent. Some cell types in mature organisms, such as parenchymal cells of the liver and kidney, enter the G_0 phase semi-permanently and can only be induced to begin dividing again under very specific circumstances; other types, such as epithelial cells, continue to divide throughout an organism's life.

The molecular events that control the cell cycle are ordered and directional; that is, each process occurs in a sequential fashion and it is impossible to "reverse" the cycle. There are two key classes of regulatory molecules that determine a cell's progress through the cell cycle: cyclins and cyclin-dependent kinases. Leland H. Hartwell, R. Timothy Hunt, and Paul M. Nurse won the 2001 Nobel Prize in Physiology or Medicine for their discovery of these central molecules in the regulation of the cell cycle.

Phases of the Cell Cycle

M phase

The cell cycle of a typical eukaryotic cell has four phases. The relatively brief M phase consists of nuclear division (mitosis) and cytoplasmic division (cytokinesis). After M phase, the daughter cells each begin interphase of a new cycle. Although the various stages of interphase are not usually morphologically distinguishable, each phase of the cell cycle has a distinct set of specialized biochemical processes that prepare the cell for initiation of cell division.

G1 phase

The first phase within interphase is called G_1 (G indicating *gap* or *growth*); during this phase the biosynthetic activities of the cell, which had been considerably slowed down during M phase, resume at a high rate.

S phase

The ensuing S phase starts when DNA synthesis commences; when it is complete, all of the chromosomes have been replicated.

G2 phase

The cell then enters the G_2 phase, which lasts until the cell enters the next round of mitosis. Metabolic activity, cell growth, and cell differentiation all occur during interphase.

The term "post-mitotic" is sometimes used to refer to both quiescent and senescent cells. Nonproliferative cells in multicellular eukaryotes generally enter the quiescent G_0 state from G_1 and may remain quiescent for long periods of time, possibly indefinitely (as is often the case for neurons). This is very common for cells that are fully differentiated. Cellular senescence is a state that occurs in response to DNA damage or degradation that would make a cell's progeny nonviable; it is often a biochemical alternative to the self-destruction of such a damaged cell by apoptosis.

Cyclins and Cyclin-Dependent Kinases

Cyclins and cyclin: Dependent kinases (CDKs) are the two critical classes of molecules in regulation of cell cycle progression. Cyclins form the regulatory subunits and CDKs the catalytic subunits of an activated heterodimer; cyclins have no catalytic activity and CDKs are inactive in the absence of a partner cyclin. When activated by a bound cyclin, CDKs perform a common biochemical reaction called phosphorylation that activates or inactivates target proteins to orchestrate coordinated entry into the next phase of the cell cycle. Different cyclin-CDK combinations determine the downstream proteins targeted.

Many of the genes encoding cyclins and CDKs are conserved among all eukaryotes, but in general more complex organisms have more elaborate cell cycle control systems that incorporate more individual components. Many of the relevant genes were first identified studying yeast, especially *Saccharomyces cerevisiae*; genetic nomenclature in yeast dubs many of these genes *cdc* (for "cell division cycle") followed by an identifying number, e.g., *cdc25*. In the following discussion generic names such as "S cyclin" will be used to maintain generality, with the understanding that this may refer to one or to several homologous molecules in any given organism, and that some organisms may combine multiple functions in one molecule.

Upon receiving a pro-mitotic extracellular signal, G_1 cyclin-CDK complexes become active to prepare the cell for S phase, promoting the expression of transcription factors that in turn promote the expression of S cyclins and of enzymes required for DNA replication. The G_1 cyclin-CDK complexes also promote the degradation of molecules that function as S phase inhibitors by targeting them for ubiquitination. Once a protein has been ubiquitinated, it is targeted for proteolytic degradation by the proteasome. Active S cyclin-CDK complexes phosphorylate proteins that make up the pre-replication complexes assembled during G_1 phase on DNA replication origins. The phosphorylation serves two purposes: to activate each already-assembled pre-replication complex, and to prevent new complexes from forming. This ensures that every portion of the cell's genome will be replicated once and only once. The reason for prevention of gaps in replication is fairly clear, because daughter cells that are missing all or part of crucial genes will die. However, for reasons related to gene copy number effects, possession of extra copies of certain genes would also prove deleterious to the daughter cells.

Mitotic cyclin: CDK complexes, which are synthesized but inactivated during S and G_2 phases, promote the initiation of mitosis by stimulating downstream proteins involved in chromosome condensation and mitotic spindle assembly. A critical complex activated during this process is a ubiquitin ligase known as the anaphase-promoting complex (APC), which promotes degradation of structural proteins associated with the chromosomal kinetochore. APC also targets the mitotic cyclins for degradation, ensuring that telophase and cytokinesis can proceed.

Plant Cell Specializations

We will learn about the cells of algae and other organisms (e.g., bacteria and fungi) traditionally covered within Botany in later chapters on those organsims. Here, we concentrate on the cells of plants.

The simplest type of plant cell is called a parenchyma cell and most of the basic metabolic and reproductive processess of the plant occur in these cells. A term for *parenchyma* cells with chloroplasts, is chlorenchyma cells. Other plant cell types that we shall be considering are:

- *Collenchyma:* Living cells with thickened walls for increased support
- *Sclerenchyma:* Lignified dead cells forming fibers for increased support
- *Epidermal:* Surface covering
- Cork
- *Xylem tracheid:* Single long (up to 1 mm) thin cells for transporting water and support
- *Xylem vessel:* Cells form individual elements in an even longer (up to 1 meter in extreme cases) tube for transporting water
- *Meristematic cells:* Growth

Chapter 3

Plant Tissues

INTRODUCTION

Most plant cells are specialized to a greater or lesser degree, and arranged together in tissues. A plant tissue can be simple or complex depending upon whether it is composed of one or more than one type of cell. The simplest tissue found in plants is called parenchyma. The cells are not very specialized, more or less rounded or angular where packed together, and thin-walled. A type of parenchyma called chlorenchyma because the cells contain chloroplasts forms tissue (usually in the leaves) responsible for most of the photosynthesis occurring in the plant. Note that in simple tissues at least (tissues comprised mostly of one cell type), the tissue name follows from the cell type. However, tissues may also have unique anatomical names related to where in the plant they occur.

MERISTEMS

The growth of a plant requires a source of undifferentiated cells located in places where growth is needed and can be initiated to further the body plan (in comparison to animals, plants are rather open in this regard). Some enlargement in size is always possible by elongation or enlargement of existing cells, or by existing cells simply dividing. But differentiation of one cell type into another is only possible if the initial cell (mother cell) is not very specialized. Tissues comprised of cells that remain undifferentiated and supply, by their divisions, cells to form new tissues and organs, are called meristems. Meristem tissue occurs in places that allow for a very orderly pattern of growth.

A meristem is a tissue in plants consisting of undifferentiated cells (meristematic cells) and found in zones of the plant where growth can take place.

Differentiated plant cells generally cannot divide or produce cells of a different type. Therefore, cell division in the meristem is required to provide new cells for expansion and differentiation of tissues and initiation of new organs, providing the basic structure of the plant body.

Meristematic cells are analogous in function to stem cells in animals, are incompletely or not at all differentiated, and are capable of continued cellular division (youthful). Furthermore, the cells are small and protoplasm fills the cell completely. The vacuoles are extremely small. The cytoplasm does not contain differentiated plastids (chloroplasts or chromoplasts), although they are present in rudimentary form (proplastids). Meristematic cells are packed closely together without intercellular cavities. The cell wall is a very thin *primary cell wall*.

Maintenance of the cells requires a balance between two antagonistic processes: organ initiation and stem cell population renewal.

Meristematic Zones

Apical meristems are the completely undifferentiated (indeterminate) meristems in a plant. These differentiate into three kinds of primary meristems. The primary meristems in turn produce the two secondary meristem types. These secondary meristems are also known as lateral meristems because they are involved in lateral growth.

Meristems located at a bud on a branch or shoot are known as a node. Tissue between nodes is known as the internode.

Apical Meristems

The apical meristem, or growing tip, is a completely undifferentiated meristematic tissue found in the buds and growing tips of root in plants. Its main function is to begin growth of new cells in young seedlings at the tips of roots and shoots (forming buds, among other things). Specifically, an active apical meristem lays down a growing root or shoot behind itself, pushing itself forward. Apical meristems are very small, compared to the cylinder-shaped lateral meristems.

Apical meristems are composed of several layers. The number of layers varies according to plant type. In general the outermost layer is called the tunica while the innermost layers are the corpus. In monocots, the tunica determine the physical characteristics of the leaf edge and margin. In dicots, layer two of the corpus determine the characteristics of the edge of the leaf. The corpus and tunica play a critical part of the plant physical appearance as all plant cells are formed from the meristems. Apical meristems are found in two locations: the root and the stem.

Shoot Apical Meristems

The source of all above-ground organs. Cells at the SAM summit serve as stem cells to the surrounding peripheral region, where they proliferate rapidly and are incorporated into differentiating leaf or flower primordia.

The shoot apical meristem is the site of most of the embryogenesis in flowering plants. Primordia of leaves, sepals, petals, stamens and ovaries are initiated here at the rate of one every time interval, called a plastochron. It is where the first indications that flower development has been evoked are manifested. One of these indications might be the loss of apical dominance and the release of otherwise dormant cells to develop as axillary shoot meristems, in some species in axils of primordia as close as two or three away from the apical dome. The SAM consists of 4 distinct cell groups -

- Stem Cells
- The immediate daughter cells of the stem cells

- A subjacent organising centre
- Founder cells for organ initiation in surrounding regions

The four distinct zones mentioned above are maintain by a complex signalling pathway. The organisation centre expresses *WUS* proteins which maintains the stem cell identity of the overlying cells. The stem cells signal back with CLAVATA3 (CLE3) which is assumed to be a ligand for the CLE1 receptor kinase. When CLV1 interacts with CLV3 it initiates a signalling pathway that results in the repression of the expression of wus. This controls the size of the organising centre.

Root Apical Meristems

The root apical meristem (RAM) is covered by the root cap, which protects the apical meristem from the rocks, dirt and pathogens.

Stem Apical Meristem

The stem apical meristem is multicellular and, unlike the root apical meristem, has no cover. Rudimentary leaves may develop as scales, hardening at the end of the growing season to protect the stem apical meristem.

Intercalary Meristem

The intercalary meristems occur only in monocot stems between mature tissues. They are cylindrical meristems located around the nodes and are an adaptation to grazing herbivores and landmowers.

Besides growing additional leaves, an apical meristem may also develop into flowers. However, once differentiated into a flower, the meristem loses its meristematic ability and thus terminate the growth of that shoot. The lateral meristems behind the flowered apical meristem will then take on the functions as the apical meristem(s). This may be observed on a fruit tree at the beginning of growing season.

Floral Meristem

As flowers derive originally from shoots, the apical part of the flower primordium is composed of meristematic tissue, referred to as the floral meristem (FM).

Primary Meristems

Apical meristems may differentiate into three kinds of primary meristem:

- Protoderm: Lies around the outside of the stem and develops into the epidermis.
- Procambium: Lies just inside of the protoderm and develops into primary xylem and primary phloem. It also produces the vascular cambium, a secondary meristem.
- Ground meristem develops into the pith. It produces the cork cambium, another secondary meristem.

These meristems are responsible for primary growth, or an increase in length or height.

Secondary Meristems

There are two types of secondary meristems:

- Vascular Cambium: Produces secondary xylem and secondary phloem, this is a process which may continue throughout the life of the plant. This is what gives rise to wood in plants. Such plants are called arborescent. This does not occur in plants which do not go through secondary growth (known as herbaceous plants).
- Cork cambium: Gives rise to the bark of a tree.

These are also called the *lateral meristems* because they surround the established stem of a plant and cause it to grow laterally (i.e. larger in diameter). Lateral growth is also known as secondary growth.

Basal Meristems

As the name implies, this type of meristem is not found at the tip of a root or shoot, but near the base. This type of meristem allows for primary growth even after the apex of the shoot has been severed. For example, the presence of basal meristem is the reason grass can continue growing after mowing.

Indeterminate Growth of Meristems

Though each plant grows according to a certain set of rules, each new root and shoot meristem can go on growing for as long as it is alive; In many plants meristematic growth is potentially indeterminate, making the overall shape of the plant not determinate in advance. This is the primary growth.

Cloning

Under appropriate conditions, each shoot meristem can develop into a complete new plant or clone. Such new plants can be grown from shoot cuttings that contain an apical meristem. Root apical meristems are not readily cloned, however. This cloning is called asexual reproduction or vegetative reproduction and is widely practiced in horticulture to mass-produce plants of a desirable genotype. This process is also known as mericloning.

Chapter 4

Plant Vegetative Organs

INTRODUCTION

As was noted in the previous chapter, most plant cells are specialized to a greater or lesser degree, and arranged together in tissues A tissue can be *simple* or *complex* depending upon whether it is composed of one or more than one type of cell. Tissues are further arranged or combined into organs that carry out life functions of the organism. Plant organs include the leaf, stem, root, and reproductive structures The first three are sometimes called the *vegetative organs* and are the subject of exploration in this chapter.

The relationships of the organs within a plant body to each other remains an unsettled subject within plant morphology. The fundamental question is whether these are truly different structures, or just modifications of one basic structure (Eames, 1936; Esau, 1965). The plant body is an integrated, functional unit, so the division of a plant into organs is largely conceptual, providing a convenient way of approaching plant form and function. A boundary between stem and leaf is particularly difficult to make, so botanists sometimes use the word shoot to refer to the stem and its appendages (Esau, 1965).

THE LEAF

The plant leaf is an organ whose shape promotes efficient gathering of light for photosynthesis, but the form of the leaf must also be balanced against the fact that most of the loss of water a plant might suffer is going to occur at its leaves. Leaves are extremely variable in details of size, shape, and adornments like hairs.

Fig. The Leaves of a Beech Tree

In botany, a leaf is an above-ground plant organ specialized for photosynthesis. For this purpose, a leaf is typically flat (laminar) and thin, to expose the cells containing chloroplast (chlorenchyma tissue, a type of parenchyma) to light over a broad area, and to allow light to penetrate fully into the tissues. Leaves are also the sites in most plants where respiration, transpiration, and guttation take place. Leaves can store food and water, and are modified in some plants for other purposes. The comparable structures of ferns are correctly referred to as fronds. Furthermore, leaves are prominent in the human diet as leaf vegetables.

Leaf Anatomy

A structurally complete leaf of an angiosperm consists of a petiole (leaf stem), a *lamina* (leaf blade), and stipules (small processes located to either side of the base of the petiole). The petiole attaches to the stem at a point that is called the "leaf axil". Not every species produces leaves with all of these structural components. In some species, paired stipules are not obvious or are absent altogether.

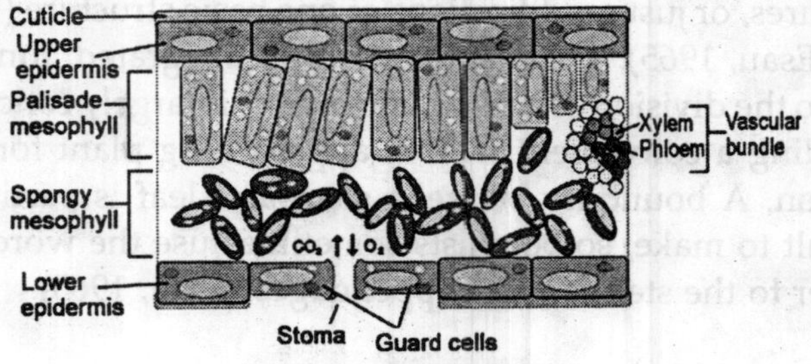

Fig. Leaf Anatomy

A petiole may be absent, or the blade may not be laminar (flattened). The tremendous variety shown in leaf structure (anatomy) from species to species is presented in detail below under Leaf morphology. After a period of time (i.e. seasonally, during the autumn), deciduous trees shed their leaves. These leaves decompose into the soil.

A leaf is considered to be a plant organ, typically consisting of the following tissues:

1. An epidermis that covers the upper and lower surfaces
2. An interior *chlorenchyma* called the mesophyll
3. An arrangement of veins (the vascular tissue).

Epidermis

The epidermis is the outer multi-layered group of cells covering the leaf. It forms the boundary separating the plant's inner cells from the external world. The epidermis serves several functions: protection against water loss, regulation of gas exchange, secretion of metabolic compounds, and (in some species) absorption of water. Most leaves show dorsoventral anatomy: the upper (adaxial) and lower (abaxial) surfaces have somewhat different construction and may serve different functions.

The epidermis is usually transparent (epidermal cells lack chloroplasts) and coated on the outer side with a waxy cuticle that prevents water loss. The cuticle is in some cases thinner on the lower epidermis than on the upper epidermis, and is thicker on leaves from dry climates as compared with those from wet climates.

The epidermis tissue includes several differentiated cell types: epidermal cells, guard cells, subsidiary cells, and epidermal hairs (trichomes). The epidermal cells are the most numerous, largest, and least specialized. These are typically more elongated in the leaves of monocots than in those of dicots.

The epidermis is covered with pores called *stomata*, part of a stoma complex consisting of a pore surrounded on each side by chloroplast-containing guard cells, and two to four subsidiary cells that lack chloroplasts. The stoma complex regulates the exchange of gases and water vapour between the outside air and the interior of the leaf. Typically, the stomata are more numerous over the abaxial (lower) epidermis than the adaxial (upper) epidermis.

Mesophyll

Most of the interior of the leaf between the upper and lower layers of epidermis is a *parenchyma* (ground tissue) or *chlorenchyma* tissue called the mesophyll (Greek for "middle leaf"). This assimilation tissue is the primary location of photosynthesis in the plant. The products of photosynthesis are called "assimilates".

In ferns and most flowering plants the mesophyll is divided into two layers:

- An upper palisade layer of tightly packed, vertically elongated cells, one to two cells thick, directly beneath the adaxial epidermis. Its cells contain many more chloroplasts than the spongy layer. These long cylindrical cells are regularly arranged in one to five rows. Cylindrical cells, with the chloroplasts close to the walls of the cell, can take optimal advantage of light. The slight separation of the cells provides maximum absorption of carbon dioxide. This separation must be minimal to afford capillary action for water distribution. In order to adapt to their different environment (such as sun or shade), plants had to adapt this structure to obtain optimal result. Sun leaves have a multi-layered palisade layer, while shade leaves or older leaves closer to the soil, are single-layered.
- Beneath the palisade layer is the spongy layer. The cells of the spongy layer are more rounded and not so tightly packed. There are large intercellular air spaces. These cells contain fewer chloroplasts than those of the palisade layer.

The pores or *stomata* of the epidermis open into substomatal chambers, connecting to air spaces between the spongy layer cells.

These two different layers of the mesophyll are absent in many aquatic and marsh plants. Even an epidermis and a mesophyll may be lacking. Instead for their gaseous exchanges they use a homogeneous aerenchyma (thin-walled cells separated by large gas-filled spaces). Their stomata are situated at the upper surface.

Leaves are normally green in colour, which comes from chlorophyll found in plastids in the chlorenchyma cells. Plants that lack chlorophyll cannot photosynthesize.

Leaves in temperate, boreal, and seasonally dry zones may be seasonally deciduous (falling off or dying for the inclement season). This mechanism to shed leaves is called abscission. After the leaf is shed, a leaf scar develops on the twig. In cold autumns they sometimes change colour, and turn yellow, bright orange or red as various accessory pigments (carotenoids and anthocyanins) are revealed when the tree responds to cold and reduced sunlight by curtailing chlorophyll production.

Veins

The veins are the vascular tissue of the leaf and are located in the spongy layer of the mesophyll. They are typical examples of pattern formation through ramification. The pattern of the veins is called venation.

The veins are made up of:

- Xylem, which brings water from the roots into the leaf.
- Phloem, which usually moves sap out, the latter containing the glucose produced by photosynthesis in the leaf.

The xylem typically lies over the phloem. Both are embedded in a dense parenchyma tissue, called "pith", with usually some structural collenchyma tissue present.

Leaf Morphology

External leaf characteristics (such as shape, margin, hairs, etc.) are important for identifying plant species, and botanists have developed a rich terminology for describing leaf characteristics. These structures are a part of what makes leaves determinant, they grow and achieve a specific pattern and shape, then stop. Other plant parts like stems or roots are non-determinant, and will usually continue to grow as long as they have the resources to do so.

Classification of leaves can occur through many different designative schema, and the type of leaf is usually characteristic of a species, although some species produce more than one type of leaf. The longest type of leaf is a leaf from palm trees, measuring at nine feet long.

Basic Leaf Types

Leaves of the White Spruce (*Picea glauca*) are needle-shaped and their arrangement is spiral

- Ferns have fronds.
- Conifer leaves are typically needle-, awl-, or scale-shaped
- Angiosperm (flowering plant) leaves: the standard form includes stipules, a petiole, and a lamina.
- Lycophytes have microphyll leaves.
- Sheath leaves (type found in most grasses).
- Other specialized leaves (such as those of *Nepenthes*)

Arrangement on the Stem

Different terms are usually used to describe leaf placement (phyllotaxis).

The leaves on this plant are arranged in pairs opposite one another, with successive pairs at right angles to each other ("decussate") along the red stem. Note developing buds in the axils of these leaves.

- *Alternate:* Leaf attachments are singular at nodes, and leaves alternate direction, to a greater or lesser degree, along the stem.

- *Opposite:* Leaf attachments are paired at each node; decussate if, as typical, each successive pair is rotated 90° progressing along the stem; or distichous if not rotated, but two-ranked (in the same geometric flat-plane)
- *Whorled:* Three or more leaves attach at each point or node on the stem. As with opposite leaves, successive whorls may or may not be decussate, rotated by half the angle between the leaves in the whorl (i.e., successive whorls of three rotated 60°, whorls of four rotated 45°, etc). Opposite leaves may appear whorled near the tip of the stem.
- *Rosulate:* Leaves form a rosette

As a *stem* grows, leaves tend to appear arranged around the stem in a way that optimizes yield of light. In essence, leaves form a helix pattern centred around the stem, either clockwise or counterclockwise, with (depending upon the species) the same angle of divergence. There is a regularity in these angles and they follow the numbers in a Fibonacci sequence: 1/2, 2/3, 3/5, 5/8, 8/13, 13/21, 21/34, 34/55, 55/89. This series tends to a limit of 360° x 34/89 = 137.52 or 137° 30', an angle known mathematically as the golden angle. In the series, the numerator indicates the number of complete turns or "gyres" until a leaf arrives at the initial position. The denominator indicates the number of leaves in the arrangement. This can be demonstrated by the following:

- Alternate leaves have an angle of 180° (or 1/2)
- 120° (or 1/3) : three leaves in one circle
- 144° (or 2/5) : five leaves in two gyres
- 135° (or 3/8) : eight leaves in three gyres.

The fact that an arrangement of anything in nature can be described by a mathematical formula is not in itself mysterious. Mathematics are the science of discovering numerical relationships and applying formulae to these relationships. The formulae themselves can provide clues to the underlying physiological processes that, in this case, determine where the next leaf bud will form in the elongating stem.

Divisions of the *Lamina* (Blade)

Two basic forms of leaves can be described considering the way the blade is divided. A simple leaf has an undivided blade. However, the leaf shape may be formed of lobes, but the gaps between lobes do not reach to the main vein. A compound leaf has a fully subdivided blade, each leaflet of the blade separated along a main or secondary vein. Because each leaflet can appear to be a simple leaf, it is important to recognize where the petiole occurs to identify a compound leaf. Compound leaves are a characteristic of some families of higher plants, such as the Fabaceae. The middle vein of a compound leaf or a frond, when it is present, is called a rachis.

- *Palmately compound* leaves have the leaflets radiating from the end of the petiole, like fingers off the palm of a hand, e.g. *Cannabis* (hemp) and *Aesculus* (buckeyes).
- *Pinnately compound* leaves have the leaflets arranged along the main or mid-vein.
-
  - Odd pinnate: With a terminal leaflet, e.g. *Fraxinus* (ash).
  - Even pinnate: Lacking a terminal leaflet, e.g. *Swietenia* (mahogany).
- *Bipinnately compound* leaves are twice divided: the leaflets are arranged along a secondary vein that is one of several branching off the rachis. Each leaflet is called a "pinnule". The pinnules on one secondary vein are called "pinna"; e.g. *Albizia* (silk tree).
- *Trifoliate*: a pinnate leaf with just three leaflets, e.g. *Trifolium* (clover), *Laburnum* (laburnum).
- *Pinnatifid*: pinnately dissected to the midrib, but with the leaflets not entirely separate, e.g. *Polypodium*, some *Sorbus* (whitebeams).

Characteristics of the *Petiole*

The overgrown petioles of Rhubarb (*Rheum rhabarbarum*) are edible

Petiolated leaves have a petiole. Sessile leaves do not: the blade attaches directly to the stem. In clasping or decurrent leaves, the blade partially or wholly surrounds the stem, often giving the impression that the shoot grows through the leaf. When this is actually the case, the leaves are called "perfoliate", such as in *Claytonia perfoliata*. In peltate leaves, the petiole attaches to the blade inside from the blade margin.

Fig. Petiole

In some *Acacia* species, such as the Koa Tree (*Acacia koa*), the petioles are expanded or broadened and function like leaf blades; these are called phyllodes. There may or may not be normal pinnate leaves at the tip of the phyllode.

A stipule, present on the leaves of many dicotyledons, is an appendage on each side at the base of the petiole resembling a small leaf. Stipules may be lasting and not be shed (a stipulate leaf, such as in roses and beans), or be shed as the leaf expands, leaving a stipule scar on the twig (an exstipulate leaf).

- The situation, arrangement, and structure of the stipules is called the "stipulation".
  - Free
  - Adnate : Fused to the petiole base
  - Ochreate : Provided with ochrea, or sheath-formed stipules, e.g. rhubarb,
  - Encircling the petiole base
  - Interpetiolar : Between the petioles of two opposite leaves.
  - Intrapetiolar : Between the petiole and the subtending stem

Venation (Arrangement of the Veins)

There are two subtypes of venation, namely, *craspedodromous,* where the major veins stretch up to the margin of the leaf, and *camptodromous,* when major veins extend close to the margin, but bend before they intersect with the margin.

- *Feather-veined, reticulate:* The veins arise pinnately from a single mid-vein and subdivide into veinlets. These, in turn, form a complicated network. This type of venation is typical for dicotyledons.
  - Pinnate-netted, penniribbed, penninerved, penniveined; the leaf has usually one main vein (called the mid-vein), with veinlets, smaller veins branching off laterally, usually somewhat parallel to each other; eg *Malus* (apples).
  - Three main veins originate from the base of the lamina, as in *Ceanothus.*
- Palmate-netted, palmate-veined, fan-veined; several main veins diverge from near the leaf base
  - where the petiole attaches, and radiate toward the edge of the leaf; e.g. most *Acer* (maples).
- Parallel-veined, parallel-ribbed, parallel-nerved, penniparallel: Veins run parallel most the length of the leaf, from the base to the apex. Commissural veins (small veins) connect the major parallel veins. Typical for most monocotyledons, such as grasses.

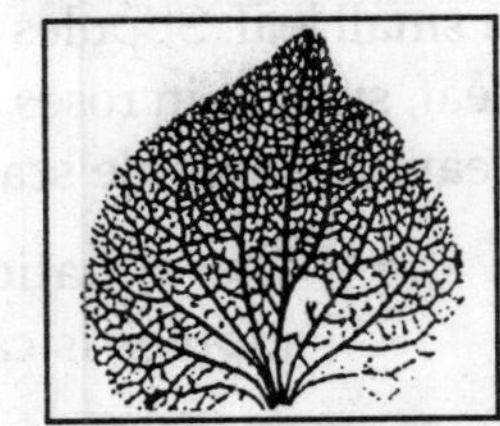

Fig. Vein Skeleton of a Hydrangea Leaf

- *Dichotomous:* There are no dominant bundles, with the veins forking regularly by pairs; found in *Ginkgo* and some pteridophytes.

Leaf Morphology Changes Within a Single Plant

- *Homoblasty:* Characteristic in which a plant has small changes in leaf size, shape, and growth habit between juvenile and adult stages.

- *Heteroblasty:* Charactistic in which a plant has marked changes in leaf size, shape, and growth habit between juvenile and adult stages.

Margins (Edge)

The leaf margin is characteristic for a genus and aids in determining the species.

- *Entire:* Even; with a smooth margin; without toothing
- *Ciliate:* Fringed with hairs
- *Crenate:* Wavy-toothed; dentate with rounded teeth, such as *Fagus* (beech)
- *Dentate:* Toothed, such as *Castanea* (chestnut)
  - Coarse-toothed: With large teeth
  - Glandular toothed: With teeth that bear glands.
- *Denticulate:* Finely toothed
- *Doubly toothed:* Each tooth bearing smaller teeth, such as *Ulmus* (elm)
- *Lobate:* Indented, with the indentations not reaching to the centre, such as many *Quercus* (oaks)
  - *Palmately lobed:* Indented with the indentations reaching to the centre, such as *Humulus* (hop).
- *Serrate:* Saw-toothed with asymmetrical teeth pointing forward, such as *Urtica* (nettle)
- *Serrulate:* Finely serrate
- *Sinuate:* With deep, wave-like indentations; coarsely crenate, such as many *Rumex* (docks)
- *Spiny:* With stiff, sharp points, such as some *Ilex* (hollies) and *Cirsium* (thistles).

Tip of the Leaf

Leaves showing various morphologies. Clockwise from upper left: tripartite lobation, elliptic with serrulate margin, peltate with palmate venation, acuminate odd-pinnate (centre), pinnatisect, lobed, elliptic with entire margin

- *Acuminate:* Long-pointed, prolonged into a narrow, tapering point in a concave manner.
- *Acute:* Ending in a sharp, but not prolonged point
- *Cuspidate:* With a sharp, elongated, rigid tip; tipped with a cusp.
- *Emarginate:* Indented, with a shallow notch at the tip.
- *Mucronate:* Abruptly tipped with a small short point, as a continuation of the midrib; tipped with a mucro.
- *Mucronulate:* Mucronate, but with a smaller spine.
- *Obcordate:* Inversely heart-shaped, deeply notched at the top.
- *Obtuse:* Rounded or blunt
- *Truncate:* Ending abruptly with a flat end, that looks cut off.

Base of the Leaf

- *Acuminate:* Coming to a sharp, narrow, prolonged point.
- *Acute:* Coming to a sharp, but not prolonged point.
- *Auriculate:* Ear-shaped
- *Cordate:* Heart-shaped with the norch away from the stem.
- *Cuneate:* Wedge-shaped.
- *Hastate:* Shaped like an halberd and with the basal lobes pointing outward.
- *Oblique:* Slanting.
- *Reniform:* Kidney-shaped but rounder and broader than long.
- *Rounded:* Curving shape.
- *Sagittate:* Shaped like an arrowhead and with the acute basal lobes pointing downward.
- *Truncate:* Ending abruptly with a flat end, that looks cut off.

Surface of the Leaf

The surface of a leaf can be described by several botanical terms:

- *Farinose:* Bearing farina; mealy, covered with a waxy, whitish powder.
- *Glabrous:* Smooth, not hairy.
- *Glaucous:* With a whitish bloom; covered with a very fine, bluish-white powder.
- *Glutinous:* Sticky, viscid.
- *Papillate, papillose:* bearing papillae (minute, nipple-shaped protuberances).
- *Pubescent:* Covered with erect hairs (especially soft and short ones)
- *Punctate:* Marked with dots; dotted with depressions or with translucent glands or colored dots.
- *Rugose:* Deeply wrinkled; with veins clearly visible.
- *Scurfy:* Covered with tiny, broad scalelike particles.
- *Tuberculate:* Covered with tubercles; covered with warty prominences.
- *Verrucose:* Warted, with warty outgrowths.
- *Viscid, viscous:* Covered with thick, sticky secretions.

The leaf surface is also host to a large variety of microorganisms; in this context it is referred to as the phyllosphere.

Hairiness (Trichomes)

Common Mullein (*Verbascum thapsus*) leaves are covered in dense, stellate trichomes.

"Hairs" on plants are properly called trichomes. Leaves can show several degrees of hairiness. The meaning of several of the following terms can overlap.

- *Glabrous:* No hairs of any kind present.
- *Arachnoid, arachnose:* With many fine, entangled hairs giving a cobwebby appearance.

- *Barbellate:* With finely barbed hairs (barbellae).
- *Bearded:* With long, stiff hairs.
- *Bristly:* With stiff hair-like prickles.
- *Canescent:* Hoary with dense grayish-white pubescence.
- *Ciliate:* Marginally fringed with short hairs (cilia).
- *Ciliolate:* Minutely ciliate.
- *Floccose:* With flocks of soft, woolly hairs, which tend to rub off.
- *Glandular:* With a gland at the tip of the hair.
- *Hirsute:* With rather rough or stiff hairs.
- *Hispid:* With rigid, bristly hairs.
- *Hispidulous:* Minutely hispid.
- *Hoary:* with a fine, close grayish-white pubescence.
- *Lanate, lanose:* with woolly hairs.
- *Pilose:* with soft, clearly separated hairs.
- *Puberulent, puberulous:* with fine, minute hairs.
- *Pubescent:* With soft, short and erect hairs.
- *Scabrous, scabrid:* Rough to the touch
- *Sericeous:* Silky appearance through fine, straight and appressed (lying close and flat) hairs.
- *Silky:* With adpressed, soft and straight pubescence.
- *Stellate, stelliform:* With star-shaped hairs.
- *Strigose:* With appressed, sharp, straight and stiff hairs.
- *Tomentose:* Densely pubescent with matted, soft white woolly hairs.
- *Cano-tomentose:* Between canescent and tomentose
-
  - *Felted-tomentose:* Woolly and matted with curly hairs.

- *Villous:* With long and soft hairs, usually curved.
- *Woolly:* With long, soft and tortuous or matted hairs.

Adaptations

The leaves of Poinsettia have evolved a red pigmentation in order to attract insects and birds to the central flowers, an adaptive function normally served by petals.

In the course of evolution, leaves adapted to different environments in the following ways:

- A certain surface structure avoids moistening by rain and contaminations (Lotus effect).
- Sliced leaves reduce wind resistance.
- Hairs on the leaf surface trap humidity in dry climates and creates a large boundary layer and reduces water loss.
- Waxy leaf surfaces reduce water loss.
- Shiny leaves deflect the sun's rays.
- Reductions of leaf sizes accompanied by a transfer of the photosynthetic functions to the stems reduces water loss.
- In more or less opaque or buried in the soil leaves translucent windows filter the light before the photosynthetis takes place at the inner leaf surfaces (e.g. Fenestraria).
- Thicker leaves store water (leaf succulents).
- Aromatic oils, poisons or pheromones produced by leaf borne glands deter herbivores (e.g. eucalypts).
- Inclusions of crystalline minerals deters herbivores.
- A transformation into petals attracts pollinators.
- A transformation into spines protects the plants (e.g. cactus).
- A transformation into insect traps helps feeding the plants (carnivorous plants).
- A transformation into bulbs helps storing food and water (e.g. onion).

- A transformation into tendrils allow the plant to climb (e.g. pea).
- A transformation into bracts and pseudanthia (*false flowers*) replaces normal flower structures if the true flowers are extremely reduced (e.g. Spurges).
  - Although the leaves of most plants carry out the same very basic functions, there is nonetheless an amazing variety of leaf sizes, shapes, margin types, forms of attachment, ornamentation (hairs), and even colour. Examine the Leaves (forms) page to learn the extensive terminology used to describe this variation. Consider that there are functional reasons for the modifications from a "basic" type.

THE STEM

The stem arises during development of the embryo as part of the *hypocotyl-root axis,* at the upper end of which are one or more cotyledons and the shoot primordium.

A stem is one of two main structural axes of a vascular plant. The stem is normally divided into nodes and internodes, the nodes hold buds which grow into one or more leaves, inflorescence (flowers), cones or other stems etc. The internodes act as spaces that distance one node from another. The term shoots is often confused with stems, shoots generally refer to new fresh plant growth and does include stems but also to other structures like leaves or flowers. The other main structural axis of plants is the root. In most plants stems are located above the soil surface but some plants have underground stems.

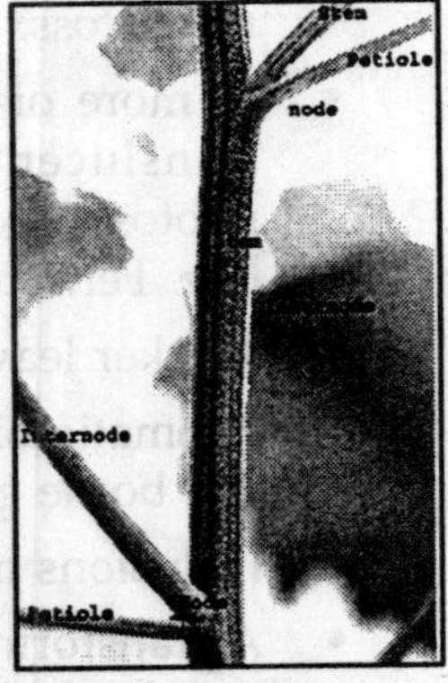

Fig. Stem Showing Internode and Nodes Plus Leaf Petiole and New Stem Rising from Node.

Stems have four main functions which are:

- Support for and the elevation of leaves, flowers and fruits. The stems keep the leaves in the light and provide a place for the plant to keep its flowers and fruits.
- Transport of fluids between the roots and the shoots in the xylem and phloem.
- Storage of nutrients.
- The production of new living tissue. The normal life span of plant cells is one to three years. Stems have cells called meristems that annually generate new living tissue.

Specialized Terms for Stems

Stems are often specialized for storage, asexual reproduction, protection or photosynthesis, including the following:

- *Acaulescent:* Plants with very short stems that appear to have no stems. The leaves appear to rise out of the ground. Some viola.
- *Arborescent:* Tree like with woody stems normally with a single trunk.
- *Bud:* An embryonic shoot with immature stem tip.
- *Bulb:* A short vertical underground stem with fleshy storage leaves attached, e.g. onion, daffodil, tulip. Bulbs often function in reproduction by splitting to form new bulbs or producing small new bulbs termed bulblets. Bulbs are a combination of stem and leaves so may better be considered as leaves because the leaves make up the greater part.
- *Caespitose:* When stems grow in a tangled mass or clump or in low growing mats.
- *Cladophyll:* A flattened stem that appears leaf like and is specialized for photosynthesis, e.g. asparagus, cactus pads.

- *Climbing:* Stems that cling or wrap around other plants or structures.
- *Corm:* A short enlarged underground, storage stem, e.g. taro, crocus, gladiolus.
- *Decumbent:* Stems that lay flat on the ground and turn upwards at the ends.
- *Fruticose:* Stems that grow shrub like with woody like habit.
- *Herbaceous:* Non woody, they die at the end of the growing season.
- *Rhizome:* A horizontal underground stem that functions mainly in reproduction but also in storage, e.g. most ferns, iris
- *Runner (plant part):* A type of stolon, horizontally growing on top of the ground and rooting at the nodes. e.g. strawberry, spider plant.
- *Scape:* A stem that holds flowers that comes out of the ground and has no normal leaves. Hosta, Lily, Iris.
- *Stolons:* A horizontal stem that produces rooted plantlets at its nodes and ends, forming near the surface of the ground.
- *Tree:* A woody stem that is longer than 5 meters with a main trunk.
- *Thorns:* A reduced stem with a sharp point and rounded shape. e.g. honeylocust, hawthorn.
- *Tuber:* A swollen, underground storage stem adapted for storage and reproduction, e.g. potato.
- *Woody:* Hard textured stems with secondary xylem

Stem Structure

Stem usually consist of three tissues, dermal tissue, ground tissue and vascular tissue. The dermal tissue covers the outer surface of the stem and usually functions to waterproof, protect and control gas exchange. The ground tissue usually consists mainly of parenchyma cells and fills in around the vascular tissue. It sometimes functions in photosynthesis. Vascular tissue provides long distance transport and structural support. Most or all ground tissue may be lost in woody stems. The dermal tissue of aquatic plants stems may lack the waterproofing found in aerial stems. The arrangement of the vascular tissues varies widely among plant species.

Dicot Stems

Dicot stems with primary growth have a pith in the centre with vascular bundles in a distinct ring visible in cross section. The outside of the stem is covered with an epidermis, which is covered by a waterproof cuticle. The epidermis also may contain stomata for gas exchange and hairs. A cortex of parenchyma cells lies between the epidermis and vascular bundles.

Woody dicots and many nonwoody dicots have secondary growth originating from their lateral or secondary meristems: the vascular cambium and the cork cambium or phellogen. The vascular cambium forms between the xylem and phloem in the vascular bundles and connects to form a continuous cylinder. The vascular cambium cells divide to produce secondary xylem to the inside and secondary phloem to the outside. As the stem increases in diameter due to production of secondary xylem and secondary phloem, the cortex and epidermis are eventually destroyed. Before the cortex is destroyed, a cork cambium develops there. The cork cambium divides to produce waterproof cork cells externally and sometimes phelloderm cells internally. Those three tissues form the periderm, which replaces the epidermis in function. Areas of loosely-packed cells in the periderm that function in gas exchange are called lenticels.

Secondary xylem is commercially important as wood. The seasonal variation in growth from the vascular cambium is what creates yearly tree rings in temperate climates. Tree rings are the basis of dendrochronology, which dates wooden objects and associated artifacts. Dendroclimatology is the use of tree rings as a record of past climates. The aerial stem of an adult tree is called a trunk. The dead, usually darker inner wood of a large diameter trunk is termed the heartwood. The outer, living wood is termed the sapwood.

Monocot Stems

Vascular bundles are present throughout the monocot stem, although concentrated towards the outside. This differs from the monocot root that has a ring of vascular bundles and often none in the centre. The shoot apex in monocot stems is more elongated. Leaf sheathes grow up around it, protecting it. This is true to some extent of almost all monocots. Monocots rarely produce secondary growth and are therefore seldom woody. However, many monocot stems increase in diameter via anamolous secondary growth.

Gymnosperm Stems

All gymnosperms are woody plants. Their stems are similar in structure to woody dicots except that most gymnosperms produce only tracheids in their xylem, not the vessels found in dicots. Gymnosperm wood also often contains resin ducts. Woody dicots are called hardwoods, e.g. oak, maple and walnut. In contrast, softwoods are gymnosperms, such as pine, spruce and fir.

Fern Stems

Most ferns have rhizomes with no vertical stem. The exception is tree ferns, with vertical stems up to about 15 meters. Stem anatomy of ferns is more complicated that dicots because fern stems often have one or more leaf gaps in cross section. A leaf gap is where the vascular tissue branches off to a frond. In cross section, the vascular tissue does not form a complete cylinder where a leaf gap occurs. Fern stems may have solenosteles or dictyosteles or variations of them. Many fern stems have phloem tissue on both sides of the xylem in cross-section.

THE ROOT

The root is the (typically) underground part of the plant axis specialized for both anchoring the plant and absorbing water and minerals.

In vascular plants, the root is that organ of a plant body that typically lies below the surface of the soil (compare with stem). However, this is not always the case, since a root can also be aerial (that is, growing above the ground) or aerating (that is, growing up above the ground or especially above water). On the other hand, a stem normally occurring below ground is not exceptional either. So, it is better to define *root* as a part of a plant body that bears no leaves, and therefore also lacks nodes. There are also important internal structural differences between stems and roots. The two major functions of roots are 1) absorption of water and inorganic nutrients and 2) anchoring the plant body to the ground. Roots also function in cytokinin synthesis, which supplies some of shoot needs. They often function in storage of food.

Root Structure

At the tip of every growing root is a conical covering of tissue called the root cap. It usually is not visible to the naked eye. It consists of undifferentiated soft tissue (parenchyma) with unthickened walls covering the apical meristem. The root cap provides mechanical protection to the meristem cells as the root advances through the soil, its cells worn away but quickly replaced by new cells generated by cell division within the meristem. The root cap is also involved in the production of mucigel, a sticky mucilage that coats the new formed cells. These cells contain statoliths, starch grains that move in response to gravity and thus control root orientation.

The outside surface of the primary root is the epidermis. Recently produced *epidermal* cells absorb water from the surrounding environment and produce outgrowths called root hairs that greatly increase the cell's absorptive surface. Root-hairs are very delicate and generally short-lived, remaining functional for only a few days. However, as the root grows, new epidermal cells emerge and these form new root hairs, replacing those that die. The process by which water is absorbed into the epidermal cells from the soil is known as *osmosis* For this reason, water that is saline is more difficult for most plant species to absorb.

Beneath the epidermis is the cortex, which comprises the bulk of the primary root. Its main function is storage of starch. Intercellular spaces in the cortex aerate cells for respiration. An endodermis is a thin layer of small cells forming the innermost part of the cortex and surrounding the vascular tissues deeper in the root. The tightly packed cells of the endodermis contain a substance known as suberin in their cell walls. This suberin layer is the Casparian strip, which creates an impermeable barrier of sorts. Mineral nutrients can only move passively within root cell walls until they reach the endodermis. At that point, they must be actively transported across a cell membrane to continue further into the root. This allows the plant to accumulate mineral nutrients in the stele.

The vascular cylinder, or stele, consists of the cells inside the endodermis. The outer part, known as the pericycle, surrounds the actual vascular tissue. In monocotyledonous plants, the xylem and phloem cells are arranged in a circle around a pith or centre, whereas in dicotyledons, the xylem cells form a central "hub" with lobes, and phloem cells fill in the spaces between the lobes.

Secondary Growth

All roots have primary growth or growth in length. Roots of many vascular plants, especially dicots and gymnosperms, often undergo secondary growth, which is an increase in diameter. A vascular cambium forms in the stele to produce secondary phloem and secondary xylem. The epidermis is replaced by a periderm. As the stele increases in diameter, the cortex, pericycle and endodermis are lost. Even nonwoody roots often undergo secondary growth, including those of tomato and alfalfa.

Root Growth

Early root growth is one of the functions of the apical meristem located near the tip of the root. The meristem cells more or less continuously divide, producing more meristem, root cap cells (these sacrificed to protect the meristem), and undifferentiated root cells. The latter will become the primary tissues of the root, first undergoing elongation, a process that pushes the root tip forward in the growing medium. Gradually these cells differentiate and mature into specialized cells of the root tissues.

Roots will generally grow in any direction where the correct environment of air, mineral nutrients and water exists to meet the plant's needs. Roots will not grow in dry soil. Over time, given the right conditions, roots can crack foundations, snap water lines, and lift sidewalks. At germination, roots grow downward due to gravitropism, the growth mechanism of plants that also causes the shoot to grow upward. In some plants (such as ivy), the "root" actually clings to walls and structures.

Growth from apical meristems is known as primary growth, which encompasses all elongation. Secondary growth encompasses all growth in diameter, a major component of woody plant tissues and many nonwoody plants. For example, storage roots of sweet potato have secondary growth but are not woody. Secondary growth occurs at the lateral meristems, namely the vascular cambium and cork cambium. The former forms secondary xylem and secondary phloem, while the latter forms the periderm.

In plants with secondary growth, the vascular cambium, originating between the xylem and the phloem, forms a cylinder of tissue along the stem and root. The cambium layer forms new cells on both the inside and outside of the cambium cylinder, with those on the inside forming secondary xylem cells, and those on the outside forming secondary phloem cells. As secondary xylem accumulates, the "girth" (lateral dimensions) of the stem and root increases. As a result, tissues beyond the secondary phloem (including the epidermis and cortex, in many cases) tend to be pushed outward and are eventually "sloughed off" (shed).

At this point, the cork cambium begins to form the periderm, consisting of protective cork cells containing suberin. In roots, the cork cambium originates in the pericycle, a component of the vascular cylinder.

The vascular cambium produces new layers of secondary xylem annually. The xylem vessels are dead at maturity but are responsible for most water transport through the vascular tissue in stems and roots.

Types of Roots

A true root system consists of a primary root and secondary roots (or lateral roots).

The primary root originates in the radicle of the seedling. During its growth it rebranches to form the lateral roots. Generally, two categories are recognized:

- The taproot system: the primary root is prominent and has a single, dominant axis; there are fibrous secondary roots running outward. Usually allows for deeper roots capable of reaching low water tables. Most common in dicots. The main function of the taproot is to store food.
- The diffuse root system: the primary root is not dominant; the whole root system is fibrous and branches in all directions. Most common in monocots. The main function of the fibrous root is to anchor the plant.

Specialized Roots

The roots, or parts of roots, of many plant species have become specialized to serve adaptive purposes besides the two primary functions described in the introduction.

- Adventitious roots arise out-of-sequence from the more usual root formation of branches of a primary root, and instead originate from the stem, branches, leaves, or old woody roots. They commonly occur in monocots and pteridophytes, but also in many dicots, such as clover (*Trifolium*), ivy (*Hedera*), strawberry (*Fragaria*) and willow (*Salix*). Most aerial roots and stilt roots are adventitious. In some conifers adventitious roots can form the largest part of the root system.
- *Aerating roots (or pneumatophores)*: Roots rising above the ground, especially above water such as in some

mangrove genera (*Avicennia, Sonneratia*). In some plants like *Avicennia* the erect roots have a large number of breathing pores for exchange of gases.

- *Aerial roots:* roots entirely above the ground, such as in ivy (*Hedera*) or in epiphytic orchids. They function as prop roots, as in maize or anchor roots or as the trunk in strangler fig.
- *Buttress roots or tabular roots:* Support roots for many tropical tree species.

Fig. Buttress Roots of Ceiba Pentandra

- *Contractile roots:* They pull bulbs or corms of monocots, such as hyacinth and lily, and some taproots, such as dandelion, deeper in the soil through expanding radially and contracting longitudinally. They have a wrinkled surface.
- *Coarse roots:* Roots that have undergone secondary thickening and have a woody structure. These roots have some ability to absorb water and nutrients, but their main function is transport and to provide a structure to connect the smaller diameter, fine roots to the rest of the plant.
- *Fine roots:* Primary roots usually <2 mm diameter that have the function of water and nutrient uptake. They are often heavily branched and support mycorrhizas. These roots may be short lived, but are replaced by the plant in an ongoing process of root 'turnover'.
- *Haustorial roots:* Roots of parasitic plants that can absorb water and nutrients from another plant, such as in mistletoe (*Viscum album*) and dodder.
- Propagative roots: Roots that form adventitious buds that develop into aboveground shoots, termed suckers, which form new plants, as in Canada thistle, cherry and many others.

- Proteoid roots or cluster roots: Dense clusters of rootlets of limited growth that develop under low phosphate or low iron conditions in Proteaceae and some plants from the following families Betulaceae, Casuarinaceae, Eleagnaceae, Moraceae, Fabaceae and Myricaceae.
- Stilt roots: These are adventitious support roots, common among mangroves. They grow down from lateral branches, branching in the soil.
- Storage roots: These roots are modified for storage of food or water, such as carrots and beets. They include some taproots and tuberous roots.
- Structural roots: Large roots that have undergone considerable secondary thickening and provide mechanical support to woody plants and trees.
- Surface roots: These proliferate close below the soil surface, exploiting water and easily available nutrients. Where conditions are close to optimum in the surface layers of soil, the growth of surface roots is encouraged and they commonly become the dominant roots.
- Tuberous roots: A portion of a root swells for food or water storage, e.g. sweet potato and dahlia. A type of storage root distinct from taproot.

Rooting Depths

The distribution of vascular plant roots within soil depends on plant form, the spatial and temporal availability of water and nutrients, and the physical properties of the soil. The deepest roots are generally found in deserts and temperate coniferous forests; the shallowest in tundra, boreal forest and temperate grasslands. The deepest observed living root, at least 60 m below the ground surface, was observed during the excavation of an open-pit mine in Arizona, USA. Some roots can grow as deep as the tree is high. The majority of roots on most plants are however found relatively close to the surface where nutrient availability and aeration are more favourable for growth. Rooting depth may be physically restricted by rock or compacted soil close below the surface, or by anaerobic soil conditions.

Chapter 5

Plant Reproduction

VEGETATIVE REPRODUCTION

Vegetative reproduction is asexual reproduction—other terms that apply are *vegetative propagation* or *vegetative multiplication*. Vegetative growth is enlargement of the individual plant; vegetative reproduction is any process that results in new plant "individuals" without production of seeds or spores. It is both a natural process in many, many species as well as one utilized or encouraged by horticulturists and farmers to obtain quantities of economically valuable plants. In this respect, it is a form of cloning that has been carried out by humankind for thousands of years and by "plants" for hundreds of millions of years.

Fig. Noni (Morinda Citrifolia)

Vegetative reproduction is a type of asexual reproduction found in plants also called vegetative propagation or vegetative multiplication. It is a process by which new plant "individuals" arise or are obtained without production of seeds or spores. It is both a natural process in many plant species (including organisms that may or may not be considered "plants", such as bacteria and fungi) and one utilized or encouraged by horticulturists to obtain quantities of economically valuable plants.

Natural vegetative reproduction is mostly a process found in herbaceous and woody perennial plants, and typically involves structural modifications of the stem, although any horizontal, underground part of a plant (whether stem or a root) can contribute to vegetative reproduction of a plant. And, in a few species (such as *Kalanchoë* shown at right), leaves are involved in vegetative reproduction. Most plant species that survive and significantly expand by vegetative reproduction would be perennial almost by definition, since specialized organs of vegetative reproduction, like seeds of annuals, serve to survive seasonally harsh conditions. A plant that persists in a location through vegetative reproduction of individuals over a long period of time constitutes a clonal colony.

In a sense, this process is not one of "reproduction" but one of survival and expansion of biomass of the individual. When an individual organism increases in size via cell multiplication and remains intact, the process is called "vegetative growth". However, in vegetative reproduction, the new plants that result are new individuals in almost every respect except genetic. And of considerable interest is how this process appears to reset the aging clock.

NATURAL VEGETATIVE STRUCTURES

The rhizome is a modified stem serving as an organ of vegetative reproduction. Prostrate aerial stems, called runners or stolons are important vegetative reproduction organs in some species, such as the strawberry, numerous grasses, and some ferns. Adventitious buds develop into above ground stems and leaves, forming on roots near the ground surface and on damaged stems (as on the stumps of cut trees). *Adventitious* roots form on stems where the latter touch the soil surface.

A form of budding called suckering is the reproduction or regeneration of a plant by shoots that arise from an existing root system. Species that characteristically produce suckers include Elm (*Ulmus*), Dandelion (*Taraxacum*), and members of the Rose Family (*Rosa*).

Another type of a vegetative reproduction is the production of bulbs. Plants like onion (*Allium cepa*), hyacinth (*Hyacinth*), narcissus (*Narcissus*) and tulips (*Tulipa*) reproduce by forming bulbs. Other plants like potatoes (*Solanum tuberosum*) and dahlia (*Dahlia*) reproduce by a similar method of producing tubers. Gladioluses and crocuses (*Crocus*) reproduce by forming a bulb-like structure called a corm.

Exceptions

Vegetative propagation is usually considered a cloning method. However, there are several cases where vegetatively propagated plants are not genetically identical. Rooted stem cuttings of thornless blackberries will revert to thorny type because the adventitious shoot develops from a cell that is genetically thorny. Thornless blackberry is a chimera, with the epidermal layers genetically thornless but the tissue beneath it genetically thorny. Leaf cutting propagation of certain chimeral variegated plants, such as snake plant, will produce mainly nonvariegated plants.

Grafting is often not a complete cloning method because sexual seedlings are used as rootstocks. In that case only the top of the plant is clonal. In some crops, particularly apples, the rootstocks are vegetatively propagated so the entire graft can be clonal if the scion and rootstock are both clones.

Apomixis is a type of asexual reproduction involving unfertilized seeds. Hawkweed (*Hieracium*), dandelion (*Taraxacum*), some Citrus (*Citrus*) and Kentucky blue grass (*Poa pratensis*) all use this form of asexual reproduction. Bulbils are sometimes formed in the flowers of garlic. The leafy crown of a pineapple fruit will root to form a new plant. These cases would not be vegetative reproduction because normally reproductive parts were involved. They would be considered asexual reproduction however. Vegetative reproduction involves only vegetative structures, i.e. roots, stems or leaves.

Horticultural Aspects

Man-made methods of vegetative reproduction are usually enhancements of natural processes, but range from simple cloning such as rooting of cuttings to grafting and artificial propagation by laboratory tissue cloning. It is very commonly practised to propagate cultivars with individual desirable characteristics. Fruit tree propagation is frequently performed by budding or grafting desirable cultivars (clones), onto rootstocks that are also clones, propagated by layering.

In horticulture, a "cutting" is a branch that has been cut off from a mother plant below an internode and then rooted, often with the help of a rooting liquid or powder containing hormones. When a full root has formed and leaves begin to sprout anew, the clone is a self-sufficient plant, genetically identical to the mother plant. Examples are cutting from the stems of blackberries (*Rubus occidentalis*), cutting from leaves of African violets (*Saintpaulia*), and cutting the stems of verbenas (*Verbena*) to create new plants. A related form of regeneration is that of grafting. This is a process of taking a bud and grafting onto a plants stem. Many nurseries now sell trees that can produce four or more varieties of apples (*Malus spp.*) from stems grafted to a common rootstock.

SEXUAL REPRODUCTION

Plant sexuality deals with the wide variety of sexual reproduction systems found across the plant kingdom. This article describes morphological aspects of sexual reproduction of plants.

That plants employ many different strategies to engage in sexual reproduction was used, from just a structural perspective, by Carolus Linnaeus (1735 and 1753) to propose a system of classification of flowering plants. Later this subject received attention from Christian Konrad Sprengel (1793) who described plant sexuality as the "revealed secret of nature" and, for the first time, understood the biotic and abiotic interactions of the pollination process. Charles Darwin's theories of natural selection are based on his work. Flowers, the reproductive structures of angiosperms, are more varied than the equivalent structures of any other group of organisms, and flowering plants also have an unrivalled diversity of sexual systems. But sexuality and the significance of sexual reproductive strategies is no less important in all of the other plant groups. The breeding system is the single most important determinant of the mating structure of nonclonal plant populations. The mating structure in turn controls the amount and distribution of genetic variation, a central element in the evolutionary process

TERMINOLOGY

The complexity of the systems and devices used by plants to achieve sexual reproduction has resulted in botanists and evolutionary biologists proposing numerous terms to describe structures and strategies. Dellaporta and Calderon-Urrea (1993) list and define a variety of terms used to describe the modes of sexuality at different levels in flowering plants. This list is reproduced here , generalized to fit more than just plants that have flowers, and expanded to include other terms and better definitions.

Individual Reproductive Unit (a Flower in Angiosperms)

- *Bisexual:* Reproductive structure with both male and female equivalent parts (stamens and pistil in angiosperms; also called a perfect or complete flower); other terms widely used are hermaphrodite, monoclinous, and synoecious.
- *Unisexual:* Reproductive structure that is either functionally male or functionally female. In angiosperms this condition is also called diclinous, imperfect or incomplete.

Individual Plant

- *Hermaphrodite:* A plant that has only hermaphrodite reproductive units (flowers, conifer cones, or functionally equivalent structures). In angiosperm terminology a synonym is monoclinous from the Greek "one bed".
- *Monoecious:* Having unisexual reproductive units (flowers, conifer cones, or functionally equivalent

structures) of both sexes appearing on the same plant; from Greek for "one household". Individuals bearing flowers of both sexes at the same time are called simultaneously or synchronously monoecious. Individuals that bear only flowers of a single sex at one time are called consecutively monoecious; "protoandrous" describes individuals that function first as males and then change to females; "protogynous" describes individuals that function first as females and then change to males.

- *Dioecious:* Having unisexual reproductive units (flowers, conifer cones, or functionally equivalent structures) occurring on different individuals; from Greek for "two households". Individual plants are not called dioecious: they are either gynoecious or androecious.
- Subdioecious, a tendency many species of dioecious conifers show towards monoecy (that is, a female plant may sometimes produce small numbers of male cones or vice versa).
- Diclinous ("two beds"), an angiosperm term, includes all species with unisexual flowers, although particularly those with *only* unisexual flowers, i.e. the monoecious and dioecious species.
- *Gynoecious:* Has only female reproductive structures; the "female" plant.
- *Androecious:* Has only male reproductive structures; the "male" plant.
- *Gynomonoecious:* Has both hermaphrodite and female structures.
- *Andromonoecious:* Has both hermaphrodite and male structures.
- *Subandroecious:* Plant has mostly male flowers, with a few female or hermaphrodite flowers.
- *Subgynoecious:* Plant has mostly female flowers, with a few male or hermaphrodite flowers.

- *Trimonoecious (Polygamous):* Male, female, and hermaphrodite structures all appear on the same plant.

Plant Population

- *Hermaphrodite:* Only hermaphrodite plants.
- *Monoecious:* Only monoecious plants.
- *Dioecious:* Only dioecious plants.
- *Gynodioecious:* Both female and hermaphrodite plants present.
- *Androdioecious:* Both male and hermaphrodite plants present.
- *Subdioecious:* Population of primarily unisexual (dioecious) plants, with a few monoecious individuals.
- *Trioecious:* Male, female, and hermaphrodite plants are all in the same population.

Some plants use a method known as self-incompatibility to promote outcrossing. In these plants, the male organs cannot fertilize the female parts of the same plant.

THE FLOWER

The flower is the reproductive organ of plants classified as *angiosperms*—that is, the flowering plants comprising the Division Magnoliophyta. All plants have the means and corresponding structures for reproducing sexually, and these other cases will be explored in later chapters. However, because flowering plants are the most conspicuous plants in almost all terrestrial environments, we justifiably devote this chapter to the flowering plants alone. You will learn how other plant groups (and non-plant groups, such as fungi) reproduce sexually in Section II of the *The Guide.*

The basic function of a flower is to produce seeds through sexual reproduction. Seeds are the next generation, and serve as the primary method in most plants by which individuals of the species are dispersed across the landscape. Actual dispersal is, in most species, a function of the fruit: structural parts that typically surround the seed. But the seed contains the germ of life of the next generation.

A flower, (<Old French *flo(u)r*<Latin *florem*<*flos*), also known as a bloom or blossom, is the reproductive structure found in flowering plants (plants of the division *Magnoliophyta*, also called angiosperms). The flower's structure contains the plant's reproductive organs, and its function is to produce seeds. After fertilization, portions of the flower develop into a fruit containing the seeds. For the higher plants, seeds are the next generation, and serve as the primary means by which individuals of a species are dispersed across the landscape. The grouping of flowers on a plant is called the inflorescence.

In addition to serving as the reproductive organs of flowering plants, flowers have long been admired and used by humans, mainly to beautify their environment but also as a source of food.

Function

The biological function of a flower is to mediate the union of male and female gametes in order to produce seeds. The process begins with pollination, is followed by fertilization, and continues with the formation and dispersal of the seed.

Morphology

Flowering plants are *heterosporangiate*, producing two types of reproductive spores. The pollen (male spores) and ovules (female spores) are produced in different organs, but the typical flower is a *bisporangiate strobilus* in that it contains both organs.

A flower is regarded as a modified stem with shortened internodes and bearing, at its nodes, structures that may be highly modified leaves. In essence, a flower structure forms on a modified shoot or *axis* with an apical meristem that does not grow continuously (growth is *determinate*). The stem is called a pedicel, the end of which is the *torus* or receptacle. The parts of a flower are arranged in whorls on the torus. The four main parts or whorls (starting from the base of the flower or lowest node and working upwards) are as follows:

Calyx is a collective term for all of the sepals, structural components of a flower.It is the outer whorl of *sepals*; typically these are green, but are petal-like in some species.

Corolla is the overall structure of the petals of a flower taken as a group within the calyx. Normally the corolla is the most conspicuous part of a flower and of a bright colour other than green. The concept of corolla description is widely used in botany as a primary determinant of vascular plant identification. Alternatively the corolla may be considered as the inner whorl of the perianth structure. The role of the corolla in plant evolution has been studied extensively since Darwin postulated a theory of the origin of elongated corollae.It is the whorl of *petals*, which are usually thin, soft and colored to attract insects that help the process of pollination.

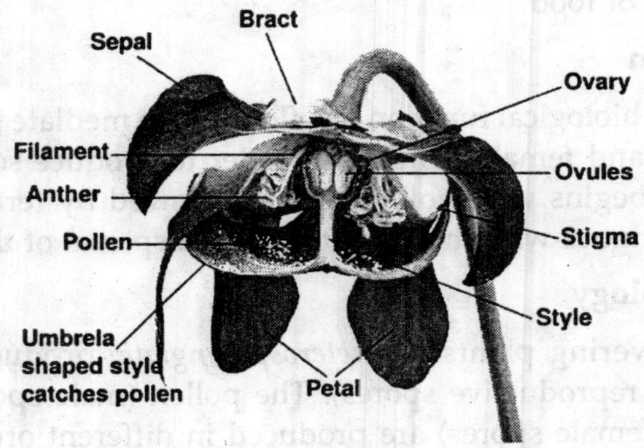

Fig. Anatomy of a *Sarracenia* flower.

The umbrella shaped style is unique to this genus, and will look different in most flowers

An androecium is a male part of a flower in a flowering plant. The androecium is composed of one or more stamina. The female part of a flower is called *gynoecium*. One or two whorls of stamens, each a filament topped by an anther where pollen is produced. Pollen contains the male gametes.

A *gynoecium* (from Ancient Greek *gyne*, "woman") is the female reproductive part of a flower. The male counterpart of is called an *androecium*. A gynoecium is composed of one or more pistils. A pistil may consist of a single free carpel, in which case the gynoecium is termed apocarpous. A pistil can also be formed from a number of carpels that are fused, and in this case the flower is synocarpous. The pistil itself is formed from the stigma, style, and ovary. One or more pistils. The female reproductive organ is the carpel: this contains an ovary with ovules (which contain female gametes). A pistil may consist of a number of carpels merged together, in which case there is only one pistil to each flower, or of a single individual carpel (the flower is then called *apocarpous*). The sticky tip of the pistil, the stigma, is the receptor of pollen. The supportive stalk, the style becomes the pathway for pollen tubes to grow from pollen grains adhering to the stigma, to the ovules, carrying the reproductive material.

Although the floral structure described above is considered the "typical" structural plan, plant species show a wide variety of modifications from this plan. These modifications have significance in the evolution of flowering plants and are used extensively by botanists to establish relationships among plant species. For example, the two subclasses of flowering plants may be distinguished by the number of floral organs in each whorl: dicotyledons typically having 4 or 5 organs (or a multiple of 4 or 5) in each whorl and monocotyledons having three or some multiple of three. The number of carpels in a compound pistil may be only two, or otherwise not related to the above generalization for monocots and dicots. In the majority of species individual flowers have both pistils and stamens as described above. These flowers are described by botanists as being *perfect*, *bisexual*, or *hermaphrodite*. However, in some species of plants the flowers are *imperfect* or *unisexual*: having only either male (stamens) or female (pistil) parts. In the latter case, if an individual plant is either male or female the species is regarded as *dioecious*. However, where unisexual male and female flowers appear on the same plant, the species is considered *monoecious*.

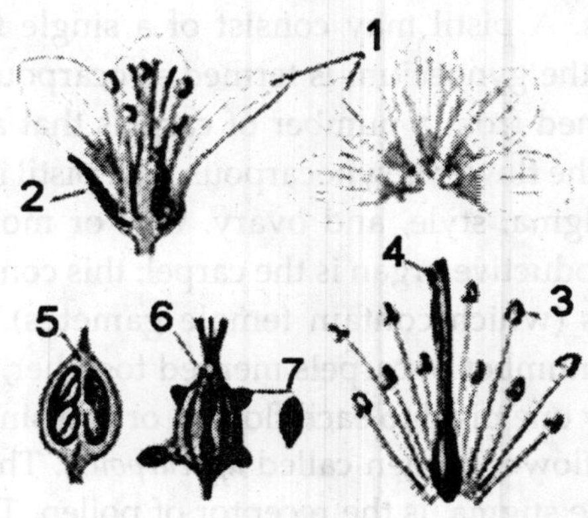

Fig. Anatomy of Oxalis Acetosella Flower.

1. Petal
2. Sepal
3. Anther
4. Stigma
5. Ovary
6. Ovary
7. Ovule.

A carpel is the outer, often visible part of the female reproductive organ of a flower; the basic unit of the *gynoecium*.

Carpel Anatomy

The parts of the carpel are:

- The stigma (from Ancient Greek *stigma* "mark, puncture"), usually the terminal (end) portion that has no epidermis and is fitted to receive pollen (male gametes); it is commonly somewhat glutinous or viscid;
- The style (from Latin *stilus* "stake, stylus"), a stalk connecting the *stigma* with the *ovary* below containing the transmitting tract, which facilitates the growth of the pollen tube and hence the movement of the male gamete to the ovule; and
- The ovary (from Latin *ovum* "egg") or *megasporophyll* containing the female reproductive cell or *ovule*.

The Pistil

A pistil (from Latin *pistillum* "pestle") is made up of a carpel (if single) or carpels (if fused). A flower with two or more fused carpels (called a *compound ovary* or *compound pistil*) is termed *syncarpous*. However, if the gynoecium consists of one or more free, simple, and distinct carpels, each carpel makes an individual *pistil* and the gynoecium is termed *apocarpous*. Fertilization of the ovule or ovules results in development of the carpel(s) into a fruit.

When two or more carpels are fused or joined together its called *syncarpy*. In a compound pistil, the carpels are fused together in one of two basic ways:

- The carpels are fused at or near their margins (parietal placentation), usually forming a single large cavity — an example would be the violet.
- The folded carpels extend in towards the centre, being fused along their outer faces (laterally concrescent), with the placentae arranged around a central column of tissue (axile placentation). There may be as many locules as there are carpels; and tissue of the receptacle may be involved in forming the axillary column. An example of axile placentation would be the lily.

A complicating factor in all of this is the fact that in some species syncarpy is present only at the base of the carpels, the pistil being apocarpous in the upper part. The manner of fusing of the carpels can also vary from one part of the pistil to another.

Inferior vs. Superior Ovaries

The *gynoecium*, the collective term for all the carpels, is the innermost whorl of the parts of a flower, and in many flowers the other parts (sepals, petals, and stamens) are attached to the receptacle beneath the gynoecium. In such cases, where the ovary lies above the attachments of the other distinct floral parts, the flower is described as *hypogynous* or as having a *superior ovary*. In some species (examples include plum, cherry, and blackberry), the other (noncarpellary) floral parts are fused to form a cup called a floral tube or hypanthium. In these flowers, the ovary lies physically lower than the lobes of the sepals and petals and below the point of attachment of the stamen filaments — the ovary is still considered to be superior but the flower is termed *perigynous*.

In those flowers in which the floral tube is fused with the ovary, the sepals, petals, and stamens appear to grow out from the top of the ovary, and the flower is said to be *epigynous* and have an *inferior ovary*. Examples of plant families with inferior ovaries include orchid, sunflower, and cactus. The position of the ovary is an important consideration in the identification and classification of plant species, as well as the kind of fruit that develops after fertilization.

The Ovule

The ovule (from Latin *ovulum* "small egg"), which represents the *megasporangium*, when mature, consists of one or two coats surrounding the central nucellus, except at the apex where an opening, the micropyle, is left. The nucellus is a cellular tissue enveloping one large cell, the embryo-sac or megaspore. The germination of the megaspore consists in the repeated division of its nucleus to form two groups of four, one group at each end of the embryo-sac. One nucleus from each group, the polar nucleus, passes to the centre of the sac, where the two fuse to form the so-called definitive nucleus. Of the three cells at the micropylar end of the sac, all naked cells (the so-called egg-apparatus), one is the egg-cell or oosphere, the other two, which may be regarded as representing abortive egg-cells (in rare cases capable of fertilization), are known as *synergidae*. The three cells at the opposite end are known as antipodal cells and become invested with a cell-wall.

The carpel of a simple *apocarpous* gynoecium appears as a folded structure, differentiated into a basal fertile part (ovary) and an upper sterile part (style). Various interpretations of the origin from a leaf-like structure have been made (Esau, 1965), but the important anatomical description is that of a variously folded tissue surrounding a cavity (called a locule) within which projects one or more ovules, attached by or along a *placenta*. Typically, a carpel has two placentae. An example of a simple carpel is that of a pea, bean or Arabidopsis: the fruit develops from the single carpel consisting of two rows of ovules aligned beside one another along the *placental* margin.

In those species that have more than one flower on an axis—so-called *composite flowers*— the collection of flowers is termed an *inflorescence*; this term can also refer to the specific arrangements of flowers on a stem. In this regard, care must be exercised in considering what a "flower" is. In botanical terminology, a single daisy or sunflower for example, is not a flower but a flower *head*—an inflorescence composed of numerous tiny flowers (sometimes called florets). Each of these flowers may be anatomically as described above. Many flowers have a symmetry, if the perianth is bisected through the central axis from any point, symmetrical halves are produced - the flower is called regular or actinomorphic e.g. rose or trillium. When flowers are bisected and produce only one line that produces symmetrical halves the flower is said to be irregular or zygomorphic. e.g. snapdragon or most orchids.

Floral Formula

A *floral formula* is a way to represent the structure of a flower using specific letters, numbers, and symbols. Typically, a general formula will be used to represent the flower structure of a plant family rather than a particular species. The following representations are used:

Ca = calyx (sepal whorl; e.g. Ca = 5 sepals)

Co = corolla (petal whorl; e.g., $Co^{(x)}$ = petals some multiple of three)

Z = add if *zygomorphic* (e.g., CoZ = zygomorphic with 6 petals)

A = *androecium* (whorl of stamens; e.g., A^{∞} = many stamens)

G = *gynoecium* (carpel or carpels; e.g., G1 = monocarpous)

x - to represent a "variable number"

" - to represent "many"

A floral formula would appear something like this:

$CaCoA^{10-''}G^{1}$

POLLINATION

Pollination is an important step in the reproduction of seed plants: the transfer of pollen grains (male gametes) to the plant carpel, the structure that contains the ovule (female gamete). The receptive part of the carpel is called a *stigma* in the flowers of angiosperms and a *micropyle* in gymnosperms. The study of pollination brings together many disciplines, such as botany, horticulture, entomology, and ecology. Pollination is important in horticulture because most plant fruits will not develop if the ovules are not fertilized. The pollination process as interaction between flower and vector was first addressed in the 18th century by Christian Konrad Sprengel.

The primary purpose of a flower is reproduction by the joining of pollen of one plant with the ovules of another (or in some cases its own ovules) in order to form seed which grows into the next generation of plants. Sexual reproduction produces genetically unique offspring, allowing for adaptation to occur. As such, each flower has a specific design which best encourages the transfer of this pollen. Many flowers are dependent upon the wind to move pollen between flowers of the same species. Others rely on animals (especially insects) to accomplish this feat. Even large animals such as birds, bats, and pygmy possums can be employed. The period of time during which this process can take place (the flower is fully expanded and functional) is called *anthesis*.

Attraction Methods

Many flowers in nature have evolved to attract animals to pollinate the flower, the movements of the pollinating agent contributing to the opportunity for genetic recombination within a dispersed plant population. Flowers that are insect-pollinated are called *entomophilous* (literally "insect-loving"). Flowers commonly have glands called *nectaries* on their various parts that attract these animals. Birds and bees are common pollinators: both having colour vision, thus opting for "colorful" flowers. Some flowers have patterns, called nectar guides, that show pollinators where to look for nectar; they may be visible to us or only under ultraviolet light, which is visible to bees and some other insects. Flowers also attract pollinators by scent. Many of their scents are pleasant to our sense of smell, but not all. Some plants, such as *Rafflesia*, the titan arum, and the North American pawpaw (*Asimina triloba*), are pollinated by flies, so they produce a scent imitating rotting meat. Flowers pollinated by night visitors such as bats or moths are especially likely to concentrate on scent - which can attract pollinators in the dark - rather than colour: most such flowers are white.

Still other flowers use mimicry to attract pollinators. Some species of orchids, for example, produce flowers resembling female bees in colour, shape, and scent. Male bees move from one such flower to another in search of a mate.

Pollination Mechanism

The pollination mechanism employed by a plant depends on what method of pollination is utilized.

Most flowers can be divided between two broad groups of pollination methods:

Entomophilous

Flowers attract and use insects, bats, birds or other animals to transfer pollen from one flower to the next. often they are specialized in shape and have an arrangement of the stamens that ensures that pollen grains are transferred to the bodies of the pollinator when it lands in search of its attractant (such as nectar, pollen, or a mate). In pursuing this attractant from many flowers of the same species, the pollinator transfers pollen to the stigmas - arranged with equally pointed precision - of all of the flowers it visits. Many flower rely on simple proximity between flower parts to ensure pollination. Others, such as the *Sarracenia* or lady-slipper orchids, have elaborate designs to ensure pollination while preventing self-pollination.

Anemophilous

Flowers use the wind to move pollen from one flower to the next, examples include the grasses, Birch trees, Ragweed and Maples. They have no need to attract pollinators and therefore tend not to be "showy" flowers. Whereas the pollen of entomophilous flowers tends to be large-grained, sticky, and rich in protein (another "reward" for pollinators), anemophilous flower pollen is usually small-grained, very light, and of little nutritional value to insects, though it may still be gathered in times of dearth. Honeybees and bumblebees actively gather anemophilous corn (maize) pollen, though it is of little value to them.

Some flowers are self pollinated and use flowers that never open or are self pollinated before the flowers open, these flowers are called clestigomous. Many Viola species and some Salvia have these types of flowers.

Types of Pollination

The process of pollination requires pollinators as agents that carry or move the pollen grains from the anther to the receptive part of the carpel. The various flower traits that attract different pollinators are known as pollination syndromes. Methods of pollination, with common pollinators or plants, are:

Biotic Pollination (by Organisms)

Entomophily: *Pollination by Insects*

Entomophily is a form of pollination whereby pollen is distributed by insects, particularly bees, Lepidoptera (e.g. butterflies and moths), flies and beetles. Entomophilous species frequently evolve mechanisms to make themselves more appealing to insects, e.g. brightly colored or scented flowers, nectar, and appealing shapes and patterns. Pollen grains of entomophilous plants are generally larger than the fine pollens of anemophilous (wind pollinated) plants. They usually are of more nutritional value to insects, who may use them for food and inadvertently spread them to other flowers.

Entomophilous species include the sunflower, orchid, and cycad.This kind of pollination include insects like Bees, wasps and occasionally ants (Hymenoptera),Beetles (Coleoptera), Moths and Butterflies (Lepidoptera),Flies (Diptera)

Zoophily: *Pollination by Vertebrates such as Birds or Bats*

Zoophily is a form of pollination whereby pollen is transferred by vertebrates, particularly by hummingbirds and other birds, and bats, but also by monkeys, marsupials, lemurs, bears, rabbits, deer, rodents, lizards and other animals. Zoomophilous species, like entomophilous species, frequently evolve mechanisms to make themselves more appealing to the particular type of pollinator, e.g. brightly colored or scented flowers, nectar, and appealing shapes and patterns. These plant animal relationships are often mutually beneficial because of the food source provided in exchange for pollination. Zoophilous species include *Arctium, Acaena,* and *Galium aparine.*

Pollination is defined as the transer of pollen from the anther to the stigma (Worldnet). There are many vectors for pollination, including abiotic (wind and water), and biotic (animal). There are different benefits and costs associated with any vector type. For instance, using animal pollination is beneficial because the process is more directed and often results in pollination. At the same time it is costly for the plant to produce rewards, such as nectar, to attract animal pollinators. Not producing such rewards is one benefit of using abiotic pollinators, but a cost associated with this approach is that the pollen may be distributed somewhat randomly. In general, pollination by vertebrates occurs when the animal reaches inside the flowers for nectar. While feeding on the nectar, the animal rubs or touches the stamens and is covered in pollen. Some of this pollen will be deposited on the stigma of the next flower it visits, pollinating the flower (Missouri Botanical Garden 2006).

Bat Pollination

Most bat species that pollinate flowers inhabit Africa, Southeast Asia, and the Pacific Islands, although bat pollination occurs over a geographically wide range. Many fruits are dependent on bats for pollination, such as mangoes, bananas, and guavas (Celebrating Wildlife 2006). Bat pollination is an integral process in tropical communities with 500 tropical plant species completely, or partially, dependent on bats for pollination (Heithaus 1974). Also, it has been noted that outcrossing (introducing unrelated genetic material into a breeding line) by bats increases genetic diversity and is important in tropical communities.

Plants pollinated by bats often have white or pale nocturnal flowers that are large and bell shaped. Many of these flowers have large amounts of nectar, and emit a smell that attracts bats, such as a strong fruity or musky odor (Gibson 2001). Bats use certain chemical cues to locate food sources. They are attracted to odors that contain esters, alcohols, aldehydes, and aliphatic acids.

The banana bat (*Musonycteris harrisoni*) is a nectarivorous species found only on the Pacific coast of Mexico. It has a very small geogrpahic range and is distinguishable by its extremely long nose. The long snout and tongue, one toungue recorded as measuring 76mm, allows this bat to feed on the nectar of long tubular flowers. This bat species is small, with the head and body length ranging from 70 to 79mm. The wild banana flower is elongated with a purple colour.

Pollination by Other Mammals

Non-flying mammals (to distinguish them from bats) have been found to feed on the nectar of several species of plant. Though some of these mammals are pollinators, others do not carry or transfer enough pollen to be considered pollinators (Johnson 2001). This group of non-flying pollinators is mainly composed of marsupials, primates, and rodents (Johnson 2001). Well-documented studies of non-flying mammal pollination now involve at least 59 species of mammal distributed among 19 families and six orders (Carthewa 1997). As of 1997, there were 85 species of plants from 43 genera and 19 families which were visited by these mammals (Carthewa 1997). In many cases, a plant species is visited by a range of mammals. Two examples of multiple mammal pollinotion are the genus *Quararibea* which is visited by 12 species and *Combretum* which is visited by 8.

Plant species that feed non-flying mammals will often exhibit similar characteristics to aide in pollination. The flowers are often large and sturdy, or are grouped together as multi-flowered inflorescences. Many non-flying mammals are nocturnal and have an acute sense of smell, so the plants tend not to have bright showy colors, but instead excrete a pungent odor. Plants will often flower profusely and produce a large amount of sugar-rich nectar. These plants also tend to produce large amounts of pollen because mammals are larger than some other pollinators, and lack the precision smaller pollinators can achieve. Animals with more precision, such as bees or other insects with a proboscis, can pollinate small flowers with less pollen necessary. This means that a plant will require more pollen for a larger mammal pollinator.

One example of a simbiotic relationship between a plant and its animal pollinators is the African Lily, *Massonia depressa,* and some rodent species of the Succulent Karoo region of South Africa. At least four rodent species, including two gerbil species, were found to be visiting *M. depressa* during the night. Traits of the *M. depressa* flowers support non-flying mammal pollination. It has dull-colored and very sturdy flowers at ground level, has a strong yeasty odor, and secretes copious amounts of sucrose-dominant nectar during the night. The nectar of *M. depressa* was also found to be 400 times as viscous, or resistant to flow, as an equivalent sugar solution. This jelly-like constistency of the nectar may discourage insect consumption while also facilitating lapping by rodents. It is assumed that *M. depressa* coevolved with its pollinators.

Bird Pollination

The term ornithophily is used to describe pollination specifically by birds. Hummingbirds, found only in North and South America, are the most recognized nectar-eating bird, but there are many other bird species throughout the world that are also important pollinators. These include: sunbirds, honeyeaters, flowerpeckers, honeycreepers, and bananaquits.

Plants pollinated by birds often have brightly colored diurnal flowers that are red, yellow, or orange, but no odor because birds have a poor sense of smell. Other characteristics of these plants are that they have suitable, sturdy places for perching, abundant nectar that is deeply nested within the flower. Often flowers are elongated or tube shaped. Also, many plants have anthers placed in the flower so that pollen rubs against the birds head/back as the bird reaches in for nectar.

The ruby-throated hummingbird (*Archilochus colubris*) is one of many species of hummingbirds. Found in North and Central America, this bird is an important pollinator for a variety of plant species. Some species, such as the trumpet creeper, are adapted specifically for ruby-throated hummingbirds. This species is quite small, measuring 7.5-9.0 cm long and weighing only 3.4-3.8g. The long narrow bill of the hummingbird is the perfect tool for extracting nectar from elongated flowers. This species is attracted to brightly colored flowers, especially those that are red in colour.

Lizard Pollination

Although lizard pollination has historically been underestimated, recent studies have shown lizard pollination to be an important part of many plant species' survival. Not only do lizards show mutualistic relationships, but these are found to occur most often on islands. This pattern of lizard pollination on islands is mainly due to their high densities, a surplus of floral food, and a relatively low predation risk when compared to lizards on the mainland.

The lizard *Hoplodactylus* is only attracted by nectar on flowers, not pollen. This means flowers pollinated by this species must produce copious nectar as a reward for *Hoplodactylus*. Scented flowers are another important adaptation to attract lizards due to their acute sense of smell. Although lizards have the ability to distinguish colors, as nocturnal feeders, it is more difficult to see bright colors.

Because *Hoplodactylus* feeds nocturnally, it is sometimes less important for flowers to allocate resources to showy inforescences. Flowers must also be robust enough to support the weight of the pollinator while feeding.

In New Zealand *Hoplodactylus* geckos visit flowers of many native plant species for nectar and pollen. The flowers of *Metrosideros excelsa* are pollinated by more than 50 types of gecko as well as birds and bees. Of the geckos visiting this species, two-thirds of them carried large amounts of pollen, suggesting a main role in pollination. However, after the arrival of humans in New Zealand, lizard populations have declined making it more difficult to witness lizard pollination.

Abiotic Pollination

Anemophily: *Pollination by Wind*

Anemophily or wind pollination is a form of pollination whereby pollen is distributed by wind. Unlike entomophilous and zoophilous species whose pollen is spread by insects and vertebrates respectively, anemophilous species do not develop scented flowers, nor do they produce nectar.

Male and female reproductive organs are generally found in separate flowers, the male flowers having a number of long filaments terminating in exposed stamens, and the female flowers having long, feather-like stigmas.

Pollen from anemophilous plants tends to be smaller and lighter in weight than pollen from entomophilous ones, with very low nutritional value to insects. However, insects sometimes gather pollen from staminate anemophilous flowers at times when higher protein pollens from entomophilous flowers are scarce. Also anemophilous pollens may also be inadvertently captured by bees' electrostatic field. This may explain why, though bees are not observed to visit ragweed flowers, its pollen is often found in honey made during the ragweed floral bloom. Other flowers that are generally anemophilous are observed to be actively worked by bees, with solitary bees often visiting grass flowers, and the larger honeybees and bumblebees frequently gathering pollen from corn tassels and other grains.

Almost all pollens that are allergens are anemophilous. Ragweed, the bane of many hayfever sufferers, is anemophilous. Its pollen has been found at sea hundreds of miles from its source. Spring hayfever often traces to pollens from birches.

Other common anemophilous plants are most grass species, conifers, sweet chestnuts, and members of the hickory family.

Hydrophily: Pollination by Water

Hydrophily is a fairly uncommon form of pollination whereby pollen is distributed by the flow of waters, particularly in rivers and streams.

Hydrophilous species fall into two categories: those that distribute their pollen to the surface of water, and those that distribute it beneath the surface. Surface pollination is rare, and appears to be a transitional phase between wind pollination and true hydrophily, where pollen is completely submerged.

Surface hydrophily has been observed in several species of pondweed and waterweed. Species exhibiting true submerged hydrophily include *Posidonia australis* and ribbonweed.

About 80% of all plant pollination is biotic. Of the 20% of abiotically pollinated species, 98% is by wind and 2% by water and sun.

THE SEED AND GERMINATION

the primary purpose of the seed is one of preserving the continuity of life—starting a new generation in a new physical location. For large plants (shrubs and trees), this can be especially important because successful germination and growth close to the parent may be difficult or impossible; the established plant monopolizes light and water resources in its immediate vicinity. Seeds can also serve the function of overwintering or surviving harsh conditions. The entire generation—every individual—may die in the Fall or the dry season. In many annual species, only the seed exists during unfavorable dry or cold conditions.

A seed is a small embryonic plant enclosed in a covering called the seed coat, usually with some stored food. It is the product of the ripened ovule of gymnosperm and angiosperm plants which occurs after fertilization and some growth within the motherplant. The formation of the seed completes the process of reproduction in seed plants (started with the development of flowers and pollination), with the embryo developed from the zygote and the seed coat from the integuments of the ovule.

Seeds have been an important development in the reproduction and spread of flowering plants, relative to more primitive plants like mosses, ferns and liverworts, which do not have seeds and use other means to propagate themselves. This can be seen by the success of seed plants (both gymnosperms and angiosperms) in dominating biological niches on land, from forests to grasslands both in hot and cold climates.

Seed also has a general meaning that predates the above - anything that can be sown i.e. "seed" potatoes, "seeds" of corn or sunflower "seeds". In the case of sunflower and corn "seeds", what is sown is the seed enclosed in a shell or hull, and the potato is a root or stem depending on what is being sowed.

Seed Formation

This process starts with double fertilization in angiosperms and it involves the fusion of the egg and sperm nuclei into a zygote. The second part of this process is the fusion of the polar nuclei with a second sperm cell nucleus, thus forming a primary endosperm. Right after fertilization the zygote is mostly inactive but the primary endosperm divides rapidly to form the endosperm tissue. This tissue becomes the food that the young plant will consume until the roots have developed after germination or it develops into a hard seed coat. The seed, which is an embryo with two points of growth (one of which forms the stems the other the roots) is enclosed in a seed coat with some food reserves. In gymnosperms the two sperm cells transferred from the pollen do not develop seed by double fertilization but instead only one sperm fertilizes the egg while the other is not used. The seed is composed of the embryo (the result of fertilization) and tissue from the mother plant, which also form a cone around the seed in coniferous plants like Pine and Spruce.

The new seed is formed in plant structures called fruits in angiosperms.

Seed Structure

A seed contains the embryo from which a new plant will grow under proper conditions. Seeds also usually contain a supply of stored energy and is wrapped in the seed coat or testa. Seeds are very diverse in size. The dust-like orchid seeds are the smallest with about one million seeds per gram. Embryotic seeds have immature embryos and no significant energy reserves. They are myco-heterotrophs, depending on mycorrhizal fungi for nutrition during germination. At over 20 kg, the largest seed is the coco de mer.

The embryo has one cotyledon or seed leaf in monocotyledons, two cotyledons in almost all dicotyledons and two or more in gymnosperms. The radicle is the embryonic root. The plumule is the embryonic shoot. The embryonic stem above the point of attachment of the cotyledon(s) is the epicotyl. The embryonic stem below the point of attachment is the hypocotyl.

In angiosperms, the stored food begins as a tissue called the endosperm, which is derived from the parent plant via double fertilization. The usually triploid endosperm is rich in oil or starch and protein. In gymnosperms, such as conifers, the food storage tissue is part of the female gametophyte, a haploid tissue. In some species, the embryo is embedded in the endosperm or female gametophyte, which the seedling will use upon germination. In others, the endosperm is absorbed by the embryo as the latter grows within the developing seed, and the cotyledons of the embryo become filled with this stored food. At maturity, seeds of these species have no endosperm and are termed exalbuminous seeds. Some exalbuminous seeds are bean, pea, oak, walnut, squash, sunflower, and radish. Seeds with an endosperm at maturity are termed albuminous seeds. Most monocots (e.g. grasses and palms) and many dicots (e.g. brazil nut and castor bean) have albuminous seeds. All gymnosperm seeds are albuminous. The seed coat develops from the tissue, the integument, originally surrounding the ovule. The seed coat in the mature seed can be a paper-thin layer (e.g. peanut) or something more substantial (e.g. thick and hard in honey locust and coconut). The seed coat helps protect the embryo from mechanical injury and from drying out.

The seeds of angiosperms are contained in a hard or fleshy (or with layers of both) structure called a fruit. Gymnosperm seeds begin their development "naked" on the bracts of cones, although the seeds do become covered by the cone scales as they develop. An example of a hard fruit layer surrounding the actual seed is that of the so-called *stone* fruits (such as the peach).

Some seeds have an appendage on the seed coat such an aril (as in yew and nutmeg) or an elaiosome (as in Corydalis) or hairs (as in cotton). The hilum is the scar on the seed coat where the seed was attached to the ovary wall by the funiculus.

In order for the seed coat to split, the embryo must imbibe (soak up water), which causes it to swell, splitting the seed coat. However, the nature of the seed coat determines how rapidly water can penetrate and subsequently initiate germination. For seeds with a very thick coat, scarification of the seed coat may be necessary before water can reach the embryo. Examples of scarification include: gnawing by animals, freezing and thawing, battering on rocks in a stream bed, or passing through an animal's digestive tract. In the latter case, the seed coat protects the seed from digestion, while perhaps weakening the seed coat such that the embryo is ready to sprout when it gets deposited (along with a bit of fertilizer) far from the parent plant. In species with thin seed coats, light may be able to penetrate into the dormant embryo. The presence of light or the absence of light may trigger the germination process, inhibiting germination in some seeds buried too deeply or in others not buried in the soil. Abscisic acid is usually the growth inhibitor in seeds.

Seed Functions

Seeds protect and nourish the embryo or baby plant. Seeds usually give a seedling a faster start than a sporling from a spore gets because of the larger food reserves in the seed.

Unlike animals, plants are limited in their ability to seek out favorable conditions for life and growth. As a consequence, plants have evolved many ways to disperse their population through their seeds (see also vegetative reproduction). A seed must somehow "arrive" at a location and be there at a time favorable for germination and growth. Seed dispersal is often attributed mainly to fruits, however many seeds aid in their own dispersal, for example:

- Many seeds (e.g. maple, pine) have a wing that aids in wind dispersal.
- The dustlike seeds of orchids are carried efficiently by the wind.
- Some seeds, (e.g. dandelion, milkweed, poplar) have hairs that aid in wind dispersal.
- Seeds with a fleshy covering (e.g. apple, cherry, juniper) are eaten by animals (birds, mammals) which then disperse these seeds in their droppings.
- Seeds (nuts) which are an attractive long-term storable food resource for animals (e.g. acorns, hazelnut, walnut); the seeds are stored some distance from the parent plant, and some escape being eaten if the animal stores hold them.
- Dock which attach to animal fur or feathers, and then drop off later.

- Seeds of some mangroves are viviparous, they begin to germinate while still attached to the parent. The large, heavy root allows the seed to penetrate into the ground when it falls.
- Some seeds have appendages called elaiosomes, e.g. bloodroot, trilliums and Acacias. Elaiosomes provide food for ants, which usually disperse such seeds.
- Some plants, such as *Mucuna* and *Dioclea,* produce buoyant seeds termed sea-beans or drift seeds because they float in rivers to the oceans and wash up on beaches .

For annuals, seeds are a way for the species to survive dry or cold seasons. Ephemeral plants are usually annuals that can go from seed to seed in as few as six weeks.

One important function of most seeds is delaying germination to allow time for dispersal and to prevent all seeds from germinating at once when conditions are favorable. Staggering germination prevents all seeds from germinating at once and being wiped out by bad weather or herbivores. Seed dormancy is defined as a seed failing to germinate under environmental conditions optimal for germination. It is often confused with seed quiescence, which is a seed failing to germinate because environmental conditions are inappropriate for germination. Many cultivated seeds lack dormancy but do not germinate in seed packets simply because there is insufficient moisture.

Germination

Germination is the process where growth emerges from a period of dormancy. The most common example of germination is the sprouting of a seedling from a seed of an angiosperm or gymnosperm. However, the growth of a sporeling from a spore, for example the growth of hyphae from fungal spores, is also germination. In a more general sense, germination can imply anything expanding into greater being from a small existence or germ.

Seed Germination

Germination is the first stage of the making of the seedling. The seed of a higher plant is a small package produced in a flower or cone containing an embryo and stored food reserves. Under favorable conditions, the seed begins to germinate, and the embryonic tissues resume growth, developing towards a seedling.

Dicot Germination

The part of the plant that emerges from the seed first is the embryonic root, termed radicle or primary root. This allows the seedling to become anchored in the ground and start absorbing water. After the root, the embryonic shoot emerges from the seed. The shoot consists of three main parts: the cotyledons (seed leaves), the section of shoot below the cotyledons (hypocotyl), and the section of shoot above the cotyledons (epicotyl). The way the shoot emerges differs between plant groups.

Epigeous

In epigeous (or epigeal) germination, the *hypocotyl* elongates and forms a hook, pulling rather than pushing the cotyledons and apical meristem through the soil. Once it reaches the surface, it straightens and pulls the cotyledons and shoot tip of the growing seedlings into the air. Beans, tamarind, papaya are examples of plant that germinate this way.

Hypogeous

Another way of germination is hypogeous (or hypogeal) where the epicotyl elongates and forms the hook. In this type of germination, the cotyledons stay underground where they eventually decompose. Peas for example germinate this way.

Monocot Germination

In monocot seeds, the embryo's radicle and cotyledon are covered by a coleorhiza and coleoptile, respectively. The coleorhiza is the first part to grow out of the seed, followed by the radicle. The coleoptile is then pushed up through the ground until it reaches the surface. There, it stops elongating and the first leaves emerge through an opening at its tip. Commonly, the primary root dies off and the plant develops shoot-borne roots.

Precocious Germination

While not a class of germination, this refers to germination of the seed occurring inside the fruit before it has begun to decay.

Requirements for Seed Germination

Seed germination depends on many factors, both internal and external. The most important external factors include: water, oxygen, temperature, light and the correct soil conditions. Every variety of seed requires a different set of variables for successful germination. This depends greatly on the individual seed variety and is closely linked to the ecological conditions in the plants' natural habitat.

Water

Germination requires moist conditions. Mature seeds are typically extremely dry and need to take up significant amounts of water before metabolism can resume. The uptake of water into seeds is called imbibition and leads to a marked swelling. The pressure caused by imbibing water aids in cracking the seed coat for germination. When seeds are formed, most plants store large amounts of food, such as starch, proteins, or oils, for the embryo inside the seed. When the seed imbibes water, hydrolytic enzymes are activated that break down these stored food resources and allow the seedling to germinate and grow non-photosynthetically until it reaches the light. Once the seedling starts growing, it requires a continuous supply of water and nutrients.

Oxygen

Most seeds respond best when water levels are enough to moisten the seeds but not soak them, and when oxygen is readily available. Once the seed coat is cracked, the germinating seedling requires oxygen for its metabolism. If the soil is waterlogged, it might cut off the necessary oxygen supply and prevent the seed from germinating as it prevents aerobic respiration, which is the main source for the seedling's energy until it starts to photosynthesize.

Temperature and Light

Seeds germinate over a wide range of temperatures, with many preferring temperatures slightly higher than room-temperature. Often, seeds have a set of temperature range for germination and will not germinate above or below a certain temperature. In addition, some seeds may require exposure to light or to cold temperature (vernalization) to break dormancy before they can germinate. As long as the seed is in its dormant state, it will not germinate even if conditions are favorable. For example, seeds requiring the cold of winter are inhibited from germinating if they never experience frost. Some seeds will only germinate when temperatures reach hundreds of degrees, as during a forest fire. Without fire, they are unable to crack their seed coats. Many seeds in forest settings will not germinate until an opening in the canopy allows them to receive sufficient light for the growing seedling. Most vegetable seeds germinate at a temperature between 50 and 75 degrees.

Stratification

Seeds must be mature and environmental factors must be favorable before germination can take place. When a mature seed is placed under favorable conditions and fails to germinate, it is said to be dormant. Some seeds will not germinate (begin to grow) until they have been dormant for a while. The length of time plant seeds remain dormant can be reduced or eliminated by a simple seed treatment called stratification. Seeds should be planted promptly after stratification.

Stratification mimics natural processes that weaken the seed coat before germination. In nature, some seeds require particular conditions to germinate, such as the heat of a fire (e.g., many Australian native plants), or soaking in a body of water for a long period of time. Others have to be passed through an animal's digestive tract to weaken the seed coat and enable germination.

Hormonal Control

Besides environmental factors, germination and dormancy in seeds are also influenced by plant hormones. The hormone abscisic acid affects seed dormancy and prevents germination, while the hormone gibberellin breaks dormancy and induces seed germination. This effect is used in brewing where barley is treated with gibberellin to ensure uniform seed germination to produce barley malt.

Seedling Establishment

In some definitions, the appearance of the radicle marks the end of germination and the beginning of "establishment", a period that ends when the seedling has exhausted the food reserves stored in the seed. Germination and establishment as an independent organism are critical phases in the life of a plant when they are the most vulnerable to injury, disease, and water stress. The germination index can be used as an indicator of phytotoxicity in soils. The mortality between dispersal of seeds and completion of establishment can be so high, that many species survive only by producing huge numbers of seeds.

Pollen Germination

Another germination event during the life cycle of gymnosperms and flowering plants is the germination of a pollen grain after pollination. Like seeds, pollen grains are severely dehydrated before being released to facilitate their dispersal from one plant to another. They consist of a protective coat containing several cells (up to 8 in gymnosperms, 2-3 in flowering plants). One of these cells is a tube cell. Once the pollen grain lands on the stigma of a receptive flower (or a female cone in gymnosperms), it takes up water and germinates. Pollen germination is facilitated by hydration on the stigma, as well as the structure and physiology of the stigma and style. Pollen can also be induced to germinate *in vitro* (in a petri dish or test tube).

During germination, the tube cell elongates into a pollen tube. In the flower, the pollen tube then grows towards the ovule where it discharges the sperm produced in the pollen grain for fertilization. The germinated pollen grain with its two sperm cells is the mature male microgametophyte of these plants.

Self-Incompatibility

Since most plants carry both male and female reproductive organs in their flowers, there is a high risk for self-pollination and thus inbreeding. Some plants use the control of pollen germination as a way to prevent this selfing. Germination and growth of the pollen tube involve molecular signaling between stigma and pollen In self-incompatibility in plants, the stigma of certain plants can molecularly recognize pollen from the same plant and prevents it from germinating.

Spore Germination

Germination can also refer to the emergence of cells from resting spores and the growth of sporeling hyphae or thalli from spores in fungi, algae, and some plants.

Resting Spores

In resting spores, germination involves cracking the thick cell wall of the dormant spore. For example, in zygomycetes the thick-walled zygosporangium cracks open and the zygospore inside gives rise to the emerging sporangiophore. In slime molds, germination refers to the emergence of amoeboid cells from the hardened spore. After cracking the spore coat, further development involves cell division, but not necessarily the development of a multicellular organism (for example in the free-living amboebas of slime molds).

Zoospores

In motile zoospores, germination frequently means a lack of motility and changes in cell shape, which allow the organism to become sessile.

Ferns and Mosses

In plants such as bryophytes, ferns, and a few others, spores germinate into independent gametophytes. In the bryophytes (e.g. mosses and liverworts), spores germinate into protonemata, similar to fungal hyphae, from which the gametophyte grows. In ferns, the gametophytes are small, heart-shaped prothalli that can often be found underneath a spore-shedding adult plant.

THE FRUIT

The fruit is the actual agent of dispersal in most flowering plants.

The term fruit has different meanings depending on context. In botany, a fruit is the ripened ovary—together with seeds—of a flowering plant. In many species, the fruit incorporates the ripened ovary and surrounding tissues. Fruits are the means by which flowering plants disseminate seeds. In cuisine, when discussing fruit as food, the term usually refers to those plant fruits that are sweet and fleshy, examples of which include plums, apples and oranges. However, a great many common vegetables, as well as nuts and grains, are the fruit of the plant species they come from. No single terminology really fits the enormous variety that is found among plant fruits. The cuisine terminology for fruits is inexact and will remain so. The term false fruit (pseudocarp, accessory fruit) is sometimes applied to a fruit like the fig or to a plant structure that resembles a fruit but is not derived from a flower or flowers. Some gymnosperms, such as yew, have fleshy arils that resemble fruits and some junipers have *berry-like*, fleshy cones. The term "fruit" has also been inaccurately applied to the seed-containing female cones of many conifers.

With most fruits pollination is a vital part of fruit culture, and the lack of knowledge of pollinators and pollenizers can contribute to poor crops or poor quality crops. In a few species, the fruit may develop in the absence of pollination/fertilization, a process known as *parthenocarpy*. Such fruits are seedless. A plant that does not produce fruit is known as *acarpous*, meaning "without fruit".

Botanic Fruit and Culinary Fruit

Many foods are botanically fruit but are treated as vegetables in cooking. These include cucurbits (e.g., squash, pumpkin, and cucumber), tomato, peas, beans, corn, eggplant (aubergine), and sweet pepper, spices, such as allspice and chillies. Occasionally, though rarely, a culinary "fruit" will not be a true fruit in the botanical sense. For example, rhubarb may be considered a fruit, though only the astringent petiole is edible. In the commercial world, European Union rules define carrot as a fruit for the purposes of measuring the proportion of "fruit" contained in carrot jam. In the culinary sense, a fruit is usually any sweet tasting plant product associated with seed(s), a vegetable is any savoury or less sweet plant product, and a nut any hard, oily, and shelled plant product.

Although a nut is a type of fruit, it is also a popular term for edible seeds, such as peanut (which is actually a legume), pistachio and walnut. Technically, a cereal grain is a fruit termed a caryopsis. However, the fruit wall is very thin and fused to the seed coat so almost all of the edible grain is actually a seed. Therefore, cereal grains, such as corn, wheat and rice are better considered edible seeds, although some references list them as fruits. Edible gymnosperms seeds are often misleadingly given fruit names, e.g. pine nuts, ginkgo nuts, and juniper berries.

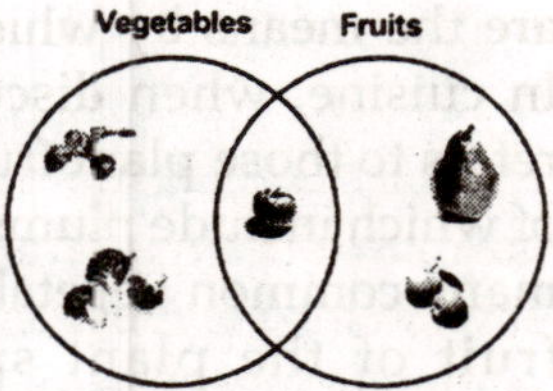

Fig. Venn Diagram

Fruit Development

A fruit is a ripened ovary. After the ovule in an ovary is fertilized in a process known as pollination, the ovary begins to ripen. The ovule develops into a seed and the ovary wall *pericarp* may become fleshy (as in berries or drupes), or form a hard outer covering (as in nuts). In some cases, the sepals, petals and/or stamens and style of the flower fall off. Fruit development continues until the seeds have matured. With some multiseeded fruits the extent to which the flesh develops is proportional to the number of fertilized ovules.

The wall of the fruit, developed from the ovary wall of the flower, is called the *pericarp*. The *pericarp* is often differentiated into two or three distinct layers called the *exocarp* (outer layer - also called epicarp), *mesocarp* (middle layer), and *endocarp* (inner layer). In some fruits, especially simple fruits derived from an inferior ovary, other parts of the flower (such as the floral tube, including the petals, sepals, and stamens), fuse with the ovary and ripen with it. The plant hormone ethylene causes ripening. When such other floral parts are a significant part of the fruit, it is called an *accessory fruit*. Since other parts of the flower may contribute to the structure of the fruit, it is important to study flower structure to understand how a particular fruit forms.

Fruits are so varied in form and development, that it is difficult to devise a classification scheme that includes all known fruits. Many common terms for seeds and fruit are incorrectly applied, a fact that complicates understanding of the terminology. *Seeds are ripened ovules; fruits are the ripened ovaries or carpels that contain the seeds.* To these two basic definitions can be added the clarification that in botanical terminology, a nut is a type of fruit and not another term for seed.

There are three basic types of fruits:

1. Simple fruit
2. Aggregate fruit
3. Multiple fruit

Simple Fruit

Simple fruits can be either dry or fleshy and result from the ripening of a simple or compound ovary with only one pistil. Dry fruits may be either dehiscent (opening to discharge seeds), or indehiscent (not opening to discharge seeds). Types of dry, simple fruits (with examples) are:

- Achene - (buttercup)
- Capsule - (Brazil nut)
- Caryopsis - (wheat)
- Fibrous drupe - (coconut, walnut)
- Follicle - (milkweed)
- Legume - (pea, bean, peanut)
- Loment
- Nut - (hazelnut, beech, oak acorn)
- Samara - (elm, ash, maple key)
- Schizocarp - (carrot)
- Silique - (radish)
- Silicle - (shepherd's purse)
- Utricle - (beet)

Fruits in which part or all of the *pericarp* (fruit wall) is fleshy at maturity are *simple fleshy fruits*. Types of fleshy, simple fruits (with examples) are:

- Berry: (tomato, avocado)
- Stone fruit or drupe (plum, cherry, peach, olive)
- False berry - accessory fruits (banana, cranberry)
- Pome: Accessory fruits (apple, pear, rosehip)

Aggregate Fruit

An aggregate fruit, or *etaerio*, develops from a flower with numerous simple pistils. An example is the raspberry, whose simple fruits are termed *drupelets* because each is like a small drupe attached to the receptacle. In some bramble fruits (such as blackberry) the receptacle is elongated and part of the ripe fruit, making the blackberry an *aggregate-accessory* fruit. The strawberry is also an aggregate-accessory fruit, only one in which the seeds are contained in achenes. In all these examples, the fruit develops from a single flower with numerous pistils.

Multiple Fruit

A multiple fruit is one formed from a cluster of flowers (called an *inflorescence*). Each flower produces a fruit, but these mature into a single mass. Examples are the pineapple, edible fig, mulberry, osage-orange, and breadfruit.

In the photograph on the right, stages of flowering and fruit development in the noni or Indian mulberry (*Morinda citrifolia*) can be observed on a single branch. First an inflorescence of white flowers called a head is produced. After fertilization, each flower develops into a drupe, and as the drupes expand, they become *connate* (merge) into a *multiple fleshy fruit* called a *syncarpet*.

There are also many dry multiple fruits, e.g.

- Tuliptree, multiple of samaras.
- Sweet gum, multiple of capsules.
- Sycamore and teasel, multiple of achenes.
- Magnolia, multiple of follicles.

Seedless Fruits

Seedlessness is an important feature of some fruits of commerce. Commercial cultivars of bananas and pineapples are examples of seedless fruits. Some cultivars of citrus fruits (especially navel oranges and mandarin oranges), table grapes, grapefruit, and watermelons are valued for their seedlessness. In some species, seedlessness is the result of *parthenocarpy*, where fruits set without fertilization. Parthenocarpic fruit set may or may not require pollination. Most seedless citrus fruits require a pollination stimulus; bananas and pineapples do not. Seedlessness in table grapes results from the abortion of the embryonic plant that is produced by fertilization, a phenomenon known as *stenospermocarpy* which requires normal pollination and fertilization.

Seed Dissemination

Variations in fruit structures largely depend on the mode of dispersal of the seeds they contain. This dispersal can be achieved by animals, wind, water, or explosive dehiscence.

Some fruits have coats covered with spikes or hooked burrs, either to prevent themselves from being eaten by animals or to stick to the hairs, feathers or legs of animals, using them as dispersal agents. Examples include cocklebur and unicorn plant.

The sweet flesh of many fruits is "deliberately" appealing to animals, so that the seeds held within are eaten and "unwittingly" carried away and deposited at a distance from the parent. Likewise, the nutritious, oily kernels of nuts are appealing to rodents (such as squirrels) who hoard them in the soil in order to avoid starving during the winter, thus giving those seeds that remain uneaten the chance to germinate and grow into a new plant away from their parent.

Other fruits are elongated and flattened out naturally and so become thin, like wings or helicopter blades, e.g. maple, tuliptree and elm. This is an evolutionary mechanism to increase dispersal distance away from the parent via wind. Other wind-dispersed fruit have tiny *parachutes*, e.g. dandelion and salsify.

Coconut fruits can float thousands of miles in the ocean to spread seeds. Some other fruits that can disperse via water are nipa palm and screw pine.

Some fruits fling seeds substantial distances (up to 100 m in sandbox tree) via explosive dehiscence or other mechanisms, e.g. impatiens and squirting cucumber.

Uses

Many hundreds of fruits, including fleshy fruits like apple, peach, pear, kiwifruit, watermelon and mango are commercially valuable as human food, eaten both fresh and as jams, marmalade and other preserves. Fruits are also found commonly in such manufactured foods as cookies, muffins, yoghurt, ice cream, cakes, and many more. Many fruits are used to make beverages, such as fruit juices (orange juice, apple juice, grape juice, etc) or alcoholic beverages, such as wine or brandy.

Many vegetables are botanical fruits, including tomato, bell pepper, eggplant, okra, squash, pumpkin, green bean, cucumber and zucchini. Olive fruit is pressed for olive oil. Apples are often used to make vinegar. The spices vanilla, paprika, allspice and black pepper are made from fruits.

Nonfood Uses

Because fruits have been such a major part of the human diet, different cultures have developed many different uses for various fruits that they do not depend on as being edible. Many dry fruits are used as decorations or in dried flower arrangements, such as unicorn plant, lotus, wheat, annual honesty and milkweed. Ornamental trees and shrubs are often cultivated for their colorful fruits, including holly, pyracantha, viburnum, skimmia, beautyberry and cotoneaster.

Fruits of opium poppy are the source of the drugs opium and morphine. Osage orange fruits are used to repel cockroaches. Bayberry fruits provide a wax often used to make candles. Many fruits provide natural dyes, e.g. walnut, sumac, cherry and mulberry. Dried gourds are used as decorations, water jugs, bird houses, musical instruments, cups and dishes. Pumpkins are carved into Jack-o'-lanterns for Halloween. The spiny fruit of burdock or cocklebur were the inspiration for the invention of Velcro.

Coir is a fiber from the fruit of coconut that is used for doormats, brushes, mattresses, floortiles, sacking, insulation and as a growing medium for container plants. The shell of the coconut fruit is used to make souvenir heads, cups, bowls, musical instruments and bird houses.

Chapter 6

Plant Morphology

INTRODUCTION

Plant morphology is the study of the external structure or form of plants. You have already covered important aspects of plant morphology in the previous chapters on plant organs (leaves, stems, roots), flowers, and fruits. Those chapters emphasized both plant anatomy (the internal structure) and plant morphology (the external form) of specific parts or organs. Here, we will be concerned especially with the various growth forms that plants take and how such morphologies contribute to the success of each species.

SIMPLE PLANT FORMS

The forms taken by simpler "plants" (e.g., algae) will be considered in greater detail in Section II of *The Guide*. However, lichens offer a good general introduction to basic plant forms in a group whose members are sometimes informally classified on the basis of morphology.

LICHEN

Lichens are symbiotic associations of a fungus (the mycobiont) with a photosynthetic partner (the photobiont also known as the phycobiont) that can produce food for the lichen from sunlight. The photobiont is usually either green algae or cyanobacteria. A few lichens are known to contain yellow-green algae or, in one case, a brown alga. Some lichens contain both green algae and cyanobacteria as photobionts; in these cases, the cyanobacteria symbiont component may specialize in fixing atmospheric nitrogen for metabolic use. The word is pronounced as though it were spelled "liken" (IPA: [lajkYn]).

The body (thallus) of most lichens is quite different from that of either the fungus or alga growing separately, and may strikingly resemble simple plants in form and growth (Sanders 2001). The fungus surrounds the algal cells, often enclosing them within complex fungal tissues unique to lichen associations; however, the algal cells are never enclosed inside the fungal cells themselves. The fungus may or may not penetrate into the algal cells with fine hyphal protrusions.

In general, the symbiosis is considered obligatory for successful growth and reproduction of the fungus; however, the significance for the algal symbiont is less clear. For some algae, the symbiosis may be obligatory for survival in a particular habitat; in other cases, the symbiosis might not be advantageous for the alga. Thus, there is some controversy as to whether the lichen symbiosis should be considered an example of mutualism or parasitism or commensalism. Nonetheless, the lichen is typically a highly stable association which probably extends the ecological range of both partners.

There is evidence that lichens might involve a controlled form of parasitism of the algal cells. In laboratory settings, cyanobacteria grow faster when they are alone rather than when they are part of a lichen. But there is also a mutualistic side to the relationship: the fungus part of the lichen provides the alga with water and minerals that the fungus absorbs from whatever the lichen is growing on, its substrate. As for the alga, it uses the minerals and water to make food for the fungus and itself.

Lichens take the external shape of the fungal partner and hence are named based on the fungus. The fungus most commonly forms the majority of a lichen's bulk, though in filamentous and gelatinous lichens this may not always be the case. The lichen fungus is typically a member of the Ascomycota—rarely a member of the Basidiomycota, and then termed basidiolichens to differentiate them from the more common ascolichens. Formerly, some lichen taxonomists placed lichens in their own division, the Mycophycophyta, but this practice is no longer accepted because the components belong to separate lineages. Neither the ascolichens nor the basidiolichens form monophyletic lineages in their respective fungal phyla, but they do form several major solely or primarily lichen-forming groups within each phylum. Even more unusual than basidiolichens is the fungus *Geosiphon pyriforme*, a member of the Glomeromycota that is unique in that it encloses a cyanobacterial symbiont inside its cells. *Geospihon* is not usually considered to be a lichen, and its peculiar symbiosis was not recognized for many years. The genus is more closely allied to endomycorrhizal genera.

The algal or cyanobacterial cells are photosynthetic, and as in higher plants they reduce atmospheric carbon dioxide into organic carbon sugars to feed both symbionts. Both partners gain water and mineral nutrients mainly from the atmosphere, through rain and dust. The fungal partner protects the alga by retaining water, serving as a larger capture area for mineral nutrients and, in some cases, provides minerals obtained from the substratum. If a cyanobacterium is present, as a primary partner or another symbiont in addition to green alga. as in certain tripartite lichens, they can fix atmospheric nitrogen, complementing the activities of the green alga.

Morphology and Structure

Lichens are often the first to settle in places lacking soil, constituting the sole vegetation in some extreme environments such as those found at high mountain elevations and at high latitudes. Some survive in the tough conditions of deserts, and others on frozen soil of the arctic regions. Recent ESA research shows that lichen can even endure extended exposure to space. Some lichens have the aspect of leaves (foliose lichens); others cover the substratum like a crust (crustose lichens); others adopt shrubby forms (fruticose lichens); and there are gelatinous lichens.

Although the form of a lichen is determined by the genetic material of the fungal partner, association with a photobiont is required for the development of that form. When grown in the laboratory in the absence of its photobiont, a lichen fungus develops as an undifferentiated mass of hyphae. If combined with its photobiont under appropriate conditions, its characteristic form emerges, in the process called morphogenesis (Brodo, Sharnoff & Sharnoff, 2001). In a few remarkable cases, a single lichen fungus can develop into two very different lichen forms when associating with either a green algal or a cyanobacterial symbiont. Quite naturally, these alternative forms were at first considered to be different species, until they were first found growing in a conjoined manner.

There is evidence to suggest that the lichen symbiosis is parasitic rather than mutualistic (Ahmadjian 1993). The photosynthetic partner can exist in nature independently of the fungal partner, but not vice versa. Furthermore, photobiont cells are routinely destroyed in the course of nutrient exchange. The association is able to continue because photobiont cells reproduce faster than they are destroyed.

Under magnification, a section through a typical foliose lichen thallus reveals four layers of interlaced fungal filaments. The uppermost layer is formed by densely agglutinated fungal hyphae building a protective outer layer called the cortex. In lichens that include both green algal and cyanobacterial symbionts, the cyanobacteria may be held on the upper or lower surface in small pustules called cephalodia/cephalodium. Beneath the upper cortex is an algal layer composed of algal cells embedded in rather densely interwoven fungal hyphae. Each cell or group of cells of the photobiont is usually individually wrapped by hyphae, and in some cases penetrated by an haustorium. Beneath this algal layer is a third layer of loosely interwoven fungal hyphae without algal cells. This layer is called the medulla. Beneath the medulla, the bottom surface resembles the upper surface and is called the lower cortex, again consisting of densely packed fungal hyphae. The lower cortex often bears rootlike fungal structures known as rhizines, which serve to attach the thallus to the substrate on which it grows. Lichens also sometimes contain structures made from fungal metabolites, for example crustose lichens sometimes have a polysaccharide layer in the cortex. Although each lichen thallus generally appears homogeneous, some evidence seems to suggest that the fungal component may consist of more than one genetic individual of that species. This seems to also be true of the photobiont species involved.

Reproduction

Many lichens reproduce asexually, either by vegetative reproduction or through the dispersal of diaspores containing algal and fungal cells. *Soredia* (singular soredium) are small groups of algal cells surrounded by fungal filaments that form in structures called *soralia,* from which the soredia can be dispersed by wind. Another form of diaspore are *isidia,* elongated outgrowths from the thallus that break off for mechanical dispersal. Fruticose lichens in particular can easily fragment. Due to the relative lack of differentiation in the thallus, the line between diaspore formation and vegetative reproduction is often blurred. Many lichens break up into fragments when they dry, dispersing themselves by wind action, to resume growth when moisture returns.

Many lichen fungi appear to reproduce sexually in a manner typical of fungi, producing spores that are presumably the result of sexual fusion and meiosis. Following dispersal, such fungal spores must meet with a compatible algal partner before a functional lichen can form. This may be a common form of reproduction in basidiolichens, which form fruitbodies resembling their nonlichenized relatives. Among the ascolichens, spores are produced in spore-producing bodies, the three most common spore body types are the apothecia, perithecia and the pycnidia.

Imagine an early earth, with most life in the oceans, but the beginnings of terrestrial life represented by primitive photoautotrophic organisms spread over rocky surfaces. In the seas, the autotrophs would be floating forms and encrusting slimes, or calcareous rich crusts.

When we think about the evolution of more complex forms, even within such groups that today appear rather primitive such as the multicellular algae, we need to consider what forces were operating on the mechanism of natural selection to drive evolution in certain directions. We must also always be mindful of realistic limitations.

When you previously read the article on algae, you may recall the following description of life forms of these simple "plants": Most algae are unicellular, flagellates or amoeboids, but colonial and non-motile forms have developed independently among several of the algal groups. The more common organizational levels beyond unicellular forms are:

Colonial: Small, regular groups of (usually) motile cells

Capsoid: Individual non-motile cells embedded in a common mucilage

Coccoid: Individual non-motile cells with adherent cell walls

Palmelloid: Non-motile cells embedded in common mucilage

Filamentous: A string of non-motile cells connected on end together and sometimes branching

Parenchymatous: Cells forming a thallus with partial differentiation of tissues.

TREE

A tree is a large perennial plant. Though there is no set definition regarding minimum size, the term generally applies to plants that grow to at least (20 ft) high at maturity and having secondary branches supported on a main stem or stems (see shrub for comparison). Most trees exhibit clear apical dominance, though this is not always the case (Mitchell, 1978). Compared with most other plant forms, trees are long-lived. A few species of trees grow to 115 m (375 ft) tall and some can live for several thousand years.

Trees are an important component of the natural landscape due to their prevention of erosion and significant elements in landscaping and agriculture, both for their aesthetic appeal and their orchard crops (such as apples). Wood from trees is a common building material. Trees also play an intimate role in many of the world's mythologies. Trees have also been found to play an important role in producing oxygen and reducing carbon dioxide in the atmosphere, as well as moderating ground temperatures.

Classification

A tree is a plant form that occurs in many different orders and families of plants. Trees show a wide variety of growth forms, leaf type and shape, bark characteristics, and reproductive organs.

The earliest trees were tree ferns and horsetails, which grew in vast forests in the Carboniferous Period; tree ferns still survive, but the only surviving horsetails are not of tree form. Later, in the Triassic Period, conifers, ginkgos, cycads and other gymnosperms appeared, and subsequently flowering plants in the Cretaceous Period. Most species of trees today are flowering plants (Angiosperms) and conifers. The listing below gives examples of many well-known trees and how they are typically classified.

A small group of trees growing together is called a grove or copse, and a landscape covered by a dense growth of trees is called a forest. Several biotopes are defined largely by the trees that inhabit them; examples are rainforest and taiga. A landscape of trees scattered or spaced across grassland (usually grazed or burned over periodically) is called a savanna.

Morphology

The basic parts of a tree are the roots, trunk(s), branches, twigs and leaves. Tree stems consist mainly of support and transport tissues (xylem and phloem). Wood consists of *xylem* cells, and bark is made of *phloem* and other tissues external to the vascular cambium. Trees may be broadly grouped into *exogenous* and *endogenous* trees according to the way in which their stem diameter increases. Exogenous trees, which comprise the great majority of modern trees (all conifers, and all broadleaf trees), grow by the addition of new wood outwards, immediately under the bark. Endogenous trees, mainly in the monocotyledons (e.g., palms), grow by addition of new material inwards.

As an exogenous tree grows, it creates growth rings. Also known as annual rings, each set of light/dark rings is equivalent to one year of growth. In temperate climates, these are commonly visible due to changes in the rate of growth with temperature variation over an annual cycle. These rings can be counted to determine the age of the tree, and used to date cores or even wood taken from trees in the past; this practice is known as the science of dendrochronology. In some tropical regions with constant year-round climate, growth is continuous and distinct rings are not formed, so age determination is impossible. Age determination is also impossible in endogenous trees.

The roots of a tree are generally embedded in earth, providing anchorage for the above-ground biomass and absorbing water and nutrients from the soil. It should be noted, however, that while ground nutrients are essential to a tree's growth the majority of its biomass comes from carbon dioxide absorbed from the atmosphere. Above ground, the trunk gives height to the leaf-bearing branches, aiding in competition with other plant species for sunlight. In many trees, the arrangement of the branches optimizes exposure of the leaves to sunlight.

Not all trees have all the plant organs or parts mentioned above. For example, most palm trees are not branched, the saguaro cactus of North America has no functional leaves, tree ferns do not produce bark, etc. Based on their general shape and size, all of these are nonetheless generally regarded as trees. Indeed, sometimes size is the more important consideration. A plant form that is similar to a tree, but generally having smaller, multiple trunks and/or branches that arise near the ground, is called a shrub. However, no sharp differentiation between shrubs and trees is possible. Given their small size, bonsai plants would not technically be 'trees', but one should not confuse reference to the form of a species with the size or shape of individual specimens. A spruce seedling does not fit the definition of a tree, but all spruces are trees.

SHRUB

A shrub or bush is a horticultural rather than strictly botanical category of woody plant, distinguished from a tree by its multiple stems and lower height, usually less than 6 m tall. A large number of plants can be either shrubs or trees, depending on the growing conditions they experience. Small, low shrubs such as lavender, periwinkle and thyme are often termed subshrubs.

A natural plant community dominated by shrubs is called a shrubland.

An area of cultivated shrubs in a park or garden is known as a shrubbery. When clipped as topiary, shrubs generally have dense foliage and many small leafy branches growing close together. Many shrubs respond well to renewal pruning, in which hard cutting back to a 'stool' results in long new stems known as "canes". Other shrubs respond better to selective pruning to reveal their structure and character.

Shrubs in common garden practice are generally broad-leaved plants, though some smaller conifers such as Mountain Pine and Common Juniper are also shrubby in structure. Shrubs can be either deciduous or evergreen.

Chapter 7

Plant Systematics

THE KINDS OF PLANTS

The diversity (number of species) of plants in the world is tremendous. Any attempt at representing this great complexity requires a system of ordering or arranging the many plant types—hopefully a system that itself contributes to our knowledge and understanding. We could take the straightforward approach of listing all plants alphabetically by their common name, or pehaps by their species name. This index approach would be handy, but would not tell us much about the plants themselves. We would have to know the plants whose names are included for the list itself to have any meaning to us. However, botanists developed a hierarchical system based on the fact—considering a variety of characteristics—that some plants are clearly more similar to each other than to other plants. Arranging all of the world's plants (and animals and minerals) by similarities to each other was an idea first promoted by Carolus Linnaeus.

LINNAEAN TAXONOMY

Linnaean taxonomy is a method of classifying living things originally devised by, and named for, Carl Linnaeus although it has changed considerably since his time. The greatest innovation of Linnaeus, and still the most important aspect of this system, is the general use of binomial nomenclature, the combination of a genus name and a single specific epithet to uniquely identify each species of organism. For example, the human species is uniquely identified by the binomial *Homo sapiens*. No other species of animal can have this binomial. Originally, animals were classified according to their mode of movement prior to Linnaean Taxonomy.

All species are classified in a ranked hierarchy, originally starting with *kingdoms* although Domains has since been added as a rank above Kingdoms. Kingdoms are divided into *phyla* (singular: *phylum*)—for animals; the term *division*, used for plants, is equivalent to the rank of phylum (and the current International Code of Botanical Nomenclature allows the use of either term). Phyla (or divisions) are divided into *classes*, and they, in turn, into *orders, families, genera* (singular: *genus*), *species* (singular: *species*).

Though the Linnaean system has proven robust, expansion of knowledge has led to an expansion of the number of hierarchical levels within the system, increasing the administrative requirements of the system though it remains the only extant working classification system at present that enjoys universal scientific acceptance. Among the later subdivisions that have arisen are such entities as Phyla (singular: phylum), Superclasses, Superorders, Infarorders, Families, Superfamilies and Tribes. Many of these extra hierarchical levels tend to arise in disciplines such as entomology, whose subject matter is replete with species requiring classification. Any biological field that is species rich, or which is subject to a revision of the state of extant knowledge concerning those species and their relationships to each other, will inevitably make use of the additional hierarchical levels, particularly if integration of living organisms with fossils is performed, and the application of newer classification tools such as cladistics to facilitate this takes place.

There are ranks below species: In zoology, *Subspecies* and *Morph*; in botany, *Variety* (varietas) and *Form* (forma). Many botanists now use "Subspecies" instead of "Variety" although the two are not, strictly speaking, of equivalent rank, and "Form" has largely fallen out of use.

Groups of organisms at any of these ranks are called *taxa* (singular: *taxon*), or *phyla*, or *taxonomic groups*.

NOMENCLATURE

A strength of Linnaean taxonomy is that it can be used to develop a simple and practical system for organizing the different kinds of living organisms. Every species is given a unique and stable name (compared with common names that are often neither unique nor consistent from place to place and language to language). This uniqueness and stability are, of course, a result of the acceptance by working systematists (biologists specializing in taxonomy); not merely of the binomial nomenclature in itself, but of much more complex codes of rules and procedures governing the use of these names.

These rules are governed by formal codes of biological nomenclature. The rules governing the nomenclature and classification of plants and fungi are contained in the International Code of Botanical Nomenclature, maintained by the International Association for Plant Taxonomy. The current code, the "Saint Louis Code" was adopted in 1999 and supersedes the "Tokyo code". The corresponding code for animals is the International Code of Zoological Nomenclature (ICZN), also last revised in 1999, and maintained by the International Commission on Zoological Nomenclature. The code for bacteria is the International Code of Nomenclature of Bacteria (ICNB), last revised in 1990, and maintained by the International Committee on Systematics of Prokaryotes (ICSP). There is also a code for virus nomenclature, the Universal Virus Database of the International Committee on Taxonomy of Viruses (ICTVdB) although it is organized on somewhat different principles, as the evolutionary history of these forms is not understood.

International Code of Botanical Nomenclature

The *International Code of Botanical Nomenclature* (*ICBN*) is the set of rules and recommendations dealing with the formal botanical names that are given to plants. Its intent is that each taxonomic group ("taxon", plural "taxa") of plants has only one correct name, accepted worldwide. The value of a scientific name is that it is a label: it is not necessarily of descriptive value, or even accurate.

- The guiding principle in botanical nomenclature is priority. The *ICBN* sets the formal starting date of plant nomenclature at 1 May 1753, the publication of *Species Plantarum* by Linnaeus (or at later dates for specified groups and ranks).
- A botanical name is fixed to a taxon by a type. This is almost invariably dried plant material and is usually deposited and preserved in a herbarium. Many type collections can be viewed online at the website of the herbarium in question.

Both these principles are regulated and limited. To avoid undesirable effects of priority, conservation of a name is possible. Above the rank of family very few hard rules apply.

The *ICBN* can only be changed by an International Botanical Congress (IBC), with the International Association for Plant Taxonomy providing the supporting infrastructure. The present edition is the *Vienna Code* (2005), based on the decisions of the XVII IBC at Vienna 2005. This was preceded by the *St Louis Code* (2000) and the *Tokyo Code* (1994), both available online. Each new edition supersedes the earlier editions and is retroactive back to 1753 (except where expressly limited).

Botanical nomenclature is independent of zoological and bacteriological nomenclature, which are governed by their own Codes.

The *ICBN* applies not only to plants, as they are now defined, but also to other organisms traditionally studied by botanists. This includes blue-green algae (*Cyanobacteria*); fungi, including chytrids, oomycetes, and slime moulds; photosynthetic protists and taxonomically related non-photosynthetic groups. There are special provisions in the *ICBN* for some of these groups, as there are for fossils.

For the naming of cultivated plants there is a separate *Code*, the *International Code of Nomenclature for Cultivated Plants*. This gives supplementary rules and recommendations.

Categorizing is an important process by which we humans gain understanding of the world around us, and something we all do to some degree as part of our observation of things and events that we encounter. In biology, as the concept of evolution was formulated, it became obvious that this concept could be the basis for catagorization. If plants that are similar in form are indeed closely related—at least more closely related than plants that are dissimilar in form—then a system of classification could be devised that reflected these relationships. This approach has important implications. Related plants have common properties, a fact that can be exploited in agriculture and other practical botanical fields.

Initially, botanists had but one approach: physical examination. The careful examination (and detailed description) of plant structures allowed for arranging each species within a system that placed all more or less similar plants (in certain "important" features) together. This approach is not as easy as it sounds, but played off of and contributed to the expansion of descriptive botany in the 18th and 19th centuries. One problem that became evident is that as species evolved, unrelated plants could come to resemble each other in many respects. After all, form and function are closely related. Within similar habitats (say deserts), species of very distantly related plants might well evolve towards a similar form. Species do not have (or certainly did not have over geological time) unrestricted access to all places on the planet and species distribution is then an important part of interpreting the evolutionary process. As species evolved, they did so within the constraints to dispersion that existed at the time. This fact provides an important clue: unrelated but similar plants are likely to be distributed far from each other on the earth's surface; and the corollary: plants that have similar structures but have widely separated distributions, may not be so closely related in an evolutionary sense.

At this point it is worthwhile to consider some examples. There are many succulant plants, as this form (typically thick, fleshy stems and/or leaves; often reduction or complete loss of leaves) incorporates adaptations necessary for a plant to survive very dry conditions. Non-botanists are tempted to classify all such plants as types of cacti. In fact, cacti evolved in the New World (the Americas), yet there are many succulants (and many plants that resemble cacti) that are not native to the New World, and evolved independently on the African continent. A large group of such plants are known as the euphorbs.

Later Developments Since Linnaeus

Over time, our understanding of the relationships between living things has changed. Linnaeus could only base his scheme on the structural similarities of the different organisms. The greatest change was the widespread acceptance of evolution as the mechanism of biological diversity and species formation. It then became generally understood that classifications ought to reflect the phylogeny of organisms, by grouping each taxon so as to include the common ancestor of the group's members (and thus to avoid polyphyly). Such taxa may be either monophyletic (including all descendants) such as genus Homo, or paraphyletic (excluding some descendants), such as genus Australopithecus.

Originally, Linnaeus established three kingdoms in his scheme, namely Plantae, Animalia and an additional group for minerals, which has long since been abandoned. Since then, various life forms have been moved into three new kingdoms: Monera, for prokaryotes (i.e., bacteria); Protista, for protozoans and most algae; and Fungi. This five kingdom scheme is still far from the phylogenetic ideal and has largely been supplanted in modern taxonomic work by a division into three domains: Bacteria and Archaea, which contain the prokaryotes, and Eukaryota, comprising the remaining forms. This change was precipitated by the discovery of the Archaea. These arrangements should not be seen as definitive. They are based on the genomes of the organisms; as knowledge on this increases, so will the categories change.

Reflecting truly evolutionary relationships, especially given the wide acceptance of cladistic methodology and numerous molecular phylogenies that have challenged long-accepted classifications, has proved problematic within the framework of Linnaean taxonomy. Therefore, some systematists have proposed a Phylocode to replace it.

Chapter 8

Phycology: The Algae

The algae (singular: alga) comprise several different groups of plant-like organisms, some of which are (and some are not) regarded as members of the Kingdom Plantae. All algae lack true leaves, roots, flowers, and other structures found in the higher plants. They are distinguished from bacteria and protozoa mainly in that they are autotrophic, obtaining energy through photosynthesis. Although no longer considered a natural group, the term *algae* is still used for convenience. The botanical discipline concerned with the study of algae is called Phycology (or sometimes, Algology); and the environment most phycologists (or algologists) focus on is the marine intertidal/shallow subtidal regions of the world oceans. It is in these environments that the diversity of structurally complex algae (called seaweeds) reaches its pinnacle.

As a grouping, the algae cut across even the prokaryote/eukaryote divide: the so-called "Blue-green algae" are cyanobacteria. All other algae are eukaryotes. Green Algae (different from Blue-green algae) are considered to be the ancestors of green plants. Other kinds of algae on the other hand are distinct from green plants and from each other in having different and unrelated accessory pigments. These pigments are responsible for the ways different algae absorb light, providing advantage to each individual type of alga to compete best at a water depth where its prefered wavelength is perhaps strongest.

Algae (singular *alga*) encompass several groups of relatively simple living aquatic organisms that capture light energy through photosynthesis, using it to convert inorganic substances into organic matter.

Algae are photosynthetic organisms that occur in most habitats. Algae varies from small, single-celled species to complex multicellular species, such as the Giant kelps that grow to 65 meters in length.

Although algae have conventionally been regarded as simple plants, they actually span more than one domain, including both Eukaryota and Bacteria as well as more than one kingdom, including plants and protists, the latter being traditionally considered more animal-like. Thus algae do not represent a single evolutionary direction or line but a level of organization that may have developed several times in the early history of life on Earth.

Algae range from single-cell organisms to multicellular organisms, some with fairly complex differentiated form and (if marine) called seaweeds. All lack leaves, roots, flowers, seeds and other organ structures that characterize higher plants (vascular plants). They are distinguished from other protozoa in that they are photoautotrophic although this is not a hard and fast distinction as some groups contain members that are mixotrophic, deriving energy both from photosynthesis and uptake of organic carbon either by osmotrophy, myzotrophy, or phagotrophy. Some unicellular species rely entirely on external energy sources and have reduced or lost their photosynthetic apparatus.

All algae have photosynthetic machinery ultimately derived from the cyanobacteria, and so produce oxygen as a byproduct of photosynthesis, unlike non-cyanobacterial photosynthetic bacteria. It is estimated that algae produce about 73 to 87 percent of the net global production of oxygen - which is available to humans and other terrestrial animals for respiration.

Algae are usually found in damp places or bodies of water and thus are common in terrestrial as well as aquatic environments. However, terrestrial algae are usually rather inconspicuous and far more common in moist, tropical regions than dry ones, because algae lack vascular tissues and other adaptations to live on land. Algae can, however, endure dryness and other conditions in symbiosis with a fungus as lichen.

The various sorts of algae play significant roles in aquatic ecology. Microscopic forms that live suspended in the water column — called phytoplankton — provide the food base for most marine food chains. In very high densities (so-called algal blooms) these algae may discolor the water and outcompete or poison other life forms. Seaweeds grow mostly in shallow marine waters. Some are used as human food or harvested for useful substances such as agar or fertilizer.

The study of marine and freshwater algae is called phycology or algology.

The US Algal Collection is represented by almost 300,000 accessioned and inventoried herbarium specimens

CLASSIFICATION

Prokaryotic Algae

Cyanobacteria have been included among the algae, referred to as the *cyanophytes* or *Blue-green algae,* (the term "algae" refers to any aquatic organisms capable of photosynthesis) though some recent treatises on algae specifically exclude them. Cyanobacteria are some of the oldest organisms to appear in the fossil record dating back to the Precambrian, possibly as far as about 3.5 billion years. Ancient cyanobacteria likely produced much of the oxygen in the Earth's atmosphere.

Cyanobacteria can be unicellular, colonial, or filamentous. They have a cockprokaryotic cell structure typical of bacteria and conduct photosynthesis on specialized cytoplasmic membranes called thylakoid membranes, rather than in organelles. Some filamentous blue-green algae have specialized cells, termed heterocysts, in which nitrogen fixation occurs.

Eukaryotic Algae

All other algae are eukaryotes and conduct photosynthesis within membrane-bound structures (organelles) called chloroplasts. Chloroplasts contain DNA and are similar in structure to cyanobacteria, presumably representing reduced cyanobacterial endosymbionts. The exact nature of the chloroplasts is different among the different lines of algae, reflecting different endosymbiotic events. The table below lists the three major groups of eukaryotic algae and their lineage relationship is shown in the figure on the left. Note many of these groups contain some members that are no longer photosynthetic. Some retain plastids, but not chloroplasts, while others have lost them entirely.

It was W.H.Harvey (1811 — 1866) who first divided the algae into four divisions based on their pigmentation. This is the first use of a biochemical criterion in plant systematics. Harvey's four divisions were: red algae (Rhodophyta), brown algae (Heteromontophyta), green algae (Chlorophyta) and Diatomaceae.

Forms of Algae

Most of the simpler algae are unicellular flagellates or amoeboids, but colonial and non-motile forms have developed independently among several of the groups. Some of the more common organizational levels, more than one of which may occur in the life cycle of a species, are:

- *Colonial:* Small, regular groups of motile cells
- *Capsoid:* Individual non-motile cells embedded in mucilage
- *Coccoid:* Individual non-motile cells with cell walls
- *Palmelloid:* Non-motile cells embedded in mucilage
- *Filamentous:* A string of non-motile cells connected together, sometimes branching

- *Parenchymatous:* cells forming a thallus with partial differentiation of tissues

In three — lines even higher levels of organization have been reached, leading to organisms with full tissue differentiation. These are the brown algae — some of which may reach 50 m in length (kelps) — the red algae [4], and the green algae [5]. The most complex forms are found among the green algae in a lineage that eventually led to the higher land plants. The point where these non-algal plants begin and algae stop is usually taken to be the presence of reproductive organs with protective cell layers, a characteristic not found in the other alga groups.

The first plants on earth were algae and these still thrive in a range of aquatic habitats today. The land plants evolved from the algae, more specifically green algae. Some 400 million years ago freshwater, green, filamentous algae invaded the land. These probably had an isomorphic alternation of generations and were probably heterotrichous. Fossils of isolated land plant spores suggest land plants may have been around as long as 475 million years ago.

Fresh-Water algae

Algae and Symbioses

Some species of algae form symbiotic relationships with other organisms. In these symbioses, the algae supply photosynthates (organic substances) to the host organism providing protection to the algal cells. The host organism derives some or all of its energy requirements from the algae. Examples include:

- *Lichens*: A fungus is the host, usually with a green alga or a cyanobacterium as its symbiont. Both fungal and algal species found in lichens are capable of living independently, although habitat requirements may be greatly different from those of the lichen pair.

- *Corals:* Algae known as zooxanthellae are symbionts with corals. Notable amongst these is the dinoflagellate *Symbiodinium*, found in many hard corals. The loss of *Symbiodinium*, or other zooxanthellae from the host is known as coral bleaching.
- *Sponges:* Green algae live close to the surface of some sponges, for example, breadcrumb sponge (*Halichondria panicea*). The alga is thus protected from predators; the sponge is provided with oxygen and sugars which can account for 50 to 80% of sponge growth in some species.

Life-Cycle

Rhodophyta, Chlorophyta and Heterokontophyta, the three main algal Phyla, have life-cycles which show tremendous variation with considerable complexity. In general there is an asexual phase where the seaweed's cells are diploid, a sexual phase where the cells are haploid followed by fusion of the male and female gametes. Asexual reproduction is advantageous in that it permits efficient population increases, but less variation is possible. Sexual reproduction allows more variation but is more costly because of the waste of gametes that fail to mate, among other things. Often there is no strict alternation between the sporophyte and gametophyte phases and also because there is often an asexual phase, which could include the fragmentation of the thallus.

Numbers and Distribution

In the British Isles the UK Biodiversity Steering Group Report estimated there to be 20,000 algal species in the UK, freshwater and marine, about 650 of these are seaweeds. Another checklist of freshwater algae reported only about 5000 species. It seems therefore that the 20,000 is an overestimate or an error (John, 2002 p.1).

World-wide it is thought that there are over 5,000 species of red algae, 1,500 — 2,000 of brown algae and 8,000 of green algae. In Australia it is estimated that there are over 1,300 species of red algae, 350 species of brown algae and approximately 2,000 species of green algae totalling 3,650 species of algae in Australia.

Around 400 species appear to be an average figure for the coastline of South African west coast.

669 marine species have been described from California (U.S.A.).

642 entities are listed in the check-list of Britain and Ireland (Hardy and Guiry, 2006).

Distribution

No publication has been found which attempts to discuss the general distribution of algae in the seas of the world.

Uses of Algae

Fertilizer

For centuries seaweed has been used as fertilizer; Orwell writing in the 16th Century referring to drift weed in South Wales: "This kind of ore they often gather and lay in heaps where it heteth and rotteth, and will have a strong and loathsome smell; when being so rotten they cast it on the land, as they do their muck, and thereof springeth good corn, especially barley" and "After spring tides or great rigs of the sea, they fetch it in sacks on horse brackets, and carry the same three, four, or five miles, and cast it on the lande, which doth very much better the ground for corn and grass".

There are also commercial uses of algae as agar.

Maerl is commonly used as a soil conditioner, it is dredged from the sea floor and crushed to form a powder. It is still harvested around the coasts of Brittany in France and off Falmouth (also extensively in western Ireland) and is a popular fertilizer in these days of organic gardening investigated

Falmouth maerl and found that *L. corallioides* predominated down to 6 m and *P. calcareum* from 6-10 m.

Chemical analysis of maerl showed that it contained 32.1% $CaCO_3$ and 3.1% $MgCO_3$ (dry weight)

Energy Source

- Algae can be used to make biodiesel and by some estimates can produce vastly superior amounts of oil, compared to terrestrial crops grown for the same purpose.
- Algae can be grown to produce hydrogen. In 1939 a German researcher named Hans Gaffron, while working at the University of Chicago, observed that the algae he was studying, *Chlamydomonas reinhardtii* (a green-alga), would sometimes switch from the production of oxygen to the production of hydrogen. Gaffron never discovered the cause for this change and for many years other scientists failed to repeat his findings. In the late 1990s professor Anastasios Melis a researcher at the University of California at Berkeley discovered that if the algae culture medium is deprived of sulfur it will switch from the production of oxygen (normal photosynthesis), to the production of hydrogen. He found that the enzyme responsible for this reaction is hydrogenase, but that the hydrogenase lost this function in the presence of oxygen. Melis found that depleting the amount of sulfur available to the algae interrupted its internal oxygen flow, allowing the hydrogenase an environment in which it can react, causing the algae to produce hydrogen. *Chlamydomonas moeweesi* is also a good strain for the production of hydrogen.
- Algae can be grown to produce biomass, which can be burned to produce heat and electricity.

Pollution Control

- Algae are used in wastewater treatment facilities, reducing the need for greater amounts of toxic chemicals than are already used.
- Algae can be used to capture fertilizers in runoff from farms. When subsequently harvested, the enriched algae itself can be used as fertilizer.
- Algae bioreactors are used by some powerplants to reduce CO_2 emissions. The CO_2 can be pumped into a pond, or some kind of tank, on which the algae feed. Alternatively, the bioreactor can be installed directly on top of a smokestack. This technology has been pioneered by Massachusetts-based GreenFuelTechnologies.

Nutrition

Algae are used by humans in many ways. They are used as fertilizers, soil conditioners and are a source of livestock feed. Because many species are aquatic and microscopic, they are cultured in clear tanks or ponds and either harvested or used to treat effluents pumped through the ponds. Algaculture on a large scale is an important type of aquaculture in some places.

Human food. Seaweeds are an important source of food, especially in Asia; They are excellent sources of many vitamins including: A, B1, B2, B6, niacin and C. They are rich in iodine, potassium, iron, magnesium and calcium.

Algae is commercially cultivated as a nutritional supplement. One of the most popular microalgal species is Spirulina *(Arthrospira platensis)*, which is a Cyanobacteria (known as blue-green algae), and has been hailed by some as a superfood. Other algal species cultivated for their nutritional value include; Chlorella (a green algae), and Dunaliella (*Dunaliella salina*), which is high in beta-carotene and is used in vitamin C supplements.

In China at least 70 species of algae are eaten as is the Chinese "vegetable" known as *fat choy* (which is actually a cyanobacterium). Roughly 20 species of algae are used in everyday cooking in Japan.

Certain species are edible; the best known, especially in Ireland is *Palmaria palmata* (Linnaeus) O. Kuntze (*Rhodymenia palmata* (Linnaeus) Kuntze, common name: dulse).This is a red alga which is dried and may be bought in the shops in Ireland. It is eaten raw, fresh or dried, or cooked like spinach. Similarly, *Durvillaea antarctica* is eaten in Chile, common name: cochayuyo.

Porphyra (common name: purple laver), is also collected and used in a variety of ways (e.g. "laver bread" in the British Isles). In Ireland it is collected and made into a jelly by stewing or boiling. Preparation also involves frying with fat or converting to a pinkish jelly by heating the fronds in a saucepan with a little water and beating with a fork. It is also collected and used by people parts of Asia, specifically China and Japan as nori and along most of the coast from California to British Columbia. The Hawaiians and the Maoris of New Zealand also use it. *Chondrus crispus*, (probably confused with *Mastocarpus stellatus*, common name: Irish moss), is also used as "carrageen" for the stiffening of milk and dairy products, such as ice-cream. One particular use is in "instant" puddings, sauces and creams. *Ulva lactuca* (common name: sea lettuce), is used locally in Scotland where it is added to soups or used in salads. *Alaria esculenta* (common name: dabberlocks), is used either fresh or cooked, in Greenland, Iceland, Scotland and Ireland.

- The oil from some algae have high levels of unsaturated fatty acids. Arachidonic acid (a polyunsaturated fatty acid), is very high in *Parietochloris incisa*, (a green alga) where it reaches up to 47% of the triglyceride pool.

Other Uses

The natural pigments produced by algae can be used as an alternative to chemical dyes and coloring agents. Many of the paper products used today are not recyclable because of the chemical inks that they use, paper recyclers have found that inks made from algae are much easier to break down. There is also much interest in the food industry into replacing the coloring agents that are currently used with coloring derived from algal pigments. In Israel, a species of green algae is grown in water tanks, then exposed to direct sunlight and heat which causes it to become bright red in colour. It is then harvested and used as a natural pigment for foods such as Salmon.

CYANOBACTERIA

The cyanobacteria comprise the structurally simplest algae, and presumably are closely related to the oldest photosynthetic organisms on the planet. Although capable of extracting energy from sunlight through photosynthesis, these algae are related to bacteria as evidenced by their prokaryotic cell structure. Yet, some Blue-greens have developed multi-cellular thalli that approach eukaryotic algal forms, and thus their traditional inclusion within the "algae."

Cyanobacteria (Greek: *κυανόs(kyanós)* = blue + bacterium) also known as Cyanophyta is a phylum (or "division") of Bacteria that obtain their energy through photosynthesis. They are often still referred to as blue-green algae, although they are in fact prokaryotes. The description is primarily used to reflect their appearance and ecological role rather than their evolutionary lineage. The name Cyanobacteria comes from the colour of the bacteria, Cyan, the bacteria do not use or produce Cyanide whose chemical prefix is cyano-. Putative fossil traces of cyanobacteria have been found from around 3.8 billion years ago (b.y.a.). They are a significant component of the marine nitrogen cycle and an important primary producer in many areas of the ocean. Their ability to perform oxygenic (plant-like) photosynthesis is thought to have converted the early reducing atmosphere into an oxidizing one, which dramatically changed the life forms on Earth and provoked an explosion of biodiversity.

Forms

Cyanobacteria are found in almost every conceivable habitat, from oceans to fresh water to bare rock to soil. Most are found in fresh water, while others are marine, occur in damp soil, or even temporarily moistened rocks in deserts. A few are endosymbionts in lichens, plants, various protists, or sponges and provide energy for the host. Some live in the fur of sloths, providing a form of camouflage.

Cyanobacteria include unicellular and colonial species. Colonies may form filaments, sheets or even hollow balls. Some filamentous colonies show the ability to differentiate into several different cell types: vegetative cells, the normal, photosynthetic cells that are formed under favorable growing conditions; akinetes, the climate-resistant spores that may form when environmental conditions become harsh; and thick-walled heterocysts, which contain the enzyme nitrogenase, vital for nitrogen fixation. Heterocysts may also form under the appropriate environmental conditions (anoxic) wherever nitrogen is necessary. Heterocyst-forming species are specialized for nitrogen fixation and are able to fix nitrogen gas, which cannot be used by plants, into ammonia (NH_3), nitrites (NO_2^-) or nitrates (NO_3^-), which can be absorbed by plants and converted to protein and nucleic acids. The rice paddies of Asia, which produce about 75% of the world's rice could not do so were it not for healthy populations of nitrogen-fixing cyanobacteria in the rice paddy fertilizer.

Many cyanobacteria also form motile filaments, called hormogonia, that travel away from the main biomass to bud and form new colonies elsewhere. The cells in a hormogonium are often thinner than in the vegetative state, and the cells on either end of the motile chain may be tapered. In order to break away from the parent colony, a hormogonium often must tear apart a weaker cell in a filament, called a necridium.

Each individual cell of a cyanobacterium typically has a thick, gelatinous cell wall, which stains gram-negative. They lack flagella, but hormogonia and some unicellular species may move about by gliding along surfaces. In water columns some cyanobacteria float by forming gas vesicles, like in archaea.

Photosynthesis

Cyanobacteria have an elaborate and highly organized system of internal membranes which function in photosynthesis. Photosynthesis in cyanobacteria generally uses water as an electron donor and produces oxygen as a by-product, though some may also use hydrogen sulfide as occurs among other photosynthetic bacteria. Carbon dioxide is reduced to form carbohydrates via the Calvin cycle. In most forms the photosynthetic machinery is embedded into folds of the cell membrane, called thylakoids. The large amounts of oxygen in the atmosphere are considered to have been first created by the activities of ancient cyanobacteria. Due to their ability to fix nitrogen in aerobic conditions they are often found as symbionts with a number of other groups of organisms such as fungi (lichens), corals, pteridophytes (Azolla), angiosperms (Gunnera) etc.

Cyanobacteria are the only group of organisms that are able to reduce nitrogen and carbon in aerobic conditions, a fact that may be responsible for their evolutionary and ecological success. The water-oxidizing photosynthesis is accomplished by coupling the activity of photosystem (PS) II and I (Z-scheme). In anaerobic conditions, they are also able to use only PS I — cyclic photophosphorylation — with electron donors other than water (hydrogen sulfide, thiosulphate, or even molecular hydrogen) just like purple photosynthetic bacteria. Furthermore, they share an archaebacterial property, which is the ability to reduce elemental sulfur by anaerobic respiration in the dark. Perhaps the most intriguing thing about these organisms is that their photosynthetic electron transport shares the same compartment as the components of respiratory electron transport. Actually, their plasma membrane contains only components of the respiratory chain, while the thylakoid membrane hosts both respiratory and photosynthetic electron transport.

Attached to thylakoid membrane, phycobilisomes act as light harvesting antennae for the photosystems . The phycobilisome components (phycobiliproteins) are responsible for the blue-green pigmentation of most cyanobacteria. The variations to this theme is mainly due to carotenoids and phycoerythrins which give the cells the red-brownish coloration. In some cyanobacteria, the colour of light influences the composition of phycobilisomes. In green light, the cells accumulate more phycoerythrin, whereas in red light they produce more phycocyanin. Thus the bacteria appear green in red light and red in green light. This process is known as complementary chromatic adaptation and is a way for the cells to maximize the use of available light for photosynthesis.

A few genera, however, lack phycobilisomes and have chlorophyll *b* instead (*Prochloron, Prochlorococcus, Prochlorothrix*). These were originally grouped together as the prochlorophytes or chloroxybacteria, but appear to have developed in several different lines of cyanobacteria. For this reason they are now considered as part of cyanobacterial group.

Relationship to Chloroplasts

Chloroplasts found in eukaryotes (algae and higher plants) likely evolved from an endosymbiotic relation with cyanobacteria. This endosymbiotic theory is supported by various structural and genetic similarities. Primary chloroplasts are found among the green plants, where they contain chlorophyll *b*, and among the red algae and glaucophytes, where they contain phycobilins. It now appears that these chloroplasts probably had a single origin, in an ancestor of the clade called Primoplantae. Other algae likely took their chloroplasts from these forms by secondary endosymbiosis or ingestion.

It was once thought that the mitochondria in eukaryotes also developed from an endosymbiotic relationship with cyanobacteria; however, we now suspect that this evolutionary event occurred when aerobic Eubacteria were engulfed by anaerobic host cells. Mitochondria are believed to have originated not from cyanobacteria but from an ancestor of Rickettsia.

Classification

The cyanobacteria were traditionally classified by morphology into five sections, referred to by the numerals I-V. The first three - Chroococcales, Pleurocapsales, and Oscillatoriales - are not supported by phylogenetic studies. However, the latter two - Nostocales and Stigonematales - are monophyletic, and make up the heterocystous cyanobacteria. The members of Chroococales are unicellular and usually aggregated in colonies.

The classic taxonomic criterion has been the cell morphology and the plane of cell division. In Pleurocapsales, the cells have the ability to form internal spores (baeocytes). The rest of the sections include filamentous species. In Oscillatorialles, the cells are uniseriately arranged and do not form specialized cells (akinets and heterocysts). In Nostocalles and Stigonematalles the cells have the ability to develop heterocysts in certain conditions. Stigonematales, unlike

Nostocalles include species with truly branched trichome. Most taxa included in the phylum or division Cyanobacteria have not yet been validly published under the Bacteriological Code. Except:

- The classes Chroobacteria, Hormogoneae and Gloeobacteria
- The orders Chroococcales, Gloeobacterales, Nostocales, Oscillatoriales, Pleurocapsales and Stigonematales
- The families Prochloraceae and Prochlorotrichaceae
- The genera Halospirulina, Planktothricoides, Prochlorococcus, Prochloron, Prochlorothrix.

Biotechnology and Applications

Certain cyanobacteria produce cyanotoxins like Anatoxin-a, Anatoxin-as, Aplysiatoxin, Cylindrospermopsin, Domoic acid, Microcystin LR, Nodularin R (from *Nodularia*), or Saxitoxin. Sometimes a mass-reproduction of cyanobacteria results in algal blooms.

The unicellular cyanobacterium *Synechocystis* sp. PCC 6803 was the first photosynthetic organism whose genome was completely sequenced. It continues to be an important model organism.

At least one secondary metabolite, cyanovirin, has shown to possess anti-HIV activity.

Some cyanobacteria are sold as food, notably *Aphanizomenon flos-aquae* (E3live) and *Arthrospira platensis* (Spirulina). It has been suggested that they could be a much more substantial part of human food supplies, as a kind of superfood.

Along with algae, some hydrogen producing cyanobacteria are being considered as an alternative energy source, notably at Oregon State University, in research supported by the U.S. Department of Energy, Princeton

University, Colorado School of Mines as well as at Uppsala University, Sweden

Health Risks

Some species of cyanobacteria produce neurotoxins, hepatotoxins, cytotoxins, and endotoxins, making them dangerous to animals and humans. Several cases of human poisoning have been documented but a lack of knowledge prevents an accurate assessment of the risks.

Chapter 9

Mycology: The Fungi

INTRODUCTION

The Fungi (singular is *fungus*) are a large group of organisms treated within the science of Botany, but not really "plants" in the usual sense of the term. The Fungi are ranked as a kingdom within the Domain Eukaryota. That is, they have eukaryotic cells with distinct nuclei, although in some species divisions between nucleated "cells" are sparse. However, they all lack chlorophyll, and the species are either saprophytic or parasitic. Included within the Fungi are the well known mushrooms, but the group also includes many microscopic forms, and fungi inhabit every environment on earth, perhaps second only to the bacteria in distribution.

MYCOLOGY

Mycology (from the Greek ìýêçò, meaning "fungus") is the study of fungi, their genetic and biochemical properties, their taxonomy, and their use to humans as a source for tinder, medicinals (e.g., penicillin), food (e.g., beer, wine, cheese, edible mushrooms) and entheogens, as well as their dangers, such as poisoning or infection. From mycology arose the field of phytopathology, the study of plant diseases, and the two disciplines remain closely related. A biologist who studies mycology is called a mycologist.

BACKGROUND

Historically, mycology was a branch of botany (though fungi are evolutionarily more closely related to animals than plants this was not recognized until recently). Pioneer *mycologists* included Elias Magnus Fries, Christian Hendrik Persoon, Anton de Bary and Lewis David von Schweinitz. The British Mycological Society was founded in 1896.

Today, the most comprehensively studied and understood fungi are the yeasts and eukaryotic model organisms *Saccharomyces cerevisiae* and *Schizosaccharomyces pombe.*

Many fungi produce toxins, antibiotics, and other secondary metabolites. For example, the cosmopolitan (worldwide) genus *Fusarium* and their toxins associated with fatal outbreaks of alimentary toxic aleukia in humans were extensively studied by Abraham Joffe. Fungi are fundamental for life on earth in their roles as symbionts, e.g. in the form of mycorrhizae, insect symbionts and lichens as well as their potency in breaking down complex organic biomolecules such as wood as well as xenobiotics, a critical step in the global carbon cycle.

Fungi and other organisms traditionally recognized as fungi (such as oomycetes and slime molds) often are economically and socially important as some cause diseases of animals (such as histoplasmosis) as well as plants (such as Dutch elm disease and Rice blast).

Field meetings to find interesting species of fungi are known as 'forays', after the first such meeting organized by the Woolhope Naturalists' Field Club in 1868 and entitled "a foray among the funguses."

FUNGUS

The Fungi (singular fungus) are a kingdom of eukaryotic organisms. They are heterotrophic and digest their food externally, absorbing nutrient molecules into their cells. Yeasts, molds, and mushrooms are examples of fungi. The branch of biology involving the study of fungi is known as mycology.

Fungi often have important symbiotic relationships with other organisms. Mycorrhizal symbiosis between plants and fungi is particularly important; over 90% of all plant species engage in some kind of mycorrhizal relationship with fungi and are dependent upon this relationship for survival. Fungi are also used extensively by humans: yeasts are responsible for fermentation of beer and bread, and mushroom farming and gathering is a large industry in many countries.

Fungi and bacteria are the primary decomposers of organic matter in almost all terrestrial ecosystems worldwide. There are an estimated 1.5 million species of fungi with around 70,000 of them having been described.

Phylogeny and Classification

Fungi were originally classified as plants, however they have since been separated as they do not fix their own carbon through chlorophyll based photosynthesis. Traditionally, they are considered heterotrophs, organisms that rely solely on carbon fixed by other organisms for metabolism. Cutting edge research at the Albert Einstein College of Medicine suggests that at least some fungi have the capability to extract energy from ionizing radiation. They apparently accomplish this by using the pigment melanin in lieu of chlorophyll. Ekaterina Dadachova has been studying certain fungi that thrive near Chernobyl in the midst of ecological catastrophe. Her experiments demonstrated the surprising result that their growth is enhanced rather than undermined by radiation. Plans are underway to extend the investigation to other species of fungi.

Fungi are now thought to be more closely related to animals than to plants, and are placed with animals in the monophyletic group of opisthokonts. For much of the Paleozoic Era, the fungi appear to have been aquatic. The first land fungi probably appeared in the Silurian, right after the first land plants appeared even though their fossils are fragmentary. For some time after the Permian-Triassic extinction event, the fungi went into a fungal spike because they were the dominant life forms - nearly 100% of the available record. Fungi absorb their food while animals ingest it; also unlike animals, the cells of fungi have cell walls. For these reasons, these organisms are placed in their own kingdom, Fungi, or Eumycota.

The Fungi are a monophyletic group, meaning all varieties of fungi come from a common ancestor. The monophyly of the fungi has been confirmed through repeated tests of molecular phylogenetics; shared ancestral traits include chitinous cell walls and heterotrophy by absorption, along with other shared characteristics.

The taxonomy of the Fungi is in a state of rapid flux at present, especially due to recent papers based on DNA comparisons, which often overturn the assumptions of the older systems of classification.

There is no unique generally accepted system at the higher taxonomic levels and there are constant name changes at every level, from species upwards. Fungal species can also have multiple scientific names depending on its life cycle. Web sites such as Index Fungorum, ITIS and Wikispecies define preferred up-to-date names (with cross-references to older synonyms), but do not always agree with each other or with names in Wikipedia in its various language variants.

Types of fungi

The major divisions (phyla) of fungi are mainly classified based on their sexual reproductive structures. Currently, five divisions are recognized:

- The Chytridiomycota are commonly known as chytrids. These fungi produce zoospores that are capable of moving on their own through liquid menstrua by simple flagella. The status of chytrids is under dispute. Some taxonomists classify them as protists on the basis of the flagellae. The genomic project should help to sort out the phylogentic relationships.
- The Zygomycota are known as zygomycetes and reproduce sexually with meiospores called

zygospores and asexually with sporangiospores. Black bread mold (*Rhizopus stolonifer*) is a common species that belongs to this group; another is *Pilobolus*, which shoots specialized structures through the air for several meters. Medically relevant genera include *Mucor*, *Rhizomucor*, and *Rhizopus*. Molecular phylogenetic investigation has shown the zygomycota to be a polyphyletic group.

- Members of the Glomeromycota are also known as the arbuscular mycorrhizal fungi. Only one species has been observed forming zygospores; all other species only reproduce asexually. This is an ancient association, with evidence dating to 400 million years ago.
- The Ascomycota, commonly known as sac fungi or ascomycetes, form meiotic spores called ascospores, which are enclosed in a special sac-like structure called an ascus. This division includes morels, some mushrooms and truffles, as well as single-celled yeasts and many species that have only been observed undergoing asexual reproduction. Because the products of meiosis are retained within the sac-like ascus, several ascomyctes have been used for elucidating principles of genetics and heredity (e.g. *Neurospora crassa*).
- Members of the Basidiomycota, commonly known as the club fungi or basidiomycetes, produce meiospores called basidiospores on club-like stalks called basidia. Most common mushrooms belong to this group, as well as rust (fungus) and smut fungi, which are major pathogens of grains.

Although the water molds and slime molds have traditionally been placed in the kingdom Fungi and those who study them are still called mycologists, they are not true fungi. Unlike true fungi, the water molds and slime molds do not have cell walls made of chitin. In the 5-kingdom system, they are currently placed in the kingdom Protista. Water moulds are descended from algae, and are placed within the phylum Oomycota, within the Kingdom Protista.

Morphology

Though fungi are part of the opisthokont clade, all phyla except for the chytrids have lost their posterior flagella. Fungi are unusual among the eukaryotes in having a cell wall of chitin. All fungi are made up of many thin thread-like structures called hyphae. These hyphae can be one of two types: septate, or coenocytic. Septate hyphae have "walls" between their cells, called septa, though these septa have holes that allow cytoplasm, organelles, and sometimes nuclei to pass through. Coenocytic hyphae have no such marked divisions between cells. Coenocytic hyphae are essentially multinucleate supercells. Parasitic fungi have special structures on their hyphae called haustoria, which penetrate directly into a host organism's cells, allowing nutrients to be taken by the fungus. All of a fungus's hyphae form a structure called the mycelium. In mushroom forming fungi, the mycelium is normally underground. In molds, the mycelium forms directly on the food source. The only fungi that do not form hyphae or mycelia are yeasts, which are unicellular.

Unlike animals and vascular plants, fungi do not spend the majority of their life cycle in a diploid condition. When a spore begins to grow into a mycelium, the organism is haploid. The haploid mycelium may or may not produce haploid spores asexually. When one haploid organism encounters another, through growth of the mycelium, since fungi are not motile, the two may merge, in a process called plasmogamy. The fungi then enter a heterokaryotic, or multinucleate stage. Usually, one nucleus from one parent fungus will pair off with one nucleus from the other parent. Some fungi spend most of their life cycle in this stage. At a given time, the paired off nuclei will merge, in a process called karyogamy, producing a diploid nucleus. This will normally happen in a separate reproductive structure; in basidiomycetes and ascomycetes, the mushroom. The diploid nucleus will then undergo meiosis to produce haploid nuclei, which are then released as spores to start the cycle once again.

Reproduction

Fungi may reproduce sexually or asexually. In asexual reproduction, the offspring are genetically identical to the "parent" organism (they are clones). During sexual reproduction, a mixing of genetic material occurs so that the offspring exhibit traits of both parents. Many species can use both strategies at different times, while others are apparently strictly sexual or strictly asexual. Sexual reproduction has not been observed in some fungi of the Glomeromycota and Ascomycota. These are commonly referred to as Fungi imperfecti or Deuteromycota.

Yeasts and other unicellular fungi can reproduce simply by budding, or "pinching off" a new cell. Many multicellular species produce a variety of different asexual spores that are easily dispersed and resistant to harsh environmental conditions. When the conditions are right, these spores will germinate and colonize new habitats.

Sexual reproduction in fungi is somewhat different from that of animals or plants, and each fungal division reproduces using different strategies. Fungi that are known to reproduce sexually all have a haploid stage and a diploid stage in their life cycles. Ascomycetes and basidiomycetes also go through a dikaryotic stage, in which the nuclei inherited by the two parents do not fuse right away, but remain separate in the hyphal cells.

In zygomycetes, the haploid hyphae of two compatible individuals fuse, forming a zygote, which becomes a resistant zygospore. When this zygospore germinates, it quickly undergoes meiosis, generating new haploid hyphae and asexual sporangiospores. These sporangiospores may then be distributed and germinate into new genetically-identical individuals, each producing their own haploid hyphae. When the hyphae of two compatible individuals come into contact with one another, they will fuse and generate new zygospores, thus completing the cycle.

In ascomycetes, when compatible haploid hyphae fuse with one another, their nuclei do not immediately fuse. The dikaryotic hyphae form structures called asci (*sing.* ascus), in which karyogamy (nuclear fusion) occurs. These asci are embedded in an ascocarp, or fruiting body, of the fungus. Karyogamy in the asci is followed immediately by meiosis and the production of ascospores. The ascospores are disseminated and germinate to form new haploid mycelium. Asexual conidia may be produced by the haploid mycelium. Many ascomycetes appear to have lost the ability to reproduce sexually and reproduce only via conidia.

Sexual reproduction in basidiomycetes is similar to that of ascomycetes. Sexually compatible haploid hyphae fuse to produce a dikaryotic mycelium. This leads to the production of a basidiocarp. The most commonly-known basidiocarps are mushrooms, but they may also take many other forms. Club-like structures known as basidia generate haploid basidiospores following karyogamy and meiosis. These basidiospores then germinate to produce new haploid mycelia.

Ecological Role

Although often inconspicuous, fungi occur in every environment on Earth and play very important roles in most ecosystems. Along with bacteria, fungi are the major decomposers in most terrestrial (and some aquatic) ecosystems, and therefore play a critical role in biogeochemical cycles and in many food webs.

Many fungi are important as partners in symbiotic relationships with other organisms, as mutualists, parasites, or commensalists, as well as in symbiotic relationships that do not fall neatly into any of these categories. One of the most important of these relationships are various types of mycorrhiza, which is a kind of mutualistic relationship between fungi and plants, in which the plant's roots are closely associated with fungal hyphae and other structures. The plant donates to the fungus sugars and other carbohydrates that it manufactures from photosynthesis, while the fungus donates water and mineral nutrients that the hyphal network is able to find much more efficiently than the plant roots alone can, particularly phosphorus. The fungi also protect against diseases and pathogens and provide other benefits to the plant. Recently, plants have been found to use mycorrhizas to deliver carbohydrates and other nutrients to other plants in the same community and in some cases can make plant species that would normally exclude each other able to coexist in the same plant community. Such mycorrhizal communities are called "common mycorrhizal networks". Over 90% of the plant species on Earth are dependent on mycorrhizae of one type or another in order to survive, and it is hypothesized that the presence of terrestrial fungi may have been necessary in order for the first plants to colonize land. Research in 2005 showed that mycorrhizal fungi facilitate significant nitrogen transfer to their plant hosts.

Lichens are formed by a symbiotic relationship between algae or cyanobacteria (referred to in lichens as "photobionts") and fungi (mostly ascomycetes of various kinds and a few basidiomycetes), in which individual photobiont cells are embedded in a complex of fungal tissue. As in mycorrhizas, the photobiont provides sugars and other carbohydrates while the fungus provides minerals and water. The functions of both symbiotic organisms are so closely intertwined that they function almost as a single organism.

Certain insects also engage in mutualistic relationships with various types of fungi. Several groups of ants cultivate various fungi in the Agaricales as their primary food source, while ambrosia beetles cultivate various kinds of fungi in the bark of trees that they infest. Termites on the African Savannah are also known to cultivate fungi.

Some fungi are parasites on plants, animals (including humans), and even other fungi. Pathogenic fungi are responsible for numerous diseases, such as athlete's foot and ringworm in humans and Dutch elm disease in plants. Some fungi are predators of nematodes, which they capture using an array of devices such as constricting rings or adhesive nets.

Human uses of fungi

Fungi have a long history of use by humans. Many types of mushrooms and other fungi are cultivated or picked from the wild eaten directly, while yeast cultures are used in many foodstuffs. The study of the historical uses and sociological impact of fungi is known as ethnomycology.

Cultured Foods

A type of single-celled yeast fungus called Saccharomyces cerevisiae yeast is used in baking bread. Yeast is also used to create alcoholic beverages through fermentation. Mycelial fungus is used to make Shoyu (soy sauce) and tempeh. Fungi are also used to produce industrial chemicals like lactic acid, antibiotics and even to make stonewashed jeans. Some types of fungi are ingested for their psychedelic properties, both recreationally and religiously.

Edible and Poisonous Fungi

Some of the most well-known types of fungi are the edible and poisonous mushrooms. Many species are commercially raised, but others must be harvested from the wild. *Agaricus bisporus*, sold as button mushrooms when small or Portobello musrooms when larger, are the most commonly eaten species, used in salads, soups, and many other dishes. Other commercially-grown mushrooms that have gained in popularity in the West and are often available fresh in grocery stores include straw mushrooms (*Volvariella volvacea*), oyster mushrooms (*Pleurotus ostreatus*), shiitakes (*Lentinula edodes*), and enokitake (*Flammulina* spp.).

There are many more mushroom species that are harvested from the wild for personal consumption or commercial sale. Milk mushrooms, morels, chanterelles, truffles, black trumpets, and porcini mushrooms (also known as king boletes) all command a high price on the market. They are often used in gourmet dishes.

It is also a common practice to permit the growth of specific species of mold in certain types of cheeses that give them their unique flavour. This mold is non-toxic and is safe for human consumption. This accounts for the blue colour in cheeses such as Stilton or Roquefort which is created using *Penicillium roqueforti* spores.

Hundreds of mushroom species are toxic to humans, causing anything from upset stomachs to hallucinations to death. Some of the most deadly belong to the genus *Amanita,* including the destroying angel *(A. virosa)* and the death cap *(A. phalloides),* the most common cause of mushroom poisoning. Stomach cramps, vomiting, and diarrhea usually occur within 6-24 hours after ingestion of these mushrooms, followed by a brief period of remission (usually 1-2 days). Patients often fail to present themselves for treatment at this time, assuming that they have recovered. However, within 2-4 weeks liver and kidney failure leads to death if untreated. There is no antidote for the toxins in these mushrooms, but kidney dialysis and administration of corticosteroids may help. In severe cases, a liver transplant may be necessary (Kaminstein 2002). It is difficult to identify a "safe" mushroom without proper training and knowledge, thus it is often advised to assume that a mushroom in the wild is poisonous and leave it alone.

Fly agaric mushrooms (*A. muscaria*) are also responsible for a large number of poisonings, but these cases rarely result in death. The most common symptoms are nausea and vomiting, drowsiness, and hallucinations. In fact, this species is used ritually and recreationally for its hallucinogenic properties. Historically Fly agaric was used by Celtic Druids in Northern Europe and the Koryak people of north-eastern Siberia for religious or shamanic purposes.

Fungi in the Biological Control of Pests

Many fungi compete with other organisms, or directly infect them. Some of these fungi are considered beneficial because they can restrict, and sometimes eliminate, the populations of noxious organisms like pest insects, mites, weeds, nematodes and other fungi, such as those that kill plants. There is much interest on the manipulation of these beneficial fungi for the biological control of pests. Some of these fungi can be used as biopesticides, like the ones that kill insects (entomopathogenic fungi). Specific examples of fungi that have been developed as bioinsecticides are *Beauveria bassiana,*

Metarhizium anisopliae, Hirsutella, *Paecilomyces fumosoroseus,* and *Verticillium lecanii.*

MUSHROOM

Mushroom(s) are the fleshy, spore-bearing fruiting bodies of fungi typically produced above ground on soil or on their food sources. The standard for the name mushroom is the cultivated white button mushroom, *Agaricus bisporus,* hence the word mushroom is most often applied to fungi (Basidiomycota, Agaricomycetes) that have a stem (called a stipe), a cap (called a pileus), and gills (each called a lamella/ pl. lamellae) on the underside of the cap just as do store-bought white mushrooms. However, mushrooms can also be a wide variety of gilled fungi, with or without stems, and the term is used even more generally to describe both fleshy fruitbodies of some Ascomycota and woody or leathery fruitbodies of some Basidiomycota, depending upon the context of the usage. Usually forms deviating from the standard form have more specific names, such as puffballs, stinkhorn, morels, etc. and gilled mushrooms themselves are often called agarics, in reference to their similarity to *Agaricus* or placement in the order Agaricales. By extension, *mushroom* can also designate the entire fungus when in culture or when referring to the whole thallus (called a mycelium) of species forming fruitbodies called mushrooms.

Mushrooms Vs. Toadstools

The terms "mushrooms" and "toadstools" go back centuries, and were never precisely defined, nor was there consensus on application, except to say that the term "toadstool" was generally, but not exclusively, applied to poisonous fungi. For an example of early usage see Badham (1863[1]). Reference was made to "tadstoles", "frogstooles", "frogge stoles", "tadstooles", "tode stoles", "toodys hatte", "toadstoole", "paddockstool", "puddockstool", "paddocstol", "toadstoole", and "paddockstooles" from 1398-1597, sometimes synonymous with "mushrom", "mushrum", "muscheron", "mousheroms", "mussheron", or "musserouns". The term "mushroom" and its variations may have been derived from the French word "Mousseron" in reference to moss (mousse). There may have been a direct connection to *toads* (in reference to poisonous properties) for toadstools. However, there is no clear-cut delimitation between edible and poisonous fungi, so that mushrooms may be edible, poisonous, or unpalatable and it makes no sense to not be able to use the term mushroom when stating there are "poisonous mushrooms" which would be an oxymoron statement if the term mushroom could not be applied to poisonous fungi. The term toadstool is nowadays used in story telling when referring to poisonous or suspect mushrooms. The classic example of a toadstool is *Amanita muscaria.*

Mushrooming

Many species of mushrooms seemingly appear overnight, growing or expanding rapidly. This phenomenon is the source of several commonly used phrases in the English language. In fact all species of mushrooms take several days to form primordial mushroom fruit bodies. The cultivated mushroom as well as the common field mushroom initially form minute fruiting body initials referred to as the pin stage because of their small size. Slightly expanded they are called buttons, once again because of the relative size and shape. Once such stages are formed, the mushroom can rapidly pull in water from its mycelium and expand, mainly by inflating preformed cells that took several days to form in the primordia. Similarly, there are even more ephemeral mushrooms, like *Parasola plicatilis* ([2] formerly *Coprinus plicatilis*) that literally appear overnight and may be gone by late afternoon on hot summer days after rainfall. The primordia form at ground level in lawns in humid spaces under the thatch of lawns and after heavy rainfall or dewy conditions, balloon to full size in a few hours, release spores, then collapse. They "mushroom" to full size. "To mushroom" means to rapidly grow in size, or to sprout up rapidly, i.e., an organization may "mushroom" from national to international almost overnight. To "pop up like mushrooms" is of similar derivation, but also has a gang slang usage.

The term "mushrooming" differs in that it generally refers to the act of gathering mushrooms, in the wild, as in the statement "I'm going mushrooming today." This is often shortened to "shrooming", which has yet another connotation, which is to "do mushrooms". To "do mushrooms" or "shrooms" often refers to taking hallucinogenic mushrooms.

Notably, not all mushrooms expand overnight. Many are very slow growing. Those types of mushrooms generally add tissue to their fruitbodies in different manners, such as growing from the edges, or inserting hyphae.

Classification of Mushrooms

Typical mushrooms are the fruitbodies of members of the order Agaricales, whose type genus is *Agaricus*, and type species is the field mushroom, *Agaricus campestris*. However, in modern molecular defined classifications, not all members of the order Agaricales produce mushroom fruitbodies, and many other gilled fungi, collectively called mushrooms, occur in other orders in the class Agaricomycetes. For example, chanterelles are in the Cantharellales, false chanterelles like *Gomphus* are in the Gomphales, milk mushrooms (*Lactarius*) and russulas (*Russula*) as well as *Lentinellus* are in the Russulales, while the tough leathery genera *Lentinus* and *Panus* are among the Polyporales, but *Neolentinus* is in the Gloeophyllales, and the little pin-mushroom genus, *Rickenella* along with similar genera are in the Hymenochaetales.

Within the main body of mushrooms, in the Agaricales, are such common fungi like the common fairy-ring mushroom (*Marasmius oreades*), shiitake, enoki, oyster mushrooms, fly agarics and other amanitas, magic mushrooms like species of *Psilocybe*, paddy straw mushrooms, shaggy manes, etc.

An atypical 'mushroom' is the Lobster mushroom, which is a deformed, cooked-lobster-colored parasitized fruitbody of a *Russula* or Lactarius colored and deformed by the mycoparasitic Ascomycete *Hypomyces lactifluorum*[3].

Other 'mushrooms' are nongilled and then the term is loosely used, so that it is difficult to give a full account of their classifications. Some 'mushrooms' have pores underneath (and are usually called boletes), others have spines, such as the hedgehog mushroom and other tooth fungi, and so on. Mushroom has been used for polypores, puffballs, jelly fungi, coral fungi, bracket fungi, stinkhorns, and cup fungi. Mushrooms and other fungi are studied by mycologists. Thus, the term *mushroom* is more one of common application to macroscopic fungal fruiting bodies than one having precise taxonomic meaning. There are approximately 14,000 described species of mushrooms.

Identification of Mushrooms

Identifying mushrooms requires a basic understanding of their macroscopic structure. Most are Basidiomycetes and gilled. Their spores, called basidiospores, are produced on the gills and fall in a fine rain of powder from under the caps as a result. At the microscopic level the basidiospores are shot off of basidia but then fall between the gills in the dead air space. As a result, for most mushrooms, if the cap is cut off and placed gill-side down, usually overnight a powdery impression reflecting the shape of the gills (or pores, or spines, etc.) is formed (when the fruit body is sporulating). The colour of the powdery print (which is called a spore print) has been used to help classify mushrooms, hence is used to help identify them. Spore print colors range from white (most common), brown, black, purple-brown, pink, yellow, cream, and almost never blue or green or red.

While modern identification of mushrooms is quickly becoming molecular, the standard methods for identification are still used by most and have developed into a fine art harking back to mediaeval and Victorian era combined with microscopic examination. The presence of juices upon breaking, bruising reactions, odors, tastes, shades of colors, and habitats and habit and season must, and are, all considered by mycologists, amateur and professional alike. Tasting and smelling mushrooms carry their own hazards because of poisons and allergens. Chemical spot tests are also used for some genera.

In general, identification to genus can often be accomplished in the field using a local mushroom guide. Identification to species, however, requires more effort; one must remember that a mushroom develops from a button stage into a mature structure and only the latter can provide certain characters needed for the identification of the species. However. over mature specimens loose features and cease producing spores. Many novices have mistaken humid water marks on paper for white spore prints, or discolored paper from oozing liquids on lamella edges for colored spored prints.

Human Use

Edible mushrooms are used extensively in cooking, in many cuisines (notably Chinese, European and Japanese). Though commonly thought to contain little nutritional value, many varieties of mushrooms are high in fiber and protein, and provide vitamins such as thiamine (B_1), riboflavin (B_2), niacin (B_3), biotin (B_7), cobalamins (B_{12}) and ascorbic acid (C), as well as minerals, including iron, selenium, potassium and phosphorus. Mushrooms have been gaining a higher profile for containing antioxidants Ergothioneine and Selenium. Research is currently being conducted as to how the mushroom may help to prevent breast cancer, prostate cancer, and other diseases such as heart disease, diabetes, high cholesterol, and obesity.

Many of the varieties of mushrooms that are sold in local supermarkets, have been commercially grown on mushroom farms. These mushrooms are safe to eat because they are grown in controlled, sterilized environments. Some of the varieties that are grown commercially include: whites, crimini, portabello, shiitake, oyster and enoki.

There are a number of species of mushrooms that are poisonous, and although some may resemble edible varieties, eating them could be fatal. Eating mushrooms gathered in the wild can be risky and a practice not to be undertaken by individuals not knowledgeable in mushroom identification. The problem is that separating edible from poisonous species depends upon the application of only a few easily recognizable traits - but there is no single trait by which all toxic mushrooms could be identified. People who collect mushrooms for consumption are known as mycophagists, and the act of collecting them for such is known as mushroom hunting, or simply "mushrooming".

Toxic Mushrooms

Of central interest with respect to chemical properties of mushrooms is the fact that many species produce secondary metabolites that render them toxic, mind-altering, or even bioluminescent. Toxicity likely plays a role in protecting the function of the basidiocarp: the mycelium has expended considerable energy and protoplasmic material to develop a structure to efficiently distribute its spores. One defence against consumption and premature destruction is the evolution of chemicals that render the mushroom inedible, either causing the consumer to vomit the meal or avoid consumption altogether.

Psychoactive Mushrooms

Psilocybin mushrooms possess psychedelic properties. They are commonly known as "magic mushrooms" or "shrooms", and are available in smart shops in many parts of the world, though some countries have outlawed their sale. A number of other mushrooms are eaten for their psychoactive effects, such as fly agaric, which is used for shamanic purposes by tribes in northeast Siberia. They have also been used in the West to potentiate, or increase, religious experiences. Because of their psychoactive properties, some mushrooms have played a role in native medicine, where they have been used to affect mental and physical healing, and to facilitate visionary states. One such ritual is the Velada ceremony. A representative figure of traditional mushroom use is the shaman, curandera (priest-healer), Maria Sabina.

Medicinal Mushrooms

Currently, many species of mushrooms and fungi utilized as folk medicines for thousands of years are under intense study by ethnobotanists and medical researchers. Maitake, shiitake, and reishi are prominent among those being researched for their potential anti-cancer, anti-viral, and/or immunity-enhancement properties. Psilocybin, originally an extract of certain psychedelic mushrooms, is being studied for its ability to help people suffering from mental disease, such as obsessive-compulsive disorder. Minute amounts have been reported to stop cluster and migraine headaches.

Other Uses

Mushrooms can be also used for dyeing wool and other natural fibers. The chromophores of mushrooms are organic compounds and produce strong and vivid colors, and all colors of the spectrum can be achieved with mushroom dyes. Before the invention of synthetic dyes the mushrooms were the primary sources on dyeing textiles. This technique has survived in Finland, and many Middle Ages re-enactors have revived the skill again. Some fungi, types of polypores, loosely called mushrooms, have been used as fire starters (known as tinder fungi). Ötzi the Iceman was found carrying such fungi. Mushrooms, and other fungi, will likely play an increasingly important role in the development of effective biological remediation and filtration technologies. The US Patent and Trademark officecan be searched for patents related to the latest developments in mycoremediation and mycofiltration.

Chapter 10

Bryology: The Liverworts and Mosses

Bryology is the study of nonvascular complex plants classified in the Division Bryophyta. The Bryophytes have several adaptations that allow them to grow on land, but are still tied to moisture for their reproduction. Their size is also restricted because they do not have vascular tissues, which limits their height and keeps them close to the nutrition and moisture of the ground. Rhizoids anchor them to the ground, but do not conduct water or nutrients as roots do. Some mosses have developed pores, but they do not open and close as stomata do.

Most mosses look like undifferentiated green mold on rocks, but while they are non vascular, they have complex forms and parts. They have what appear to be leaves coming off their stalks, but are not leaves because they have no veins. Most notable in their anatomy is the peculiarity of their life cycle. The haploid organism is actually most of what we would call the moss. That green mat is the gametecphyte contains half of the DNA, or one full set of chromosomes. When the female gametophyte is fertilized with motile sperm from the male gametophyte, the diploid sporophyte grows directly out of the gametophyte. So out of an ordinary spiky green moss stalk, a tall curved stalk grows, with a large cap on it. That structure is the sporophyte. It will produce spores (that are haploid) that will propagate more haploid mosses. Each successive evolutionary adaptation of plants has reduced the haploid stage in size, and increased the diploid, or sporophyte stage. In ferns, the next evolutionary innovation based in bryophytes, the gametophyte stage functions the same way as here, the haploid spores make a haploid organism, which after fertilization grows a full diploid sporophyte from it. However, what we know as a fern is the diploid organism, the gametophyte is a very small heart shaped organism at the base of the fern.

MOSS

Mosses are small and soft plants that are typically 1–10 cm tall, though some species are much larger. They commonly grow close together in clumps or mats in damp or shady locations. They do not have flowers or seeds, and their simple leaves cover the thin wiry stems. At certain times mosses produce spore capsules which may appear as beak-like capsules borne aloft on thin stalks.

Overview

Botanically, mosses are bryophytes, or non-vascular plants. They can be distinguished from the apparently similar liverworts (Marchantiophyta or Hepaticae) by their multi-cellular rhizoids. Other differences are not universal for all mosses and all liverworts, but the presence of clearly differentiated "stem" and "leaves", the lack of deeply lobed or segmented leaves, and the absence of leaves arranged in three ranks, all point to the plant being a moss.

There are approximately 10,000 species of moss classified in the Bryophyta.The division *Bryophyta* formerly included not only mosses, but also liverworts and hornworts. These other two groups of bryophytes now are often placed in their own divisions.

In addition to lacking a vascular system, mosses have a gametophyte-dominant life cycle, i.e. the plant's cells are haploid for most of its life cycle. Sporophytes (i.e. the diploid body) are short-lived and dependent on the gametophyte. This is in contrast to the pattern exhibited by most "higher" plants and by most animals. In vascular plants, for example, the haploid generation is represented by the pollen and the ovule, whilst the diploid generation is the familiar flowering plant.

Life Cycle

Most kinds of plants have a double portion of chromosomes in their cells (diploid, i.e. each chromosome exists with a partner that contains the same genetic information) whilst mosses (and other bryophytes) have only a single set of chromosomes (haploid, i.e. each chromosome exists in a unique copy within the cell). There are periods in the moss lifecycle when they do have a full, paired set of chromosomes but this is only during the sporophyte stage.

The life of a moss starts from a haploid spore, which germinates to produce a protonema, which is either a mass of filaments or thalloid (flat and thallus-like). This is a transitory stage in the life of a moss. From the protonema grows the gametophore ("gamete-bearer") that is differentiated into stems and leaves ('microphylls'). From the tips of stems or branches develop the sex organs of the mosses. The female organs are known as archegonia (sing. archegonium) and are protected by a group of modified leaves known as the perichaetum (plural, perichaeta). The archegonia have necks called venters which the male sperm swim down. The male organs are known as antheridia (singular antheridium) and are enclosed by modified leaves called the perigonium (plural, perigonia).

Mosses can be either dioicous (compare dioecious in seed plants) or monoicous (compare monoecious). In dioicous mosses, both male and female sex organs are borne on different gametophyte plants. In monoicous (also called autoicous) mosses, they are borne on the same plant. In the presence of water, sperm from the antheridia swim to the archegonia and fertilisation occurs, leading to the production of a diploid sporophyte. The sperm of mosses is biflagellate, i.e. they have two flagella that aid in propulsion. Without water, fertilisation cannot occur. After fertilisation, the immature sporophyte pushes its way out of the archegonial venter. It takes about a quarter to half a year for the sporophyte to mature. The sporophyte body comprises a long stalk, called a seta, and a capsule capped by a cap called the operculum. The capsule and operculum are in turn sheathed by a haploid calyptra which is the remains of the archegonial venter. The calyptra usually falls off when the capsule is mature. Within the capsule, spore-producing cells undergo meiosis to form haploid spores, upon which the cycle can start again. The mouth of the capsule is usually ringed by a set of teeth called peristome. This may be absent in some mosses.

In some mosses, green vegetative structures called gemmae are produced on leaves or branches, which can break off and form new plants without the need to go through the cycle of fertilization. This is a means of asexual reproduction.

Classification of Mosses

Mosses were traditionally grouped with the liverworts and hornworts in the Division Bryophyta (bryophytes), within which the mosses made up the class Musci. This group, however, is paraphyletic and now tends to be split up. In such system, the Division Bryophyta refers specifically to mosses. They appear to be the closest living relatives of the vascular plants.

Fig. Red Moss Stalks, A Winter Native of the Yorkshire Dales Moorland.

The mosses are grouped as a single class, now named Bryopsida, and divided into seven subclasses:

- Andreaeidae
- Sphagnidae
- Tetraphidae
- Polytrichidae
- Buxbaumiidae
- Bryidae
- Archidiidae

Andreaeidae are distinguished by the biseriate (two rows of cells) rhizoids, multiseriate (many rows of cells) protonema, and sporangium that splits along longitudinal lines. Most mosses have capsules that open at the top.

The Sphagnidae, the peat-mosses, comprise the single genus *Sphagnum*. These large mosses form extensive acidic bogs in peat swamps. The leaves of *Sphagnum* have large dead cells alternating with living photosynthetic cells. The dead cells help to store water. Aside from this character, the unique branching, thallose (flat and expanded) protonema, and explosively rupturing sporangium place it apart from other mosses.

The Tetraphidae are unique (as their name implies) in having only four large peristome teeth surrounding the opening of the capsule.

Polytrichidae have leaves with lamellae, which are flaps on the leaves that look like the fins on a heat sink. These help it retain moisture. They differ from other mosses in other details of their development and anatomy too, and can also become larger than most other mosses, with e.g. *Polytrichum commune* forming cushions up to 40 cm high. The tallest land moss, a member of the Polytrichidae is supposed to be *Dawsonia superba*, a native to New Zealand and Australia.

The Buxbaumiidae are called 'bug mosses' because they usually have a very small and reduced gametophore and the whole plant is mostly the sporophyte capsule. The shape reminds one of a bug, which is the reason for its common name.

The Bryidae are the most diverse group; over 95% of moss species belong to this subclass.

The Archidiidae are distinguished by their extremely large spores and the way the sporangium develops

Habitat

Mosses are found chiefly in areas of low light and dampness. Mosses are common in wooded areas and at the edges of streams. Mosses are also found in cracks between paving stones in damp city streets. Some types have adapted to urban conditions and are found only in cities. A few species are wholly aquatic, such as *Fontinalis antipyretica*, and others such as *Sphagnum* inhabit bogs, marshes and very slow-moving waterways. Such aquatic or semi-aquatic mosses can greatly exceed the normal range of lengths seen in terestial mosses. Individual plants 20–30 cm or more long are common in *Sphagnum* species for example.

Wherever they occur, mosses require moisture to survive because of the small size and thinness of tissues, lack of cuticle (waxy covering to prevent water loss), and the need for liquid water to complete fertilisation. Some mosses can survive desiccation, returning to life within a few hours of rehydration.

In northern latitudes, the north side of trees and rocks will generally have more moss on average than other sides (though south-side outcroppings are not unknown). This is assumed to be because of the lack of sufficient water for reproduction on the sun-facing side of trees. South of the equator the reverse is true. In deep forests where sunlight does not penetrate, mosses grow equally well on all sides of the tree trunk.

Cultivation

Moss is considered a weed in grass lawns, but is deliberately encouraged to grow under aesthetic principles exemplified by Japanese gardening. In old temple gardens, moss can carpet a forest scene. Moss is thought to add a sense of calm, age, and stillness to a garden scene. Rules of cultivation are not widely established. Moss collections are quite often begun using samples transplanted from the wild in a water-retaining bag. However, specific species of moss can be extremely difficult to maintain away from their natural sites with their unique combinations of light, humidity, shelter from wind, etc.

Growing moss from spores is even less controlled. Moss spores fall in a constant rain on exposed surfaces; those surfaces which are hospitable to a certain species of moss will typically be colonised by that moss within a few years of exposure to wind and rain. Materials which are porous and moisture retentive, such as brick, wood, and certain coarse concrete mixtures are hospitable to moss. Surfaces can also be prepared with acidic substances, including buttermilk, yogurt, urine, and gently pureed mixtures of moss samples, water and ericaceous compost.

Inhibiting Moss Growth

Moss growth can be inhibited by a number of methods:

- Decreasing availability of water through drainage or direct application changes.
- Increasing direct sunlight.
- Increasing number and resources available for competitive plants like grasses.
- Increasing the soil pH with the application of lime.

Heavy traffic or manually disturbing the moss bed with a rake will also inhibit moss growth.

The application of products containing ferrous sulfate or ferrous ammonium sulfate will kill moss, these ingredients are typically in commercial moss control products and fertilizers. Sulfur and Iron are essential nutrients for some competing plants like grasses. Killing moss will not prevent regrowth unless conditions favourable to their growth are changed.

Mossery

A passing fad for moss-collecting in the late 19th century led to the establishment of mosseries in many British and American gardens. The mossery is typically constructed out of slatted wood, with a flat roof, open to the north side (maintaining shade). Samples of moss were installed in the cracks between wood slats. The whole mossery would then be regularly moistened to maintain growth.

Commercial Use

There is a substantial market in mosses gathered from the wild. The uses for intact moss are principally in the florist trade and for home decoration. Decaying moss in the genus *Sphagnum* is also the major component of peat, which is "mined" for use as a fuel, as a horticultural soil additive, and in smoking malt in the production of Scotch whisky. There are growing concerns in parts of the world where this trade is growing that significant environmental damage may be caused by commercial peat harvesting. In World War II, *Sphagnum* mosses were used as first-aid dressings on soldiers' wounds, as these mosses are highly absorbent and have mild antibacterial properties. Some early people used it as a diaper due to its high absorbency.

In rural UK, *Fontinalis antipyretica* was traditionally used to extinguish fires as it could be found in substantial quantities in slow-moving rivers and the moss retained large volumes of water which helped extinguish the flames. This historical use is reflected in its specific Latin/Greek name, the approximate meaning of which is "against fire".

In Finland, peat mosses have been used to make bread during famines.

HEPATICA

Hepatica is a genus of herbaceous perennial plants belonging to the buttercup family, Ranunculaceae. A native of central and northern Europe, Asia and northeastern North America, *Hepatica* is sometimes called liverleaf or "liverwort". It should not be confused with liverworts, which may also be called "Hepaticae". A few botanists include *Hepatica* within a wider interpretation of *Anemone*, as *Anemone hepatica* .

Between two and ten species of *Hepatica* are recognised, with some of the taxa more often treated as varieties:

- *Hepatica nobilis*: Common Hepatica
  - *H. nobilis* var. *pyrenaica* (*H. pyrenaica*) - Pyrenees
  - *H. nobilis* var. *japonica* (*H. japonica*) - Japan
  - *H. nobilis* var. *nobilis* - European Hepatica - Alps north to Scandinavia
  - *H. nobilis* var. *pubescens* (*H. pubescens*) - Japan
  - *H. nobilis* var. *acuta* (*H. acutiloba*) - Sharp-lobed Hepatica - North America
  - *H. nobilis* var. *obtusa* (*H. americana*) - Round-lobed Hepatica - North America
- *Hepatica transsilvanica*: Carpathian Mountains and Transylvania

Hepatica cultivation has been popular in Japan since the 18th Century (mid-Edo period), where flowers with doubled petals and a range of colour patterns have been developed .

Noted for their tolerance of alkaline limestone-derived soils, *Hepatica* may grow in a wide range of conditions; it can be found either in deeply shaded deciduous (especially beech) woodland and scrub or grassland in full sun. *Hepatica* will also grow in both sandy and clay-rich substrates, being associated with limestone. Moist soil and winter snowfall is a requirement; *Hepatica* is tolerant of winter snow cover, but less so of dry frost.

Hepatica reaches a height of 10 cm and produces hermaphroditic flowers from February to May. The leaves are basal and dark leathery green, each with three lobes. The flowers may be white bluish purple or pink; they are supported singly on hairy, largely leafless stems. Butterflies, moths, bees, flies and beetles are known to act as pollinators for *Hepatica*.

Hepatica is named from its leaves, which, like the human liver (Greek *hepar*), have three lobes. It was once used as a medicinal herb. Owing to the doctrine of signatures, the plant was thought an effective treatment for liver disorders. Although poisonous in large doses, the leaves and flowers may be used as an astringent, demulcent for slow-healing injuries and as a diuretic .

MARCHANTIOPHYTA

Liverworts are a division of plants commonly called hepatics, Marchantiophyta or liverworts. They are typically small plants that are often overlooked. They frequently have the appearance of small irregular leaf-like plaques, often covering large areas of the ground but they may also occur on rocks, trees or any other reasonably firm substrate. They can also take on a form very much like flattened mosses. They most often occur in damp locations and are typically found in moderate to deep shade. Some species can be a nuisance in shady green-houses. They do not have flowers or seeds.

Overview

Botanically, liverworts are bryophytes, or non-vascular plants. They can most reliably be distinguished from the apparently similar mosses by their single-celled rhizoids. Other differences are not universal for all mosses and all liverworts, but the occurrence of leaves arranged in three ranks, the presence of deep lobes or segmented leaves, or a lack of clearly differentiated stem and leaves all point to the plant being a liverwort.

Aside from lacking a vascular system, liverworts have a gametophyte-dominant life cycle, i.e. the plant's cells are haploid for most of its life cycle. Sporophytes (i.e. the diploid body) are short-lived and dependent on the gametophyte. This is in contrast to the pattern exhibited by vascular plants and by most animals. In vascular plants, for example, the haploid generation is represented by the pollen and the ovule, whilst the diploid generation is the familiar flowering plant.

Originally, the Marchantiophyta were grouped as class Hepaticae alongside the mosses in the Division Bryophyta, but the liverworts are now usually given their own division with two classes: Jungermanniopsida (simple thalloids and leafy liverworts) and the Marchantiopsida (complex-thallus liverworts and bottle hepatics). It is estimated that there are 6000 to 8000 species of liverworts, at least 85% of which belong to the leafy group.

Today, liverworts can be found in many ecosystems across the planet except the sea and dry environments or those exposed to high levels of direct solar radiation. As with most groups of living plants, they are most common (both in numbers and species) in moist tropical areas.

Description

The most familiar liverworts consist of a prostrate, flattened, branching structure called a thallus (plant body). These liverworts are termed *thallose liverworts*. However, most liverworts produce flattened stems with overlapping scales or leaves in three or more ranks, the middle rank being conspicously different from the outer ranks. These are called *leafy liverworts* or *scale liverworts*.

They can be distinguished from the apparently similar mosses by their single celled rhizoids. Other differences are not universal for all mosses and all liverworts, but the lack of clearly differentiated stem and leaves, the presence of deeply lobed or segmented leaves, and the presence of leaves arranged in three ranks all point to the plant being a liverwort. Confirmation of the identifiaction of a moss or a leafy liverwort can only be performed with certainty by microscopical investigation.

Aside from lacking a vascular system, liverworts have a gametophyte-dominant life cycle, i.e. the plant's cells are haploid for most of its life cycle. Sporophytes (i.e. the diploid body) are short-lived and dependent on the gametophyte. This is in contrast to the pattern exhibited by most higher plants and animals. In higher plants, for example, the haploid generation is represented by the pollen and the ovule while the diplod generation is the familiar flowering plant.

Life Cycle

The life of a liverwort starts from a haploid spore, which germinates to produce a protonema, which is either a mass of filaments or thalloid (flat and thallus-like). This is a transitory stage in the life of a liverwort. From the protonema grows the gametophore ("gamete-bearer") that produces the sex organs of the liverworts. The female organs are known as archegonia (singular archegonium) and are protected by the perichaetum (plural perichaeta). The archegonia have necks called venters which the male sperm swim down. The male organs are known as antheridia (singular antheridium) and are enclosed by the perigonium (plural perigonia).

Liverworts can be either dioicous or monoicous. In dioecious liverworts, female and male sex organs are borne on different plants. In monoecious liverworts, they are borne on the same plant. In the presence of water, sperm from the antheridia swim to the archegonia and fertilisation occurs, leading to the production of a diploid sporophyte. The sperm of liverworts is biflagellate, i.e. they have two flagellae that aid in propulsion. Without water, fertilisation cannot occur. After fertilisation, the immature sporophyte elongates, pushing its way out of the archegonial venter. The sporophyte body comprises a long stalk, called a seta, and a spherical or ellipsoidal capsule. Within the capsule, cells divide to produce elater cells and spore-producing cells that will undergo meiosis to form haploid spores, upon which the cycle can start again.

Classification

In ancient times, it was believed that liverworts cured diseases of the liver, hence the name. In Old English, the word liverwort literally means *liver plant*. This probably stemed from the superficial appearance of some thalloid liverworts (which resemble a liver in outline), and led to the common name of the group as *hepatics*, from the Latin word for liver. An unrelated flowering plant, *Hepatica*, is sometimes also referred to as liverwort because it was once also used in treating diseases of the liver. This archaic relationship of plant form to function was based in the "Doctrine of Signatures".

Bryologists classify liverworts in the division Marchantiophyta. This divisional name is based on the name of the type species *Marchantia polymorpha.* In addition to this taxon-based name, the liverworts are often called Hepaticophyta. This name is derived from their common Latin name as Latin was the language in which botanists published their descriptions of species. This name has led to some confusion, partly because it appears to be a taxon-based name derived from the genus *Hepatica* which is actually a flowering plant of the buttercup family Ranunculaceae. In addition, the name Hepaticophyta is frequently misspelled in textbooks as Hepatophyta, which only adds to the confusion.

Traditionally, the liverworts were grouped together with other bryophytes (mosses and hornworts) in the Division Bryophyta, within which the liverworts made up the class Hepaticae (also called Marchantiopsida). However, since this grouping makes the Bryophyta paraphyletic, the liverworts are now usually given their own division. The use of the division name Bryophyta *sensu latu* is still found in the literature, but more frequently the Bryophyta now is used in a restricted sense to include only the mosses.

Another reason that liverworts are now classified separately is that liverworts appear to have diverged from all other embryophyte plants near the beginning of their evolution. The strongest line of supporting evidence is that liverworts are the only living group of land plants that do not have stomata on the sporophyte generation. The earliest fossils believed to be liverworts are compression fossils of *Pallaviciniites* from the Upper Devonian of New York. These fossils resemble modern species in the Metzgeriales. Another Devonian fossil called *Protosalvinia* also looks like a liverwort, but its relationship to other plants is still uncertain, so it may not belong to the Marchantiophyta.

The Marchantiophyta is subdivided into two classes. The Jungermanniopsida includes primarily the two orders Metzgeriales (simple thalloids) and Jungermanniales (leafy liverworts), as well as a smaller order Haplomitriales. The Marchantiopsida includes primarily the orders Marchantiales (complex-thallus liverworts) and Sphaerocarpales (bottle hepatics), as well as the problematic genus *Monoclea,* which is sometimes placed in its own order Monocleales.

HORNWORT

Hornworts are a group of bryophytes, or non-vascular plants, comprising the division Anthocerotophyta. The common name refers to the elongated horn-like structure, which is the sporophyte. The flattened, green plant body of a hornwort is the gametophyte plant.

Hornworts may be found world-wide, though they tend to grow only in places that are damp or humid. Some species grow in large numbers as tiny weeds in the soil of gardens and cultivated fields. Large tropical and sub-tropical species of *Dendroceros* may be found growing on the bark of trees.

Description

The plant body of a hornwort is a haploid gametophyte stage. This stage usually grows as a thin rosette or ribbon-like thallus between one and five centimeters in diameter. Each cell of the thallus usually contains just one chloroplast per cell. In most species, this chloroplast is fused with other organelles to form a large pyrenoid that both manufactures and stores food. This particular feature is very unusual in land plants, but is common among algae.

Many hornworts develop internal mucilage-filled cavities when groups of cells break down. These cavities are invaded by photosynthetic cyanobacteria, especially species of *Nostoc*. Such colonies of bacteria growing inside the thallus give the hornwort a distinctive blue-green colour. There may also be small *slime pores* on the underside of the thallus. These pores superficially resemble the stomata of other plants.

The horn-shaped sporophyte grows from an archegonium embedded deep in the gametophyte. Hornworts sporophytes are unusual in that the sporophyte grows from a meristem near its base, instead of from its tip the way other plants do. Unlike liverworts, most hornworts have true stomata on the sporophyte as mosses do. The exceptions are the genera *Notothylas* and *Megaceros*, which do not have stomata.

When the sporophyte is mature, it has a multicellular outer layer, a central rod-like columella running up the centre, and a layer of tissue in between that produces spores and pseudo-elaters. The pseudo-elaters are multi-cellular, unlike the elaters of liverworts. They have helical thickenings that change shape in response to drying out, and thereby twist in and thereby help to disperse the spores. Hornwort spores are relatively large for bryophytes, measuring between 30 and 80 μm in diameter or more. The spores are polar, usually with a distinctive Y-shaped tri-radiate ridge on the proximal surface, and with a distal surface ornamented with bumps or spines.

Life cycle

The life of a hornwort starts from a haploid spore. In most species, there is a single cell inside the spore, and a slender extension of this cell called the *germ tube* germinates from the proximal side of the spore. The tip of the germ tube divides to form an octant of cells, and the first rhizoid grows as an extension of the original germ cell. The tip continues to divide new cells, which produces a thalloid protonema. By contrast, species of the family Dendrocerotaceae may begin dividing within the spore, becoming multicellular and even photosynthetic before the spore germinates. In either case, the protonema is a transitory stage in the life of a hornwort.

From the protonema grows the adult gametophyte, which is the persistent and independent stage in the life cycle. This stage usually grows as a thin rosette or ribbon-like thallus between one and five centimeters in diameter, and several layers of cells in thickness. It is green or yellow-green from the chlorophyll in its cells, or bluish-green when colonies of cyanobacteria grow inside the plant.

When the gametophyte has grown to its adult size, it produces the sex organs of the hornwort. Most plants are monoicous, with both sex organs on the same plant, but some plants (even within the same species) are dioicous, with separate male and female gametophytes. The female organs are known as archegonia (singular archegonium) and the male organs are known as antheridia (singular antheridium). Both kinds of organs develop just below the surface of the plant and are only later exposed by disintegration of the overlying cells.

The biflagellate sperm must swim from the antheridia, or else be splashed to the archegonia. When this happens, the sperm and egg cell fuse to form a zygote, the cell from which the sporophyte stage of the life cycle will develop. Unlike all other bryophytes, the first cell division of the zygote is longitudinal. Further divisions produce three basic regions of the sporophyte.

At the bottom of the sporophyte (closest to the interior of the gametophyte), is a foot. This is a globular group of cells that receives nutrients from the parent gametophyte, on which the sporophyte will spend its entire existence. In the middle of the sporophyte (just above the foot), is a meristem that will continue to divide and produce new cells for the third region. This third region is the capsule. Both the central and surface cells of the capsule are sterile, but between them is a layer of cells that will divide to produce pseudo-elaters and spores. These are released from the capsule when it splits lengthwise from the tip.

Classification of Hornworts

Hornworts were traditionally considered a class within the Division Bryophyta (bryophytes). However, it now appears that this group is paraphyletic, so the hornworts tend to be given their own division, called Anthocerotophyta. The Bryophyta is now restricted to include only mosses.

There is a single class of hornworts, called Anthocerotopsida, or traditionally Anthocerotae. This class includes a single order of hornworts (Anthocerotales) in this classification scheme. In some other classification schemes, a second order Notothyladales (containing only the genus *Notothylas*) is recognized because of the unique and unusual features present in that group.

Among land plants, hornworts appear to be one of the oldest surviving groups. There are only about 100 species known, but new species are still being discovered. The number and names of genera are a current matter of investigation, and several competing classification schemes have been published since 1988.

Families and Genera

Anthocerotaceae

- *Anthoceros*
- *Folioceros*
- *Leiosporoceros*

- *Phaeoceros*
- *Sphaerosporoceros*

Dendrocerotaceae

- *Dendroceros*
- *Megaceros*
- *Notoceros*

Notothyladaceae

- *Notothylas*

Chapter 11

Pteridology: The Ferns

Pteridology is the study of ferns—plants classified in the Division Pterophyta (or Filicophyta). Ferns do not have seeds the way trees and flowering plants do. Rather, they have spores the way mosses do. The haploid spores grow small haploid organisms, which then undergo fertilization and grow the diploid fern plant directly out of the haploid gametophyte, similar to the sporophyte stalk growing out of the moss. The larger part, what we think of as the fern, is the sporophyte. The gametophyte is a small green prothallus that the sporophyte grows out of. Ferns are still tied to an aquatic environment, in that once a spore grows into a prothallus, there must be moisture enough for the egg in the prothallus to be fertilized by swimming, flagellated fern sperm.

Having a large sporophyte allows ferns to produce many more spores than a moss could- recall that each sporophyte on a moss only carried one sporangia. Producing many more propagules increased fern presence and dominance. Besides having a larger sporophyte generation, ferns have many important adaptations that increase their capabilities above the mosses. Ferns have roots, which, unlike moss rhizoids, not only anchor, but take up nutrients. Ferns are vascular plants, with lignified vascular tissues. These allow active water transport. That water transport along with the strength of the ligified cells allow ferns to be much larger than their moss ancestors. At one point, ferns and fern trees were the most advanced plant life, and grew even larger than ferns today do, with great size and variety of ferns. There were no flowering plants in the early cretaceous- the first forests of the dinosaurs were composed of fern trees.

FERN

A fern is any one of a group of about 20,000 species of plants classified in the phylum or division Pteridophyta, also known as Filicophyta. The group is also referred to as polypodiophyta, or polypodiopsida when treated as a subdivision of tracheophyta (vascular plants). The study of ferns is called pteridology; one who studies ferns is called a pteridologist. The term pteridophytes has traditionally been used to describe all seedless vascular plants so is synonymous with "ferns and fern allies". This can be confusing given that the fern phylum Pteridophyta is also sometimes referred to as pteridophytes.

A fern is a vascular plant that differs from the more primitive lycophytes in having true leaves (megaphylls), and from the more advanced seed plants (gymnosperms and angiosperms) in lacking seeds. Like all vascular plants, it has a life cycle, often referred to as alternation of generations, characterized by a diploid sporophytic and a haploid gametophytic phase. Unlike the gymnosperms and angiosperms, in ferns the gametophyte is a free-living organism. The life cycle of a typical fern is as follows:

1. A sporophyte (diploid) phase produces haploid spores by meiosis;
2. A spore grows by cell division into a gametophyte, which typically consists of a photosynthetic prothallus
3. The gametophyte produces gametes (often both sperm and eggs on the same prothallus) by mitosis
4. A mobile, flagellate sperm fertilizes an egg that remains attached to the prothallus
5. The fertilized egg is now a diploid zygote and grows by mitosis into a sporophyte (the typical "fern" plant).

FERN ECOLOGY

Ferns have a popular image of growing in moist, shady woodland nooks, but the reality is far more complex. Ferns grow in a wide variety of habitats, ranging from remote mountain elevations to dry desert rock faces to bodies of water to open fields. Ferns in general may be thought of as largely being specialists in marginal habitats, often succeeding in places where various environmental delimiters limit the success of flowering plants. On the other hand, some ferns are among the world's most serious weed species, such as the bracken growing in the British highlands, or the mosquito fern (*Azolla*) growing in tropical lakes. There are four particular types of habitats that are often key places to find ferns: the afore-mentioned moist, shady forest cove; the sheltered rock face, especially when sheltered from the full sun; acid bogs and swamps; and tropical trees, where many species are epiphytes.

Many ferns depend on associations with mycorrhizal fungi. Many ferns only grow within specific pH ranges; for instance, the climbing fern (*Lygodium*) of eastern North America will only grow in moist, intensely acid soils, while the bulblet bladder fern (*Cystopteris bulbifera*) with overlapping range is only ever found on limestone

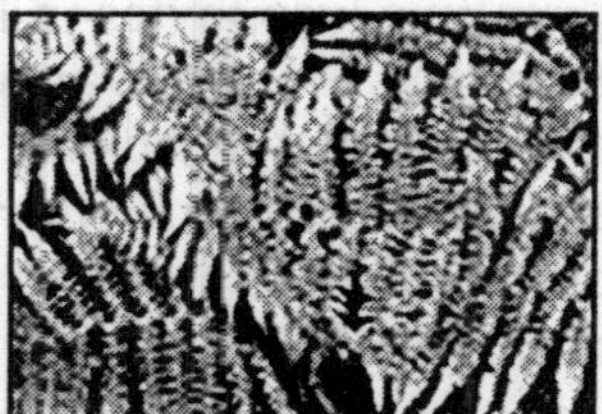

Fig. Ferns

Fern Structure

Like the sporophytes of seed plants, those of ferns consist of:

- *Stems:* Most often an underground creeping rhizome, but sometimes an above-ground creeping stolon (e.g., Polypodiaceae), or an above-ground erect semi-woody trunk (e.g., Cyatheaceae) reaching up to 20 m in a few species (e.g., *Cyathea brownii* on Norfolk Island and *Cyathea medullaris* in New Zealand).
- *Leaf:* The green, photosynthetic part of the plant. In ferns, it is often referred to as a frond, but this is

because of the historical division between people who study ferns and people who study seed plants, rather than because of differences in structure. New leaves typically expand by the unrolling of a tight spiral called a crozier or fiddlehead. This uncurling of the leaf is termed circinate vernation. Leaves are divided into two types:

- *Trophophyll:* A leaf that does not produce spores, instead only producing sugars by photosynthesis. Analogous to the typical green leaves of seed plants.
- *Sporophyll:* A leaf that produces spores. These leaves are analogous to the scales of pine cones or to stamens and pistil in gymnosperms and angiosperms respectively. Unlike the seed plants, however, the sporophylls of ferns are typically not very specialized, looking similar to trophophylls and producing sugars by photosynthesis as the trophophylls do.

- *Roots:* The underground non-photosynthetic structures that take up water and nutrients from soil. They are always fibrous and are structurally very similar to the roots of seed plants.

The gametophytes of ferns, however, are very different from those of seed plants. They typically consist of:

- *Prothallus:* A green, photosynthetic structure that is one cell thick, usually heart or kidney shaped, 3-10 mm long and 2-8 mm broad. The thallus produces gametes by means of:
  - *Antheridia:* Small spherical structures that produce flagellate sperm.
  - *Archegonia:* A flask-shaped structure that produces a single egg at the bottom, reached by the sperm by swimming down the neck.

- *Rhizoids:* Root-like structures (not true roots) that consist of single greatly-elongated cells, water and mineral salts are absorbed over the whole structure. Rhizoids anchor the prothallus to the soil.

Evolution and Classification

Ferns first appear in the fossil record in the early-Carboniferous period. By the Triassic, the first evidence of ferns related to several modern families appeared. The "great fern radiation" occurred in the late-Cretaceous, when many modern families of ferns first appeared.

Ferns have traditionally been grouped in the Class Filices, but modern classifications assign them their own division in the plant kingdom, called Pteridophyta.

Traditionally, three discrete groups of plants have been considered ferns: two groups of eusporangiate ferns—families Ophioglossaceae (adders-tongues, moonworts, and grape-ferns) and Marattiaceae—and the leptosporangiate ferns. The Marattiaceae are a primitive group of tropical ferns with a large, fleshy rhizome, and are now thought to be a sibling taxon to the main group of ferns, the leptosporangiate ferns. Several other groups of plants were considered "fern allies": the clubmosses, spikemosses, and quillworts in the Lycopodiophyta, the whisk ferns in Psilotaceae, and the horsetails in the Equisetaceae. More recent genetic studies have shown that the Lycopodiophyta are only distantly related to any other vascular plants, having radiated evolutionarily at the base of the vascular plant clade, while both the whisk ferns and horsetails are as much "true" ferns as are the Ophioglossoids and Marattiaceae. In fact, the whisk ferns and Ophioglossoids are demonstrably a clade, and the horsetails and Marattiaceae are arguably another clade.

One possible means of treating this situation is to consider only the leptosporangiate ferns as "true" ferns, while considering the other three groups as "fern allies". In practice, numerous classification schemes have been proposed for ferns and fern allies, and there has been little consensus among them. A new classification by Smith et al. (2006) is based on recent molecular systematic studies, in addition to morphological data. This classification divides ferns into four classes:

- Psilotopsida
- Equisetopsida

- Marattiopsida
- Polypodiopsida

The last group includes most plants familiarly known as ferns. Modern research supports older ideas based on morphology that the Osmundaceae diverged early in the evolutionary history of the leptosporangiate ferns; in certain ways this family is intermediate between the eusporangiate ferns and the leptosporangiate ferns.

FROND

A frond is the leaf- like structure of a fern or alga. The term is colloquially applied to the leaves of palms, cycads, and plants with pinnately compound leaves. A significant difference is that, unlike the leaves of the latter, fern fronds bear the reproductive structures (spore-bearing structures) of the sporophyte plant. Because many ferns grow fronds that are held more vertical than horizontal, the "upper" and "lower" surfaces of a frond are more correctly referred to as the adaxial and abaxial surfaces, respectively.

A fern frond consists of a stipe, the stem supporting the blade, and the blade consists of both a laminar (flattened) photosynthetic tissue and a rachis—that portion of the stem to which the laminar tissue is attached. The blades of fern fronds may vary from being simple (undivided) to being highly dissected, even "lace-like". If the leaf tissue is undissected, or the dissections do not reach to the *rachis,* the frond may be described as lobed or *pinnatifid*. Otherwise, the blade is compound and each large division of the laminar tissue arising from the rachis is called a pinna (pl., pinnae). The main vein or mid-rib of a pinna is known as a costa (pl., costae). Pinnae may be arranged along the *rachis* either directly opposite one another or alternating up the stem. The arrangement may change from the base of a blade to the tip, as in the example of *Blechnum* shown below (from base to tip: pinnae opposite to alternate, and pinnatisect to pinnatifid).

Many ferns have *pinnae* that are divided two or more times, and the level of division of the fronds is termed *pinnate* (or 1-pinnate), or *twice-pinnate* (2-pinnate), or the like. Each secondary division (division of a pinna) is termed a pinnule, and its mid-vein, a costule. A few species of ferns with divided fronds are not pinnate, but are palmate or bifurcate.

On some or all mature blades (usually on the *abaxial* surface) occur sporangia, which bear the spores. The sporangia are clustered in a sporus (pl., spori) or "fruit dot". Associated with each sporus in many species is a membranous structure called an indusium: an outgrowth of the blade surface that may partly cover the sporangial cluster. Fronds also may bear hairs or scales, glands, and, in some species, bulblets for vegetative reproduction.

Each frond arises from the stem or rhizome, which in most species is concealed in the ground or creeps along the ground (or branch or rock) surface. Growth of a fern frond differs from that of a leaf of a flowering plant. The fern frond unrolls from a tightly-coiled structure called a "fiddle-head".

Some fern species feature frond dimorphism, in which fertile and sterile fronds differ in appearance and structure.

Evolution and Ferns

Ferns have a big advantage over the mosses in their vascular tissue. They can grow taller, and can exist in more diverse environments. This is a trend that will continue in evolution, eventually leading to the rise of such large sporophyte generations as the great sequoia trees. But if ferns are so much more fit for survival, why are there still mosses? And if a larger sporophyte generation is more fit, why haven't sequoias become dominant enough to eliminate the ferns? While there are clear benefits to a larger sporophyte generation, in some recurring natural situations, natural selection favors mosses over ferns or ferns over trees. Spores are better at spreading by wind than many seeds are, for instance. So while in the long term, the protection of a seed allows seed plants to be dominant on the planet, in many situations the lightness and transport of a spore is still efficient in spreading ferns.

Chapter 12

Equisetophyta: The Club Mosses and Horsetails

The Equisetophyta is a very old division—the plants of this division were among the first plants to grow on land. Equisetophyta are sometimes reffered to as fern allies, as they have the same life cycles as ferns, and have similar developments that allow them to grow taller than the bryophytes.

EQUISETUM

Equisetum is a genus of vascular plants that reproduce by spores rather than seeds. The genus includes 15 species, commonly known as horsetails and scouring rushes. *Equisetum* is the only one in the family Equisetaceae, which in turn is the only family in the order Equisetales and the class Equisetopsida This class is often placed as the sole member of the Division Equisetophyta (also called Arthrophyta in older works), though some recent molecular analyses place the genus within the ferns (Pteridophyta), related to Marattiales. Other classes and orders of Equisetophyta are known from the fossil record, where they were important members of the world flora during the Carboniferous period.

The name horsetail, often used for the entire group, arose because the branched species somewhat resemble a horse's tail, the name *Equisetum* being from the Latin *equus*, "horse", and *seta*, "bristle". Other names include candock (applied to branching species only), and scouring-rush (applied to the unbranched or sparsely branched species). The latter name refers to the plants' rush-like appearance; the stems were used for scouring cooking pots in the past (due to them being coated with abrasive silica).

The genus is near-cosmopolitan, being absent only from Australasia and Antarctica. They are perennial plants, either herbaceous, dying back in winter (most temperate species) or evergreen (some tropical species, and the temperate species *Equisetum hyemale, E. scirpoides, E. variegatum* and *E. ramoissimum*). They mostly grow 0.2-1.5 m tall, though *E. telmateia* can exceptionally reach 2.5 m, and the tropical American species *E. giganteum* 5 m, and *E. myriochaetum* 8 m.

In these plants the leaves are greatly reduced, in whorls of small, segments fused into nodal sheaths. The stems are green and photosynthetic, also distinctive in being hollow, jointed, and ridged (with (3-) 6-40 ridges). There may or may not be whorls of branches at the nodes; when present, these branches are identical to the main stem except smaller.

The spores are borne under sporangiophores in cone-like structures (*strobilus*, pl. *strobili*) at the tips of some of the stems. In many species the cone-bearing stems are unbranched, and in some (e.g. *E. arvense*) they are non-photosynthetic, produced early in spring separately from photosynthetic sterile stems. In some other species (e.g. *E. palustre*) they are very similar to sterile stems, photosynthetic and with whorls of branches.

Horsetails are mostly homosporous, though in *E. arvense*, smaller spores give rise to male prothalli. The spores have four elaters that act as moisture-sensitive springs, assisting spore dispersal after the sporangia have split open longitudinally.

Many plants in this genus prefer wet sandy soils, though some are aquatic and others adapted to wet clay soils. One horsetail, *E. arvense*, can be a nuisance weed because it readily regrows after being pulled out. The stalks arise from rhizomes that are deep underground and almost impossible to dig out. It is also unaffected by many herbicides designed to kill seed plants. The foliage of some species is poisonous to grazing animals if eaten in large quantities. *Equisetum* is cooked and eaten in Japan.

The horsetails were a much larger and more diverse group in the distant past before seed plants became dominant across the Earth. Some species were large trees reaching to 30 m tall. The genus *Calamites* (family Calamitaceae) is abundant in coal deposits from the Carboniferous period.

Chapter 13

Pinophyta

The Division Pinophyta in the Kingdom Plantae comprises those species of plants that were formerly classified as the "modern" gymnosperms of the Class Coniferales—that is the *conifers*. Unlike many of the gymnosperm groups covered in the previous chapter, *pinophytes* are today still broadly represented on the landscape, forming extensive forests in both the Northern and Southern hemispheres. This does not mean this is a recent group in the paleontological record. Pinophytes are found as fossils as far back as the Upper Carboniferous (Paleozoic Era). Thus, it is a very ancient group, but one still having significant ecological importance on the planet. Pinophytes are mostly evergreen trees (some are shrubs) and many have great commercial value for their wood.

GYMNOSPERM

Gymnosperms (Gymnospermae) are a group of Spermatophyte seed-bearing plants with ovules on the edge or blade of an open sporophyll, the sporophylls usually arranged in cone-like structures. The other major group of seed-bearing plants, the angiosperms, have ovules enclosed in a carpel, a sporophyll with fused margins. The term gymnosperm comes from the Greek word *gumnospermos*, meaning "naked seeds" and referring to the unenclosed condition of the seeds, as when they are produced they are found naked on the scales of a cone or similar structure.

Gymnosperms are heterosporous, producing *microspores* that develop into pollen grains and *megaspores* that are retained in an ovule. After fertilization (joining of the micro- and megaspore), the resulting embryo, along with other cells comprising the ovule, develops into a seed. The seed is a sporophyte resting stage.

In early classification schemes, the gymnosperms (Gymnospermae) "naked seed" plants were regarded as a "natural" group. However, certain fossil discoveries suggest that the angiosperms evolved from a gymnosperm ancestor, which would make the gymnosperms a paraphyletic group if all extinct taxa are included. Modern cladistics only accepts taxa that are monophyletic, traceable to a common ancestor and inclusive of all descendants of that common ancestor. So, while the term gymnosperm is still widely used for non-angiosperm seed-bearing plants, the plant species once treated as gymnosperms are usually distributed among four groups, which can be given equal rank as divisions within the Kingdom Plantae.

With regard to extant gymnosperms, molecular phylogenies of living taxa have conflicted with morphological datasets with regard to whether they comprise a monophyletic or paraphyletic group with respect to angiosperms. At issue is whether the Gnetophyta are the sister taxon to angiosperms, or whether they are sister to, or nested within, other extant gymnosperms.

The conifers, division Pinophyta, also known as division Coniferaeor looser frwak dummy eiiot, are one of 13 or 14 division level taxa within the Kingdom Plantae. They are cone-bearing seed plants with vascular tissue; all extant conifers are woody plants, the great majority being trees with just a few being shrubs. Typical examples of conifers include cedars, cypresses, douglas-firs, firs, junipers, kauris, larches, pines, redwoods, spruces, and yews. Species of conifers can be found growing naturally in almost all parts of the world, and are frequently dominant plants in their habitats, as in the taiga, for example. Conifers are of immense economic value, primarily for timber and paper production; the wood of conifers is known as softwood. The division contains approximately 700 living species.

TAXONOMY AND NAMING

The division name Pinophyta conforms with the rules of the *ICBN*, which state (Art 16.1) that the names of higher taxa in plants (above the rank of family) are either formed from the name of an included family, in this case Pinaceae (the pine family), or are descriptive. In the latter case the name for the conifers (at whatever rank is chosen) is Coniferae (Art 16 Ex 2), which is also in widespread use. Older scientific names (no longer allowed) are Coniferophyta and Coniferales.

According to Article 16 of the *ICBN* such "descriptive botanical names" may be used at any rank above family. Alternatively, it is also possible to use a name formed by replacing the termination *-aceae* in the name of an included family, preferably Pinaceae, by the appropriate termination. Both are allowed.

This means that if the conifers are regarded to be an order they may be called Coniferae or Pinales if regarded as a class they may be called Coniferae or Pinopsida; if regarded as a division they may be called Coniferae or Pinophyta.

Commonly the conifers are considered equivalent to the Gymnosperms, particularly in areas with a temperate climate where they may be the only commonly occurring gymnosperms. However, these are two different groupings; conifers are the largest and economically most important component group of the gymnosperms, but nevertheless they comprise only one of the four groups.

The division Pinophyta consists of just one class, Pinopsida, which includes both living and fossil taxa. Subdivision of the living conifers into two or more orders has been proposed from time to time. The most commonly seen in the past was a split into two orders, Taxales (Taxaceae only) and Pinales (the rest), but recent research into DNA sequences suggests that this interpretation leaves the Pinales without

Taxales as paraphyletic, and the latter order is no longer regarded as distinct. A more accurate subdivision would be to split the class into three orders, Pinales containing only Pinaceae, Araucariales containing Araucariaceae and Podocarpaceae, and Cupressales containing the remaining families (including Taxaceae), but there has not been any significant support for such a split, with the majority of opinion preferring retention of all the families within a single order Pinales, despite their antiquity and diverse morphology.

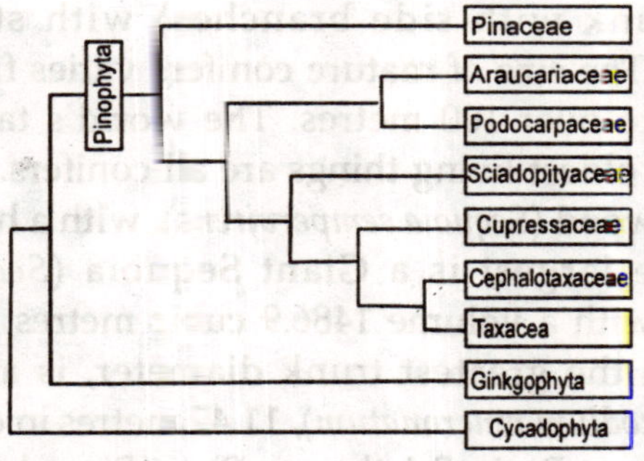

Fig. Phylogeny of the Pinophyta Based on Cladistic Analysis of Molecular Data.

The conifers are now accepted as comprising six to eight families, with a total of 65-70 genera and 600-650 species. The seven most distinct families are linked in the box above right and phylogenetic diagram left. In other interpretations, the Cephalotaxaceae may be better included within the Taxaceae, and some authors additionally recognize Phyllocladaceae as distinct from Podocarpaceae (in which it is included here). The family Taxodiaceae is here included in family Cupressaceae, but was widely recognized in the past and can still be found in many field guides.

The conifers are an ancient group, with a fossil record extending back about 300 million years to the Paleozoic in the late Carboniferous period; even many of the modern genera are recognizable from fossils 60-120 million years old. Other classes and orders, now long extinct, also occur as fossils, particularly from the late Paleozoic and Mesozoic eras. Fossil conifers included many diverse forms, the most dramatically distinct from modern conifers being some herbaceous conifers with no woody stems. Major fossil orders of conifers or conifer-like plants include the Cordaitales, Vojnovskyales, Voltziales and perhaps also the Czekanowskiales (possibly more closely related to the Ginkgophyta).

MORPHOLOGY

All living conifers are woody plants, and most are trees, the majority having monopodial growth form (a single, straight trunk with side branches) with strong apical dominance. The size of mature conifers varies from less than one metre, to over 100 metres. The world's tallest, largest, thickest and oldest living things are all conifers. The tallest is a Coast Redwood (*Sequoia sempervirens*), with a height of 115.2 metres. The largest is a Giant Sequoia (*Sequoiadendron giganteum*), with a volume 1486.9 cubic metres. The thickest, or tree with the greatest trunk diameter, is a Montezuma Cypress (*Taxodium mucronatum*), 11.42 metres in diameter. The oldest is a Great Basin Bristlecone Pine (*Pinus longaeva*), 4,700 years old.

Foliage

The leaves of many conifers are long, thin and needle-like, but others, including most of the Cupressaceae and some of the Podocarpaceae, have flat, triangular scale-like leaves. Some, notably *Agathis* in Araucariaceae and *Nageia* in Podocarpaceae, have broad, flat strap-shaped leaves. In the majority of conifers, the leaves are arranged spirally, exceptions being most of Cupressaceae and one genus in Podocarpaceae, where they are arranged in decussate opposite pairs or whorls of 3. In many species with spirally arranged leaves, the leaf bases are twisted to present the leaves in a flat plane for maximum light capture. Leaf size varies from 2 mm in many scale-leaved species, up to 400 mm long in the needles of some pines (e.g. Apache Pine *Pinus engelmannii*). The stomata are in lines or patches on the leaves, and can be closed when it is very dry or cold. The leaves are often dark green in colour which may help absorb a maximum of energy from weak sunshine at high latitudes or under forest canopy shade. Conifers from hotter areas with high sunlight levels (e.g. Turkish Pine *Pinus brutia*) often have yellower-green leaves, while others (e.g. Blue Spruce *Picea pungens*) have a very strong glaucous wax bloom to reflect ultraviolet light. In the great majority of genera the leaves are evergreen, usually remaining on the plant for several (2-40) years before falling, but five genera (*Larix, Pseudolarix, Glyptostrobus, Metasequoia* and *Taxodium*) are deciduous, shedding the leaves in autumn and leafless through the winter. The seedlings of many conifers, including most of the Cupressaceae, and *Pinus* in Pinaceae, have a distinct juvenile foliage period where the leaves are different, often markedly so, from the typical adult leaves.

Reproduction

Most conifers are monoecious, but some are subdioecious or dioecious; all are wind-pollinated. Conifer seeds develop inside a protective cone called a *strobilus* (or, very loosely, "pine cones", which technically occur only on pines, not other conifers!). The cones take from four months to three years to reach maturity, and vary in size from 2 mm to 600 mm long. In Pinaceae, Araucariaceae, Sciadopityaceae and most Cupressaceae, the cones are woody, and when mature the scales usually spread open allowing the seeds to fall out and be dispersed by the wind. In some (e.g. firs and cedars), the cones disintegrate to release the seeds, and in others (e.g. the pines that produce pine nuts) the nut-like seeds are dispersed by birds (mainly nutcrackers and jays) which break up the specially adapted softer cones. Ripe cones may remain on the plant for a varied amount of time before falling to the ground; in some fire-adapted pines, the seeds may be stored in closed cones for up to 60-80 years, being released only when a fire kills the parent tree.

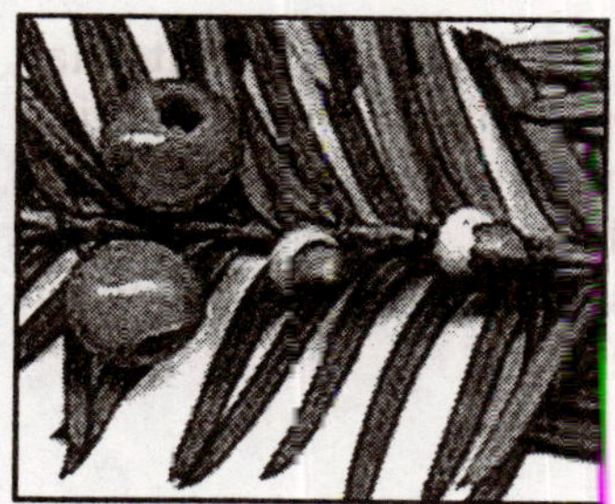

Fig. Taxaceae

In the families Podocarpaceae, Cephalotaxaceae, Taxaceae, and one Cupressaceae genus (*Juniperus*), the scales are soft, fleshy, sweet and brightly coloured, and are eaten by fruit-eating birds, which then pass the seeds in their droppings. These fleshy scales are (except in *Juniperus*) known as arils. In some of these conifers (e.g. most Podocarpaceae), the cone consists of several fused scales, while in others (e.g. Taxaceae), the cone is reduced to just one seed scale or (e.g. Cephalotaxaceae) the several scales of a cone develop into individual arils, giving the appearance of a cluster of berries.

The male cones have structures called microsporangia which produce yellowish pollen. Pollen is released and carried by the wind to female cones. Pollen grains from living pinophyte species produce pollen tubes, much like those of angiosperms. When a pollen grain lands near a female gametophyte, it undergoes meiosis and fertilizes the female gametophyte. The resulting zygote develops into an embryo, which along with its surrounding integument, becomes a seed. Eventually the seed may fall to the ground and, if conditions permit, grows into a new plant.

In forestry, the terminology of flowering plants has commonly though inaccurately been applied to cone-bearing trees as well. The male cone and unfertilized female cone are called "male flower" and "female flower", respectively. After fertilization, the female cone is termed "fruit", which undergoes "ripening" (maturation).

Life Cycle

1. To fertilize the ovum, the male cone releases pollen that is carried on the wind to the female cone.
2. A fertilized female gamete (called a zygote) develops into an embryo.
3. Along with integument cells surrounding the embryo, a seed develops containing the embryo.
4. Mature seed drops out of cone onto the ground.
5. Seed germinates and seedling grows into a mature plant.
6. When mature, the adult plant produces cones.

Other Facts

Although the total number of species is relatively small, conifers are of immense ecological importance. They are the dominant plants over huge areas of land, most notably the boreal forests of the northern hemisphere, but also in similar cool climates in mountains further south.

Many conifers have distinctly scented resin, secreted to protect the tree against insect infestation and fungal infection of wounds. Fossilised resin hardens into amber.

Evolution of the Gymnosperms

Gymnosperms are very different from the earliest vascular plants. Gymnosperms have very reduced gametophyte generations- female ovules and male pollen. Many gymnosperms, like the Pinophytes, have secondary growth from a vascular cambium. While tree ferns are tall and have a trunk resembling a tree, they are only very superficially similar. They do not have woody growth the way trees with secondary growth from a vascular cambium do. A single mutation of a fern or a horsetail could not produce a functioning seed plant So gymnosperms must be descended from some progymnosperm ancestor that evolved these adaptations from the ferns, but was not fit enough to remain on earth. One possible example of a progymnosperm is the spore bearing woody tree Archaeopteris, which was at one point probably prevalent on earth.

Chapter 14

Magnoliophyta

The Division Magnoliophyta in the Kingdom Plantae comprises those species of plants that were formerly classified as angiosperms and are known widely as the *flowering plants*. You have already studied flowers so now understand that the Division Magnoliophyta comprises all those species of plants that have flowers. For perspective: all of the plants we have read about in Section II of the *Botany Study Guide* up to this point do not have flowers, but certainly do have reproductive structures. We also know that flowers are the reproductive structures of the plants that bear them, and that reproductive structures are not limited to flowering plants. Thus, "flowers" are structures that distinguish plants in the Division Magnoliophyta from plants in the other divisions of the Kingdom Plantae. Observe a flower, and you know you are examining an angiosperm. However, not all angiosperms have obvious, showy flowers. You will need to consider that the structure of a flower is quite variable across all the many species of angiosperms (about a quarter million have been identified) and a few flowering plants actually seldom flower. However, angiosperm botanists put great stock in the structure of flowers as a way of classifying plants—more than any other part of a plant, the flower provides the basis for placement of a species in subtaxa (classes, orders, and families) of the Division Magnoliophyta.

The Division Magnoliophyta is split into two large classes: the Magnoliopsida and the Liliopsida.

DICOTYLEDON

Dicotyledons or "dicots" is a name for a group of flowering plants whose seed typically contains two embryonic leaves or cotyledons. There are around 199,350 species within this group Flowering plants that are not dicotyledons are monocotyledons, typically having one embryonic leaf.

The dicotyledons no longer are regarded as a "good" group, and the names "dicotyledons" and "dicots" are no longer to be used at least in a taxonomic sense. The vast majority of the former dicots, however, form a monophyletic group called the eudicots or tricolpates. These may be distinguished from all other flowering plants by the structure of their pollen. Other dicotyledons and monocotyledons have monosulcate pollen, or forms derived from it, whereas eudicots have tricolpate pollen, or derived forms, the pollen having three or more pores set in furrows called colpi.

Traditionally the dicots have been called the Dicotyledones (or Dicotyledoneae), at any rank. If treated as a class, as in the Cronquist system, they may be called the Magnoliopsida after the type genus *Magnolia*. In some schemes, the eudicots are treated as a separate class, the Rosopsida (type genus *Rosa*), or as several separate classes. The remaining dicots (palaeodicots) may be kept in a single paraphyletic class, called Magnoliopsida, or further divided.

The following lists are of the orders formerly placed in the dicots, giving their new placement in the APG-system and that under the older Cronquist system, which is still in wide use.

COMPARED TO MONOCOTYLEDONS

Aside from cotyledon number, other broad differences have been noted between monocots and dicots, although these have proven to be differences primarily between monocots and eudicots. Many early-diverging dicot groups have "monocot" characteristics such as scattered vascular bundles, trimerous flowers, and non-tricolpate pollen. In addition, some monocots have "dicot" characteristics such as reticulated leaf veins.

Seeds: The embryo of the monocot has one cotyledon while the embryo of the dicot has two.

Flowers: The flower parts in monocots are multiples of three while in dicots are multiples of four or five.

Stems: In monocots, the stem vascular bundles are scattered, while in dicots they are in a ring.

Secondary growth: In monccots, stems rarely show secondary growth; in dicots, stems frequently have secondary growth.

Pollen: In monocots, pollen has one furrow or pore while in dicots they have three.

Roots: The roots are adventitious in monocots, while in dicots they develop from the radicle.

Leaves: In monocots, the major leaf veins are parallel, while in dicots they are reticulated.

MONOCOTYLEDON

Monocotyledons or monocots are one of two major groups of flowering plants (angiosperms) that are traditionally recognized, Dicotyledons or dicots being the other. Monocots have been recognized at various taxonomic ranks, and under various names. The APG II system recognizes a clade called "monocots" but does not assign it to a taxonomic rank.

Monocots comprise the majority of agricultural plants in terms of biomass produced. There are between 50,000 and 60,000 species within this group; according to IUCN there are 59,300 species. The largest family in this group (and in the flowering plants as a whole) by number of species are the orchids (family Orchidaceae), with about twenty thousand species. The economically most important family in this group (and in the flowering plants) are the grasses, family Poaceae (Gramineae). These include all the true grains (rice, wheat, maize, etc.), the pasture grasses and the bamboos. This family of the true grasses have evolved in another direction, becoming highly specialized for wind pollination. Grasses produce much smaller flowers, which are gathered in highly visible plumes (inflorescences). Other economically important monocot families are the palm family (Arecaceae), banana family (Musaceae), ginger family (Zingiberaceae) and the Alliaceae, which includes such ubiquitously used vegetables as onions and garlic.

NAME, CHARACTERS

The name monocotyledons is derived from the traditional botanical name *Monocotyledones*, which derives from the fact that most members of this group have one cotyledon, or embryonic leaf, in their seeds. This as opposed to the traditional Dicotyledones, which typically have two cotyledons. From a diagnostic point of view the number of cotyledons is neither a particularly handy (as they are only present for a very short period in a plant's life), nor totally reliable character.

Nevertheless, monocots are a distinctive group. One of the most noticeable traits is that a monocot's flower is trimerous, with the flower parts in threes or in multiples of three. For example, a monocotyledon's flower typically has three, six, or nine petals. Many monocots also have leaves with parallel veins.

MORPHOLOGY

The traditionally listed differences between monocotyledons and dicotyledons are as follows. This is a broad sketch only, not invariably applicable, as there are a number of exceptions. The differences indicated are more true for monocots versus eudicots, as per the APG II system:

Flowers: In monocots, flowers are trimerous (number of flower parts in a whorl in threes) while in dicots the flowers are tetramerous or pentamerous (flower parts are in fours or fives).

Pollen: In monocots, pollen has one furrow or pore while dicots have three.

Seeds: In monocots, the embryo has one cotyledon while the embryo of the dicot has two.

Stems: In monocots, vascular bundles in the stem are scattered, in dicots arranged in a ring.

Roots: In monocots, roots are advertitious, while in dicots they develop from the radicle.

Leaves: In monocots, the major leaf veins are parallel, while in dicots they are reticulate.

However, these differences are not hard and fast: some monocots have characteristics more typical of dicots, and vice-versa. This is in part because "dicots" are a paraphyletic group with respect to monocots, and some dicots may be more closely related to monocots than to other dicots. In particular, several early-branching lineages of "dicots" share "monocot" characteristics, suggesting that these are not defining characters of monocots. When monocots are compared to eudicots, the differences are more concrete.

Fig. Slice of Onion, Showing Parallel Veins

TAXONOMY

The monocots are considered to form a monophyletic group arising early in the history of the flowering plants. The earliest fossils presumed to be monocot remains date from the early Cretaceous period.

Taxonomists have considerable latitude in naming this group, as the monocots are a group above the rank of family. Article 16 of the *ICBN* allows either a descriptive name or a name formed from the name of an included family.

Historically, the monocotyledons were named:

- Monocotyledoneae in the de Candolle system and the Engler system.
- Monocotyledones in the Bentham & Hooker system and the Wettstein system
- Class Liliopsida in the Takhtajan system and the Cronquist system.

- Subclass Liliidae in the Dahlgren system and the Thorne system (1992).
- Clade monocots in the APG system and the APG II system.

Each of the systems mentioned above use their own internal taxonomy for the group. The monocotyledons are famous as a group that is extremely stable in its outer borders (it is a well-defined, coherent group), while in its internal taxonomy is extremely unstable (historically no two authoritative systems have agreed with each other on how the monocotyledons are related to each other).

Recent molecular studies have both confirmed the monophyly of the monocots and helped elucidate relationships within this group. The APG II system does not assign the monocots to a taxonomic rank, instead recognizing a monocots clade.

FLOWER EVOLUTION

Conifers are pollinated by wind, meaning they must produce a large amount of pollen grains for only a few to arrive at the female megaspore, resulting in fertilization. An advancement that allowed for higher rates of fertilization per energy expended in making pollen would surely result in that new plant type being prolific and even dominant. However, a single mutation would never result in a flower adequately adapted to spread pollen using animals. Flowers are the result of a special kind of evolution called co-evolution.

Suppose one wind pollinated plant began to have its' pollen eaten by an animal, for example a beetle. Then suppose the beetle spreads the pollen from the male cones to the female cones while looking for pollen because it can't tell the difference between the male and female cones; it searches both, spreading pollen from the male to the female in doing so. This plant has an advantage over all wind pollinated plants, because of a higher rate of fertilization. Any mutation that leads to pollen being spread by an animal soon becomes prevalent within a species, because the plant pollinated by an animal is more efficient. The plant and the animal rely on each other, one for food, the other for reproduction. Any mutation in the animal that helps pollinate the plant will become prolific through natural selection, just as mutations that favour feeding the insect or making the insect come to the flower will be dominant. This back and forth evolution results in such seemingly improbable structures as flowers that look like female moths that are pollinated by amorous male moths, or flowers and bird beaks that complement the feeding of the bird and the pollination of the flowers.

CLASS MAGNOLIOPSIDA

Members of the Class Magnoliopsida are defined partly on the basis of the seed or seedling having two cotyledons, most obvious at germination. But the differences between dicots and monocots are many, and we will be able to recognize most flowering plants that we encounter as belonging to one or the other class without having to dissect the seed or observe the seedling.

MAGNOLIOPHYTA (II)

The Class Liliopsida constitutes the monocotyledonous angiosperms and includes some of the largest plant families such as the orchids with some 20,000 species and the grasses with perhaps 15,000 species. In the previous chapter we learned how to separate the two major flowering plant groups: the dicotyledons and the monocotyledons, and covered a number of dicot families. In this chapter, we shall do the same by considering representative monocot families.

The liliopsids are considered to form a monophyletic group evolved from an early dicot. The oldest fossils presumed to be monocot remains date from the Early Cretaceous. Features that are generally common to monocots include vascular bundles that are irregularly distributed (in cross-section of the stem), leaves with parallel venation, and flower parts in multiples of three. Although true secondary growth is absent, most growth habits are found in the group including floating and submerged aquatics, lianas, trees, epiphytes, and forbs of all sizes.

GRASSES

The largest and most successful of all families of monocotyledons is the grass family, classified as the Family Poaceae (or Gramineae) and comprising nearly 10,000 species distributed more widely than any other angiosperm family. This family is also the most important economically, providing species that are the world's staple food supply. The grasses have reduced floral structures compared with most angiosperms for the reason that grasses are almost exclusively pollinated by wind. Therefore, these plants have had no cause to evolve floral structures that are attractive to insect (or other animal) pollinators. The grasses also have a fairly specific body plan that is immediately recognizable and very successful for colonizing seasonally dry landscapes, yet modifiable to suit a wide range of ecological conditions. Bor (1960) stated:

> The amazing diversity of shapes, sizes, texture and so on in the various parts that make up the vegetative and the reproductive shoot is alike the admiration and exasperation of the taxonmist. It can be said... that in no other family of monocotyledons, or dicotyledons for that matter, has a relatively simple structure been so altered in the process of evolution as to produce the amazing wealth of forms known to us. It is as if Evolution, in an excess of exuberance, had added this, subtracted that, enlarged this, suppressed that, in fact tried out every possible combination and permutation of characters to produce a family as complicated and as difficult [taxonomically] as, if not more difficult than any other.

Poaceae

The family Poaceae, in the Class Liliopsida of the flowering plants, is also known as Gramineae. Plants of this family are usually called grasses. There are about 600 genera and between 9,000-10,000 species of grasses (Kew Index of World Grass Species). Plant communities dominated by Poaceae are called grasslands; it is estimated that grasslands comprise 20% of the vegetation cover of the earth. This family is the most important of all plant families to human economies: it includes the staple food grains grown around the world, lawn and forage grasses, and bamboo, widely used for construction throughout Asia.

The term "grass" is also applied to many grass-like plants not in the Poaceae, leading to plants of the Poaceae often being called "true grasses".

Structure and Growth

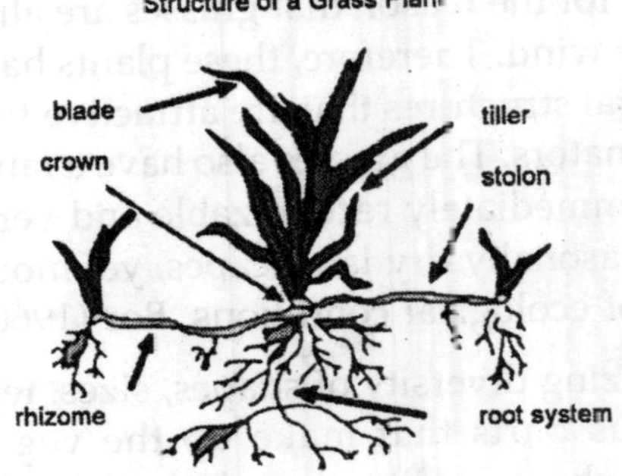

Fig. Structure of a Grass Plant.

Poaceae have hollow stems called *culms,* plugged at intervals called *nodes.* Leaves are alternate, *distichous* (in one plane) or rarely spiral, parallel-veined and arise at the nodes. Each leaf is differentiated into a lower *sheath* hugging the stem for a distance and a *blade* with margin usually entire. The leaf blades of many grasses are hardened with silica phytoliths, which helps discourage grazing animals. In some grasses (such as sword grass) this makes the grass blades sharp enough to cut human skin. A membranous appendage or ring of hairs, called the *ligule,* lies at the junction between sheath and blade, preventing water or insects to penetrate into the sheath.

Grass blades grow at the base of the blade and not from growing tips. This location of the grass growing point near the ground allows it to be grazed regularly without damage to the growing point.

Reproduction

Flowers of Poaceae are peculiar. They are typically arranged in a terminal *spike* made of many smaller *spikelets*, each spikelet having more than one flower. The flowers are usually hermaphroditic (maize, monoecious, is an exception) and pollination is always anemophilous. The perianth is reduced. Each spikelet is usually protected by a couple of bracts called the *glumes* and each single flower is held into other two bracts called the *lemma* (the external one) and the *palea* (the internal). This complex structure can be seen in the image on the left, portraying a wheat (*Triticum aestivum*) spike.

The fruit of Poaceae is a *caryopsis*.

Grass plants also spread out from a parent plant. Growth habit describes the type of shoot growth present in particular grass plants and is directly related to their ability to spread out from the parent plant and ultimately form a clonal colony. There are three general classifications of growth habit present in grasses; bunch-type, stoloniferous, and rhizomatous.

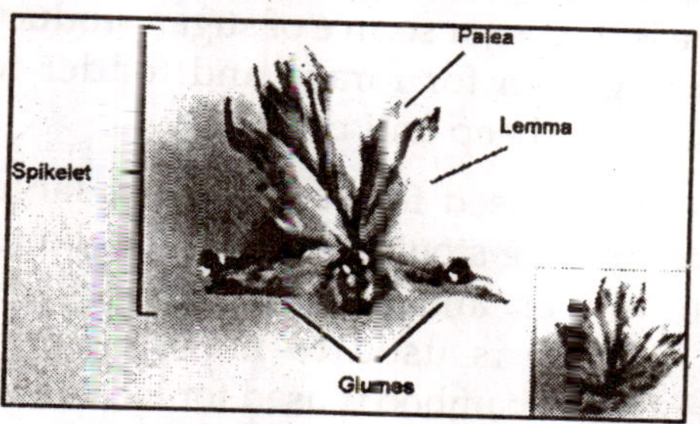

Fig. Parts of a Spikelet

The success of the grasses lies in part in their morphology and growth processes, and in part in their physiological diversity. The grasses divide into two physiological groups, using the C3 and C4 carbon fixation processes. The C4 grasses have a photosynthetic pathway linked to specialised leaf anatomy that particularly adapts them to hot climates.

Grass Evolution

Until recently grasses were thought to have evolved around 55 million years ago, based on fossil records. However, recent findings of 65-million-year-old phytoliths resembling grass phytoliths (including ancestors of rice and bamboo) in Cretaceous dinosaur coprolites, may place the diversification of grasses to an earlier date.

The flowers of grass are reduced from the general monocotyledon type. The immediate ancestor of the first grass may have been a small Liliaceous plant with rhizomes and many small flowers, growing in dense patches, which changed over to wind pollination to get round limitations caused by shortage of insects to pollinate the flowers.

Cultivation and Uses

Agricultural grasses grown for seed for human food production are called *cereals*. Cereals constitute the major source of food energy for humans and perhaps the major source of protein, and include rice in southern and eastern Asia, maize in Central and South America, and wheat and barley in Europe, northern Asia and the Americas. Some other grasses are of major importance for foliage production. Sugarcane is the major source of sugar production. Many other grasses are grown for forage and fodder for animal food, particularly for sheep and cattle.

Grasses are used for construction; larger bamboos and *Arundo donax* have stout culms that can be used in a manner similar to timber, and grass roots stabilize the sod of sod houses. *Arundo* is used to make reeds for woodwind instruments, and bamboo is used for innumerable implements.

Grass fibre can be used for making paper, and for biofuel production. Grasses are the primary plant used in lawns, which themselves derive from grazed grasslands in Europe. *Phragmites australis* is important in water treatment, wetland habitat preservation and land reclamation in the Old World.

Grasses are used as food plants by many species of butterflies and moths. see List of Lepidoptera which feed on grasses.

Anatomy

Grasses have fibrous roots and three kinds of stems: culms, rhizomes, and stolons. The culm is the main aerial shoot to which leaves and flower head are attached. The culm is a rounded or slightly flattened stem with one or more solid joints known as *nodes*. The leaves are attached at the nodes and if the stem is not simple but branched, branches arise only at nodes. Roots may also develop from a node where the node comes into contact with the ground (as in decumbent and prostrate stems). The portion of the stem between the nodes is called the *internode*, and is usually hollow in temperate zone grasses and solid in tropical grasses. All or a portion of an internode may be surrounded by the basal part of the leaf known as the *sheath*.

A rhizome is a modified stem that grows underground. Rhizomes are jointed (thus distinguishable from roots) with bladeless leaves (*scales*) arising from the joints. Rhizomes enable the grass plant to spread horizontally as new culms develp vertically from the joints. Thus, grasses with extensive rhizome development will form a turf rather than distinct tufts or bunches.

A stolon is a stem that creeps across the surface of the ground, and is really a basal branch of the culm that will develop roots and shoots from some or all of its nodes. Like a rhizome, a stolon results in a spreading or turf forming grass plant.

Grasses display two types of leaves:

1. Green leaves consisting of a sheath and blade, and
2. Reduced leaves consisting of only a sheath. With but a few exceptions, the green leaves arise at nodes alternately up the culm. Leaves that are concentrated near the base where the internodes are very short are termed *basal leaves*; leaves arising at nodes along an elongated culm are *cauline leaves*. These vegetative leaves typically surround the culm as a *sheath*, then diverge outward (at the "collar") as a long narrow *blade* with longitudinal parallel venation. If the veins are conspicuous, the leaf is striate; if the veins are raised, the leaf is ribbed.

The sheath of the leaf surrounds and protects the shoot. In some species, the sheath extends beyond the next node, so that consecutive leaf sheaths overlap, hiding the nodes. The longitudinal edges of the sheath may overlap, completely surrounding the culm, or the sheath may be tubular (the margins connate). The upper end of the sheath, known as the sheath mouth is the collar on the lower (outer) surface that may be produced into short appendages called *auricles*. On the inner, upper surface of the leaf, between the sheath and the blade, is an outgrowth called a *ligule*. This may be a flap of membranous tissue or simply a fringe of hairs, an inconspicuous rim, or even absent all together, marked only by dark tissue.

Although there is variation in leaf blade shape, most grasses have linear-shaped leaves that are many times longer than wide, with margins that are parallel then taper to a point at the apex. Grasses that grow in shady places may have lanceolate or even ovate leaf blades.

Flowers of grasses are borne in an inflorescence or flower head which terminates the culm and other branches of the stem. Smaller units of the inflorescence are called spikelets and these are arranged on one or more branches in a wide variety of different ways to which the standard terminology for inflorescences can be applied, but using the spikelet instead of the individual flower. Thus:

- In a panicle, spikelets are carried on branches of the axis.

The panicle can be either open or dense and crowded, with the branches so short that they are concealed by the spikelets, forming a *false spike*.

- In a spike, the spikelets are carried, unstalked, directly on an unbranched axis, called a *rachis*.
- In a raceme some or all of the spikelets are attached to the rachis by stalks called *pedicels*.